Biopolymers –
New Materials for
Sustainable Films
and Coatings

Biopolymers – New Materials for Sustainable Films and Coatings

Editor

David Plackett
Risø National Laboratory for Sustainable Energy
Technical University of Denmark
Roskilde
Denmark

A John Wiley and Sons, Ltd., Publication

Library of Congress Cataloging-in-Publication Data

A catalogue record for this book is available from the British Library.

Print ISBN: 9780470683415
ePDF ISBN 9781119994329
oBook ISBN: 9781119994312
ePub ISBN: 9781119995791
eMobi ISBN: 9781119995807

Set in 9/11 Sabon by Thomson Digital, Noida, India

Front cover images kindly supplied by Lluís Martín-Closas. Bottom right image: Biodegradable
Mater-Bi® fruit protecting bag applied to table grapes. Courtesy of S. Guerrini, Novamont S.p.A.
Bottom image: Biodegradable mulching in a commercial organic tomato field in North-East Spain.

Contents

Preface   xiii

About the Editor   xv

List of Contributors   xvii

PART I   1

1 Introductory Overview   3
David Plackett

1.1 Introduction   3
1.2 Worldwide Markets for Films and Coatings   4
     1.2.1 Total Polymer Production and Use   4
     1.2.2 Total Production and Use of Plastic Films   5
     1.2.3 Coatings   7
1.3 Sustainability   8
1.4 Bio-Derived Polymers   9
1.5 Other Topics   12
References   13

2 Production, Chemistry and Degradation of Starch-Based Polymers   15
Analía Vázquez, María Laura Foresti and Viviana Cyras

2.1 Introduction   15
2.2 Gelatinization   18
2.3 Effect of Gelatinization Process and Plasticizer on Starch Properties 19
2.4 Retrogradation   22
2.5 Production of Starch–Polymer Blends   23
2.6 Biodegradation of Starch-Based Polymers   27
2.7 Concluding Remarks   35
2.8 Acknowledgement   39
References   39

3 Production, Chemistry and Properties of Polylactides   43
Anders Södergård and Saara Inkinen

3.1 Introduction   43
3.2 Production of Polylactides   44
     3.2.1 Lactic Acid and its Production   44
     3.2.2 Production Methods for Polylactide   45
3.3 Polylactide Chemistry   49
     3.3.1 Tacticity   49
     3.3.2 Molecular Weight and its Distribution   50
     3.3.3 Conversion and Yield   50
     3.3.4 Copolymerization   51
     3.3.5 Characterization of Lactic Acid Derivatives and Polymers   52
3.4 Properties of Polylactides   54
     3.4.1 Processability   54
     3.4.2 Thermal Stability   55
     3.4.3 Hydrolytic Stability   56
     3.4.4 Thermal Transitions and Crystallinity of PLA   56
     3.4.5 Barrier and Other Properties   57
3.5 Concluding Remarks   59
References   60

4 Production, Chemistry and Properties of Polyhydroxyalkanoates   65
Eric Pollet and Luc Avérous

4.1 Introduction   65
4.2 Polyhydroxyalkanoate Synthesis   67
     4.2.1 Background   67
     4.2.2 Bacterial Biosynthesis of Polyhydroxyalkanoates   68
     4.2.3 Production of Polyhydroxyalkanoates by Genetically
            Modified Organisms   71
     4.2.4 Chemical Synthesis of Polyhydroxyalkanoates   72
4.3 Properties of Polyhydroxyalkanoates   72
     4.3.1 Polyhydroxyalkanoate Structure and Mechanical Properties   72
     4.3.2 Polyhydroxyalkanoate Crystallinity and Characteristic
            Temperatures   74
4.4 Polyhydroxyalkanoate Degradation   74
     4.4.1 Hydrolytic Degradation of PHAs   74
     4.4.2 Biodegradation of PHAs   75
     4.4.3 Thermal Degradation of PHAs   76
4.5 PHA-Based Multiphase Materials   77
     4.5.1 Generalities   77
     4.5.2 PHA Plasticization   77
     4.5.3 PHA Blends   78
     4.5.4 PHA-Based Multilayers   78
     4.5.5 PHA Biocomposites   79
     4.5.6 PHA-Based Nano-Biocomposites   80
4.6 Production and Commercial Products   81
References   82

5   Chitosan for Film and Coating Applications   87
Patricia Fernandez-Saiz and José M. Lagaron

5.1   Introduction   87
5.2   Physical and Chemical Characterization of Chitosan   88
    5.2.1   Degree of N-acetylation   88
    5.2.2   Molecular Weight   89
    5.2.3   Solvent and Solution Properties   89
5.3   Properties and Applications of Chitosan   89
    5.3.1   Waste/Effluent Water Purification   90
    5.3.2   Cosmetics   90
    5.3.3   Fat Trapping Agent   90
    5.3.4   Pharmaceutical and Biomedical Applications: Controlled
        Drug Release, Tissue Engineering   90
    5.3.5   Antimicrobial Properties and Active Packaging Applications   91
    5.3.6   Agriculture   94
    5.3.7   Biosensors – Industrial Membrane Bioreactors and Functional
        Food Processes   95
    5.3.8   Other Applications of Chitosan-Based Materials in the
        Food Industry   95
5.4   Processing of Chitosan   97
5.5   Concluding Remarks   98
References   99

6   Production, Chemistry and Properties of Proteins   107
Mikael Gällstedt, Mikael S. Hedenqvist and Hasan Ture

6.1   Introduction   107
6.2   Plant-Based Proteins   108
    6.2.1   Rapeseed   108
    6.2.2   Wheat Gluten   109
    6.2.3   Corn Zein   109
    6.2.4   Soy Protein   110
    6.2.5   Kafirin (Grain Sorghum)   111
    6.2.6   Oat Avenin   111
    6.2.7   Rice Bran Protein (RBP)   111
    6.2.8   Lupin   111
    6.2.9   Cottonseed Proteins   112
    6.2.10 Peanut Protein   113
6.3   Animal-Based Proteins   113
    6.3.1   Whey Protein   113
    6.3.2   Casein   114
    6.3.3   Egg White   115
    6.3.4   Keratin   115
    6.3.5   Collagen   115
    6.3.6   Gelatin   115
    6.3.7   Myofibrillar Proteins   116

6.4   Solution Casting of Proteins – an Overview   117
      6.4.1   Solvent Casting Procedures   117
      6.4.2   Importance of pH   118
      6.4.3   Drying Conditions   118
      6.4.4   Viscosity   119
      6.4.5   Importance of Temperature   119
      6.4.6   Selection of Solvent   119
      6.4.7   Plasticizers for Protein Films and Coatings   119
      6.4.8   Proteins as Coatings and in Composites   120
      6.4.9   Water Sensitivity of Protein Films   121
6.5   Dry Forming of Protein Films   121
      6.5.1   Compression Moulding   121
      6.5.2   Properties of Compression-Moulded Protein-Based Films   122
      6.5.3   Extrusion and Injection Moulding   125
6.6   Concluding Remarks   128
References   129

7   Synthesis, Chemistry and Properties of Hemicelluloses   133
Ann-Christine Albertsson, Ulrica Edlund and Indra K. Varma

7.1   Introduction   133
7.2   Structure   134
7.3   Sources   137
      7.3.1   Species   138
      7.3.2   Distribution   140
      7.3.3   Co-Constituents   141
7.4   Extraction Methodology   141
7.5   Modifications   143
      7.5.1   Esterification   143
      7.5.2   Etherification   144
      7.5.3   Miscellaneous Treatments   145
7.6   Applications   146
7.7   Concluding Remarks   147
References   148

8   Production, Chemistry and Properties of Cellulose-Based Materials   151
Mohamed Naceur Belgacem and Alessandro Gandini

8.1   Introduction   151
8.2   Pristine Cellulose as a Source of New Materials   154
      8.2.1   All-Cellulose Composites   154
      8.2.2   Cellulose Nano-Objects   154
      8.2.3   Model Cellulose Films   159
8.3   Novel Cellulose Solvents   160
8.4   Cellulose-Based Composites and Superficial Fiber Modification   162
      8.4.1   Composites with Pristine Fibers   162
      8.4.2   Superficial Fiber Modification   165
8.5   Cellulose Coupled with Nanoparticles   172
8.6   Electronic Applications   172

8.7   Biomedical Applications   173
8.8   Cellulose Derivatives   173
8.9   Concluding Remarks   174
References   174

9   Furan Monomers and their Polymers: Synthesis, Properties
    and Applications   179
    Alessandro Gandini

9.1   Introduction   179
9.2   Precursors and Monomers   181
9.3   Polymers   183
       9.3.1   Chain-Growth Systems   186
       9.3.2   Step-Growth Systems   189
       9.3.3   The Application of the Diels–Alder Reaction to Furan Polymers   201
9.4   Biodegradability of Furan Polymers   205
9.5   Concluding Remarks   206
References   206

PART II   211

10   Food Packaging Applications of Biopolymer-Based Films   213
     N. Gontard, H. Angellier-Coussy, P. Chalier, E. Gastaldi, V. Guillard,
     C. Guillaume and S. Peyron

10.1   Introduction   213
10.2   Food Packaging Material Specifications   214
        10.2.1   Functional Properties   214
        10.2.2   Safety Issues   216
        10.2.3   Environmental Aspects   218
10.3   Examples of Biopolymer Applications for Food Packaging
        Materials   220
        10.3.1   Short Shelf-Life Fresh Food Packaging   220
        10.3.2   Long Shelf-Life Dry or Liquid Food Packaging   221
10.4   Research Directions and Perspectives   222
        10.4.1   Improving/Modulating Functional Properties   222
        10.4.2   Active Biopolymer Packaging   223
        10.4.3   Improving Safety and Stability   224
        10.4.4   Towards an Integrated Approach for Biopolymer-Based
                  Food Packaging Development   225
10.5   Concluding Remarks   227
References   228

11   Biopolymers for Edible Films and Coatings in Food Applications   233
     Idoya Fernández-Pan and Juan Ignacio Maté Caballero

11.1   Introduction   233
11.2   Materials for Edible Films and Coatings   236
        11.2.1   Protein–Based Films and Coatings   237

11.2.2   Polysaccharide-Based Films and Coatings   239
11.2.3   Lipid-Based Films and Coatings   241
11.2.4   Composite/Multilayer Films   242
11.2.5   Additives   243
11.3   Edible Films and Coatings for Food Applications   243
11.3.1   Edible Coatings on Fruit and Vegetables   243
11.3.2   Edible Films and Coatings on Meat and Poultry   244
11.3.3   Edible Films and Coatings on Foods with Low Water
Content   246
11.3.4   Edible Coatings on Deep Fat Frying Foods   246
11.4   Concluding Remarks   247
References   250

12   Biopolymer Coatings for Paper and Paperboard   255
Christian Aulin and Tom Lindström

12.1   Introduction   255
12.2   Biopolymer Films and Coatings   257
12.2.1   Starches   258
12.2.2   Chitosan   259
12.2.3   Hemicelluloses   263
12.2.4   Cellulose Derivatives   265
12.3   Bio-Nanocomposite Films and Coatings   267
12.3.1   Nano-Sized Clay   267
12.3.2   Nanocellulose   269
12.4   Concluding Remarks   272
12.5   Acknowledgement   272
References   273

13   Agronomic Potential of Biopolymer Films   277
Lluís Martín-Closas and Ana M. Pelacho

13.1   Introduction   277
13.2   The Potential Role of Biodegradable Materials in Agricultural Films   278
13.3   Presently Available Biopolymers and Biocomposites   279
13.4   Past and Current International Projects on Biodegradable
Agricultural Films   282
13.5   Present Applications of Biopolymer Films in Agriculture 288
13.5.1   Overview   288
13.5.2   Biodegradable Mulching   289
13.6   Potential Uses: Current Limitations and Future Applications   293
13.6.1   Solarization with Biodegradable Films   293
13.6.2   Biodegradable Low Tunnels   293
13.6.3   Fruit Protecting Bags   295
13.6.4   Future Biodegradable Film Applications   295
13.7   Concluding Remarks   296
13.8   Acknowledgements   296
References   297

14 Functionalized Biopolymer Films and Coatings for Advanced Applications   301

David Plackett and Vimal Katiyar

14.1 Introduction   301
14.2 Optoelectronics   303
    14.2.1 Photovoltaics   303
    14.2.2 Other Optoelectronic Devices   306
14.3 Sensors   308
    14.3.1 Chemical Sensors   308
    14.3.2 Biosensors   310
14.4 Miscellaneous Applications   311
14.5 Concluding Remarks   312
References   313

15 Summary and Future Perspectives   317

David Plackett

15.1 Introduction   317
15.2 Bioplastics   318
15.3 Bio-Thermoset Resins   320
15.4 Nanocomposites Based on Inorganic Nanofillers   320
15.5 Nanocomposites Based on Cellulose Nanofillers   321
15.6 Concluding Remarks   322
References   322

Index   325

Preface

Since you have picked up this book and are now reading the preface, chances are that you have a special interest in the science and technology of new materials from bioresources. Perhaps you are a university student, research scientist, university professor or are involved in technology development in industry. Whatever the case, I hope that this volume will have some content of real value to you. As a reader of this book, you are likely to be keenly interested in the future prospects for a world in which we can become less dependent on fossil fuels for energy and materials, take steps to dramatically reduce greenhouse gas emissions and efficiently address the significant challenges associated with plastic waste in the global environment. The bio-derived polymers discussed in this book and their applications in packaging products and beyond can, as suggested by the various authors, provide part of the solution to these problems.

The invitation to prepare this book came from the publisher after a colleague and I had produced a journal article reviewing research on films and coatings from hemicelluloses, a widely available but relatively underutilized component of most biomass. After accepting the offer to edit this book, I decided to aim for chapters combining state-of-the-art summaries of our knowledge about individual biopolymers and complementary chapters written from the perspective of key applications in films and coatings. With the help of all the chapter authors, that structure is basically what you find here. There were a few challenges along the way, but not too many! At least one author remarked that various chapters might overlap and there is some truth to this comment; however, I think this has not turned out to be a significant problem. Where overlapping comments do occur, this is mostly related to the topic of packaging, which is easily the largest commodity market for bio-derived plastics. Of the original 13 authors invited to contribute, I am pleased to say that 12 of them stayed with the process to the very end and were relatively prompt in sending me their input, for which I am indeed grateful. One of the authors did thank me for my "infinite patience" – an indicator that scientists do naturally have other priorities as well as writing book chapters!

There are already a number of fine texts that comprehensively cover the subject of polymers from renewable resources in great detail, but it is hoped that this update covering key bio-derived polymers, combined in a unique way with a discussion on their use in films and coatings, will also make a contribution to this rapidly growing scientific field. It has been a privilege to work with the other chapter authors on this book and a real pleasure to read and edit their contributions, which I think have captured the essence of each topic. As a final tribute, I would like to express my sincere gratitude to Sarah Tilley and her colleagues at Wiley as well as Sarah's predecessors in managing this book.

David Plackett
Risø DTU
Roskilde
Denmark

13 December 2010

About the Editor

David Plackett holds a PhD in Chemistry from the University of British Columbia in Canada and has held research and research management positions in various companies, research institutes and universities in the UK, Canada, New Zealand and Denmark. He has a career background in bio-based materials research and since 2002 he has been Senior Scientist and Biopolymers group leader at Risø National Laboratory for Sustainable Energy, part of the Technical University of Denmark (DTU) located near Roskilde. Dr Plackett has more than 60 peer-reviewed publications and his research interests currently include the production and characterization of bio-derived polymers and their property enhancement through the use of nanotechnology.

List of Contributors

Ann-Christine Albertsson, Fiber and Polymer Technology, School of Chemical Science and Engineering, Royal Institute of Technology (KTH), Stockholm, Sweden.

H. Angellier-Coussy, Joint Research Unit Agro-polymers Engineering and Emerging Technologies, Université Montpellier II, Montpellier, France.

Christian Aulin, Innventia AB, Stockholm, Sweden, and Wallenberg Wood Science Center, Royal Institute of Technology, Stockholm, Sweden.

Luc Avérous, LIPHT-ECPM, Université de Strasbourg, Strasbourg, France.

Mohamed Naceur Belgacem, Grenoble INP-Pagora, St. Martin d'Hères, France.

Juan Ignacio Maté Caballero, Department of Food Technology, College of Agricultural Engineering, Universidad Pública de Navarra, Pamplona, Spain.

P. Chalier, Joint Research Unit Agro-polymers Engineering and Emerging Technologies, Université Montpellier II, Montpellier, France.

Viviana Cyras, Research Institute of Materials Science and Technology, Faculty of Engineering, National University of Mar del Plata, Argentina.

Ulrica Edlund, Fiber and Polymer Technology, School of Chemical Science and Engineering, Royal Institute of Technology (KTH), Stockholm, Sweden.

Idoya Fernández-Pan, Department of Food Technology, College of Agricultural Engineering, Universidad Pública de Navarra, Pamplona, Spain.

Patricia Fernández-Saiz, Novel Materials and Nanotechnology Group, IATA-CSIC, Paterna, Spain.

María Laura Foresti, Institute of Engineering in Technology and Science, Faculty of Engineering, University of Buenos Aires, Argentina.

Mikael Gällstedt, Innventia AB, Packaging Solutions, Stockholm, Sweden.

Alessandro Gandini, CICECO and Chemistry Department, University of Aveiro, Aveiro, Portugal.

E. Gastaldi, Joint Research Unit Agro-polymers Engineering and Emerging Technologies, Université Montpellier II, Montpellier, France.

N. Gontard, Joint Research Unit Agro-polymers Engineering and Emerging Technologies, Université Montpellier II, Montpellier, France.

V. Guillard, Joint Research Unit Agro-polymers Engineering and Emerging Technologies, Université Montpellier II, Montpellier, France.

C. Guillaume, Joint Research Unit Agro-polymers Engineering and Emerging Technologies, Université Montpellier II, Montpellier, France.

Mikael S. Hedenqvist, Royal Institute of Technology, Department of Fiber and Polymer Technology, Stockholm, Sweden.

Saara Inkinen, Laboratory of Polymer Technology, Åbo Akademi University, Turku, Finland.

Vimal Katiyar, Risø National Laboratory for Sustainable Energy, Technical University of Denmark, Roskilde, Denmark.

José M. Lagaron, Novel Materials and Nanotechnology Group, IATA-CSIC, Paterna, Spain.

Tom Lindström, Innventia AB, Stockholm, Sweden, and Wallenberg Wood Science Center, Royal Institute of Technology, Stockholm, Sweden.

Lluís Martín-Closas, Department of Horticulture, Botany and Gardening, University of Lleida, Lleida, Spain.

Ana M. Pelacho, Department of Horticulture, Botany and Gardening, University of Lleida, Lleida, Spain.

S. Peyron, Joint Research Unit Agro-polymers Engineering and Emerging Technologies, Université Montpellier II, Montpellier, France.

David Plackett, Risø National Laboratory for Sustainable Energy, Technical University of Denmark, Roskilde, Denmark.

Eric Pollet, LIPHT-ECPM, Université de Strasbourg, Strasbourg, France.

Anders Södergård, Laboratory of Polymer Technology, Åbo Akademi University, Turku, Finland.

Hasan Ture, Royal Institute of Technology, Department of Fiber and Polymer Technology, Stockholm, Sweden.

Indra K. Varma, Centre for Polymer Science and Engineering, Indian Institute of Technology, Delhi, India.

Analía Vázquez, Institute of Engineering in Technology and Science, Faculty of Engineering, University of Buenos Aires, Argentina.

Part I

1

Introductory Overview

David Plackett

Risø National Laboratory for Sustainable Energy, Technical University of Denmark, Roskilde, Denmark

1.1 INTRODUCTION

This book presents the latest knowledge about the synthesis and properties of bio-derived polymers in terms of practical film and coating applications. These biopolymers include starch, polyactides (PLAs), polyhydroxyalkanoates (PHAs), chitosan, proteins, hemicelluloses, cellulose and furan polymers. The contents are divided into two parts: Part 1 in which the synthesis or production and properties of biopolymers are discussed and Part 2 in which specific applications are covered. With this structure, the objective of the book is to provide the reader with an up-to-date summary of current knowledge concerning individual biopolymers and then to discuss the state of their development and uptake in a number of key fields, including packaging, edible films and coatings, paper and paperboard, and agronomy. A further chapter deals with specialized uses of biopolymers in optoelectronics and sensor technologies, a field which is still very much in its infancy. In this introductory overview, biopolymers and the question of sustainability are discussed in the context of total world markets for plastics and resins.

Biopolymers – New Materials for Sustainable Films and Coatings, First Edition.
Edited by David Plackett.
© 2011 John Wiley & Sons, Ltd. Published 2011 by John Wiley & Sons, Ltd.

1.2 WORLDWIDE MARKETS FOR FILMS AND COATINGS

1.2.1 Total Polymer Production and Use

In order to set the scene for this book, it is useful to get a sense of the total volumes and types of polymer used worldwide as well as in use specifically in films and coatings. This includes such everyday materials as kitchen films, grocery carrier bags and garbage bags, and many types of industrial and specialty films, as well as industrial or architectural paints and coatings. Since polymer production and consumption is reported in various ways, it can be difficult to separate out uses in films and coatings from total polymer usage; however, some insights can be gained through reports from industry associations and other sources.

A RAPRA report from 2004 [1] states that in 2003 the total global consumption of polymers in solid form (i.e., not adhesives, paints, binders) was 160 million tonnes and, of this amount, about 40 million tonnes was in the form of films. Information presented by Ambekar *et al.* [2] illustrates the steady increase in total plastics consumption over the past 50–60 years and indicates that this figure reached 260 million tonnes in 2007, of which about one-third was used in packaging. Volumes of plastics produced worldwide according to figures for 2007 are shown in Figure 1.1. According to the American Chemistry Council (http://www.americanchemistry. com), US production of thermoset resins in 2009 was 12.7 billion pounds on a dry weight basis, down nearly 16% from 2008, probably as a result of the impact of the economic crisis on construction activity. The corresponding figure for thermoplastic production in 2009 was 86 billion pounds which reflected only a 0.5% decrease over 2008, possibly because of the greater diversity of markets for thermoplastics as compared with thermosets.

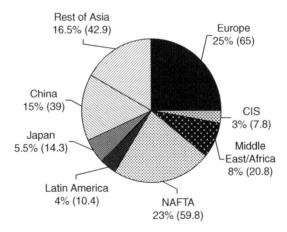

Figure 1.1 Volumes of plastics produced worldwide in 2007. Figures in brackets are millions of tonnes (*Source*: Plastics Europe Market Research Group)

Table 1.1 World polymer consumption in 2006 and projected consumption for 2016 (thousands of metric tonnes) [3]. Adapted from Accenture report "Trends in manufacturing polymers: Achieving high performance in a multi-polar world", 2008

Market sector	2006	2016
Construction	45 886	72 919
Plastic products	43 500	78 361
Food	42 025	71 774
Textiles	32 176	51 630
Electrical/electronic	13 810	25 499
Furniture	13 687	22 993
Vehicles and parts	10 746	15 625
Machinery	2397	3658
Fabricated metals	1519	2259
Printing	780	1220
Other transportation	9330	16 181
Other equipment	3852	6334
Other manufacturing	21 238	33 569
Total	240 947	402 022

A report by the consulting company Accenture discusses polymer consumption by market sector in 2006 and consumption figures predicted for 2016 [3]. A summary of the relevant data is shown in Table 1.1.

1.2.2 Total Production and Use of Plastic Films

The world market for plastic films is dominated by polyethylene (PE) and polypropylene (PP), which together comprise some 34 million tonnes per annum. These polyolefins are subject to increasing demand as the main materials used in packaging films, particularly in the developing areas of the world. Besides PE and PP, polyethylene terephthalate (PET) film is used in packaging and in a wide range of industrial and specialty products, such as in electrical (e.g., transformer insulation films, thermal printing tapes) and imaging products (e.g., microfilm, x-ray films, business graphics). Polyvinyl chloride (PVC) films are found in consumer goods and medical applications and polyvinyl butyrate (PVB), because of its optical clarity, toughness, flexibility and ability to bind to many surfaces, is mainly used in automotive and construction applications as glazing protection. Polystyrene (PS) films are also used in packaging and a variety of other medical, commercial and consumer goods. The primary types of plastics used in films, their properties and various applications are shown in Table 1.2. Polymer films as a whole are a massive market sector with Europe and North America each consuming about 30% of total world production and increasing volumes being consumed in the growing economies.

Regardless of polymer type, packaging is the main end use for plastic films. In this context, films are generally defined as being planar materials less than 10 mils or ~250 μm in thickness (i.e., thick enough to be self-supporting but thin enough to be flexed, folded and/or creased without cracking). Above this thickness, the term sheet is frequently used instead of film. Data from Plastics Europe (http://www.plasticseurope.org) confirm that packaging is the biggest end-use for

Table 1.2   Commodity thermoplastic films – their properties and applications (* in addition, trays produced by thermoforming films of PET, PE and other thermoplastics are widely used in the packaging industry). PET = polyethylene terephthalate, HDPE = high-density polyethylene, LDPE = low-density polyethylene, PP = polypropylene, PS = polystyrene, PVC = polyvinyl chloride, PVB = polyvinyl butyral

Polymer type	Properties	Applications*
PET	Clear and optically smooth surfaces, barrier to oxygen, water and carbon dioxide, heat resistance for hot filling, chemical resistance	Oveneable films and microwave trays, packaging films, industrial and specialty films
HDPE	Solvent resistance, higher tensile strength than other PEs	Grocery bags, cereal box liners, wire and cable coverings
LDPE	Resistance to acids, bases and vegetable oils, good properties for heat-sealing packaging	Bags for dry cleaning, newspapers, frozen bread, fresh produce and household garbage, shrink wrap and stretch film, coatings for paper milk cartons, and hot and cold beverage cups, wire and cable coverings
PP	Excellent optical clarity in BOPP films, low water vapour transmission, inert to acids, bases and most solvents	Packaging, electronics, kitchen laminates, furniture, ceiling and wall panels
PS	Excellent water barrier for short shelf-life products, good optical clarity, hard wearing	Packaging, electronic housings, medical products, interior furnishing panels
PVC	Biologically and chemically resistant	Packaging films, wire and cable coverings, waterproof clothing, roofing membranes
PVB	Adheres well to various surfaces, optically clear, tough and flexible	Laminated safety glass for use in automotive and architectural applications

plastics (38%), followed by building and construction (21%), automotive (7%), and electrical and electronic (6%). Other applications for plastics, which include medical and leisure, use 28% of the total production volume (Figure 1.2).

In addition to packaging, the myriad applications of polymer films include decorative wrap, form-fill-seal, blood bags, flexible printed circuits, bed sheeting, diapers, and in-mould decorating of car parts (to replace painting and provide a more durable surface coating) to name just a few. Carrier bags and garbage bags are big markets with significant imports into Europe. In construction, films are used in glazing, damp proofing, tarpaulins, and geomembranes.

Multi-material or mutli-layer films account for around seven million tonnes annually, with about 95% of this volume going into packaging. Multi-layer materials

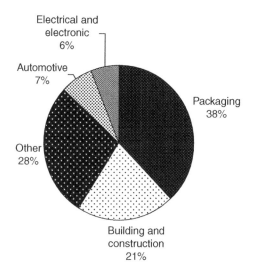

Figure 1.2 Plastics demand by converters in Europe in 2008. Total demand was 48.5 million tonnes (*Source*: Plastics Europe Market Research Group)

are attractive because they permit custom adaptation of properties such as barrier and strength. As well as the possibility to colour or print, and produce single or multi-layered products, films are also often combined with other materials such as aluminum or paper. An example is the aseptic packaging manufactured by Tetra Pak (Lund, Sweden) which contains layers of PE, paper and a very thin layer of aluminum acting as a water vapour and gas barrier. The introduction of technology such as orientation of PP films has also contributed significantly to the availability of more valuable film materials. In particular, biaxially oriented PP films (BOPP) have grown strongly on world markets because of their improved shrinkage, stiffness, transparency, seal-ability, twist retention and barrier properties.

According to Andrew Reynolds of Applied Market Information Ltd (http://www.amiplastics.com), the global agricultural film market consumed 3.6 million tonnes of plastic in 2007, of which roughly 40% was devoted to mulch films and a similar volume to greenhouse films. The remaining 20% was used in silage films.

Plastic films can be made via a number of well-established converting processes: extrusion, co-extrusion, casting, extrusion coating, extrusion laminating and metal-lizing. Blown extrusion was the first process used to make PE films. These processes have advantages and disadvantages, depending on the material in use, as well as the width and thickness of film required. Readers interested in the fundamentals of polymer processing are referred to recent textbooks on this subject [4, 5].

1.2.3 Coatings

In terms of industrial coatings, the world market for architectural paints is forecast to grow to 22.8 million metric tonnes, worth $51 billion, by 2013, according to a new report by The Freedonia Group, a Cleveland, Ohio-based industry consulting firm (http://www.freedoniagroup.com). On the basis of the large German industry, Europe is expected to remain the world's leading regional net exporter of architectural paint. The paint and coatings industry in North America, Europe and Japan is mature and generally correlates with the health of the economy, especially housing and construction and transportation. According to an industry overview published early in 2009 [6], annual growth rates of 1.5–2% were expected; however, the prolonged economic downturn in the West and in Japan will probably result in lower growth figures in the short term. The same report indicated good prospects for growth in the paint and coatings sector in the Asia-Pacific, Eastern Europe and Latin America. Growth of coatings in China was expected to continue at 8–10% per year and most multi-national paint companies have established production there.

Paper and paperboard represent large markets for polymer coatings in respect to printing, conversion and uses in packaging. Extruded coatings on paper and paper-board include those based on PE, PP, and PET, providing whiteness, smoothness and gloss. Functional coatings may also be used on paper and paperboard for grease resistance, water repellency, non-slip and release characteristics.

1.3 SUSTAINABILITY

In today's world, manufacturers are under pressure to satisfy varied, and often conflicting, demands such as lower costs, improved performance and enhanced environmental attributes. Material selection for a particular purpose can therefore be critical in a number of ways. As in many other industries, manufacturers and converters of polymer films and coatings now recognize the need to meet environmental demands and understand the need for sustainable manufacturing processes and products. In this context, there are now more and more efforts made to recycle and reuse plastics, even though total recycling is still far from adequate in many locations.

Concerning plastics recycling and disposal, with Europe as an example, converters used 48.5 million tonnes of plastics in 2008, down 7.5% on 2007, probably as a result of the first stages of the 2008–2009 financial crisis. Of all plastics used by consumers, 24.9 million tonnes ended up as post-consumer waste, up from 24.6 million tonnes in 2006, 51.3% of post-consumer used plastic was recovered and the remainder (12.1 million tonnes) went to disposal. Of the 51.3% recovered, 5.3 million tonnes was recycled as material and feedstock and 7.5 million tonnes was recovered as energy. The total material recycling rate of post-consumer plastics in 2008 was reported to be 21.3%. Mechanical recycling was 21% (up 0.9 percentage points over 2007) and feedstock recycling was at 0.3% (unchanged from 2007). The energy recovery rate increased from 29.2 to 30%.

The use of synthetic polymers has grown rapidly over the past few decades and is forecast to roughly quadruple by the year 2100 as a result of growing human population and prosperity. If this growth were to occur, we would need to use one-quarter of the world's current oil production just to manufacture plastics. Something – or a number of things – clearly must change. In this book we look at the issue of sustainability through opportunities to adopt increased use of bio-derived polymers from renewable sources as an alternative to traditional petroleum-based polymers. Biopolymers are now playing a significant, if still relatively small role, across a number of industries and can be viewed as sustainable in terms of variables such as raw material supply, water and energy use and waste product generation. Product viability, human resources and technology development also need to be viewed through the lens of sustainability. Since most biopolymers are either biodegradable or compostable, it can also be argued that bio-based polymers generally fit with the 'cradle-to-cradle' (C2C) concept in that, on disposal, they can become 'food' for the next generation of materials [7].

The achievement of true sustainability in any field requires a fine balance between environmental, economic and social concerns. In this respect, the life cycle assessment (LCA) approach can be especially useful [8, 9]. For the polymer industry as well as for consumers, a key question is the difference between 'bio-sourced' and 'biodegradable' and which is the more important. This question is complicated since most, but not all, bio-sourced polymers are biodegradable and some petroleum-sourced polymers are biodegradable. As a property, biodegradation can be viewed as an added feature for plastics, suited to certain specific uses and environments. Even if biodegradability is not necessarily a key market driver, the sourcing of polymers in a renewable way from biomass is increasingly in focus for industry and governments alike. As well as broad legislation in some countries requiring the introduction of bioplastics, particularly in grocery bags, cities such as San Francisco have taken steps to prohibit plastic carry bags completely, and similar steps are being considered elsewhere. In the food and beverage industry, McDonalds is using bioplastics packaging for the Big Mac sandwich and companies such as Biota (Telluride, Colorado) are making biodegradable water bottles.

Since bio-derived polymers are mostly sourced from biomass in one form or the other, it is useful to consider raw material supply. Unlike petroleum, biomass as a raw material for biopolymer production is widely available on a renewable basis. World biomass production is estimated at 170×10^9 metric tonnes per annum, of which 6×10^9 metric tonnes is said to be used by humans, with 2% for food, 33% in the form of wood for energy, paper, furniture and construction and 5% for non-food uses in areas such as clothing and chemicals [10].

The Sustainable Biomaterials Collaborative (SBC) based in Washington, DC has, in its 2009 guidelines for sustainable bioplastics, defined sustainable biomaterials as those that are: (1) sourced from sustainably grown and harvested croplands and forests; (2) manufactured without hazardous inputs and impacts; (3) healthy and safe for the environment during use; (4) designed to be reutilized at the end of their intended use; and (5) provide living wages and do not exploit workers or communities throughout the product lifecycle. The SBC is working to implement a new market-based approach connecting use of agriculturally derived biopolymers with best environmental practices on the agricultural land on which biopolymer production may essentially be based. The sustainability concept is then extended through biopolymer manufacturing (e.g., process safety, minimized emissions), the establishment of suitable infrastructure to compost, recycle or reuse products, and the development of appropriate new technologies for various markets (http://www.sustainablebiomaterials.org).

1.4 BIO-DERIVED POLYMERS

Current discussion on practical applications of bio-derived polymers generally focuses on bio-based thermoplastics, which are not only biodegradable or compostable, but are considered advantageous to the environment. Points that may be made in support of using bio-derived polymers can be based on: (1) the opportunity to close the carbon cycle by eventually returning plant-based carbon to the soil through biodegradation or composting and thereby reducing environmental impacts; (2) less use of fossil energy and reduced carbon dioxide emissions over manufactured product life cycles.

Market reports from the European Bioplastics Association (http://www.european-bioplastics.org) point to strong growth for bioplastics and suggest that this industry has been relatively resilient during the recent economic crisis. Evidence for this conclusion comes from investment in new plants, in new inter-company cooperation and, most importantly, from new innovations. Biopolymer industry growth has been strong in areas which have prospered relatively well during the economic crisis, but more mature markets in Europe and North America are also now predicted to grow. Outside of economic issues, challenges exist in some areas because some biopolymers, including PLA and PHA bioplastics, may not have the required technical properties to compete with conventional plastics in some applications. However, solutions are being developed and adopted by the industry, as in the use of heat-tolerant bio-based additives to improve thermal stability (e.g., collaboration between the US Agricultural Service and the plastics company Lapol based in Santa Barbara, California) [11]. This type of development may provide PLA films that can withstand high temperatures, to overcome one of the present drawbacks of the material. The impact resistance of PLA can also be enhanced through the use of impact modifiers and there are now developments in the use of non-bio-derived additives which can enhance both PLA processability and properties (e.g., Joncryl® ADR chain extender for PLA from BASF, Biostrength® 280 impact modifier from Arkema). In addition to the technical challenges associated with the processing and properties of biopolymers, commercial bioplastics are generally higher in cost than commodity petroleum-derived polymers. However, it is reasonable to expect that continued investment in R & D should ensure that bioplastics become technologically and economically more competitive over time.

The world's largest producer of PLA, NatureWorks LLC, has a production capacity of 140 000 tonnes per year and is in a strong market position in terms of the world's most widely used bioplastic with its Ingeo™ trade name. Ingeo™ biopolymers are already in successful commercial use in fiber and non-wovens, extruded and thermoformed containers, and extrusion and emulsion coatings. Among the material advantages, packaging products from Ingeo™ can be clear, opaque, flexible or rigid and offer gloss, clarity and mechanical properties which are similar to those of polystyrene. In an interesting example of customer feedback, the acoustic properties of PLA films used in packaging of snacks such as potato chips has recently been identified as a drawback and alternative less noisy materials are being developed. Other PLA manufacturers are Purac (Netherlands), Teijin (Japan), Galactic (Belgium), Pyramid Bioplastics (Germany), Zhejiang HiSun Chemical Co., Ltd (China), and Tong-Jie-Liang Biomaterials Co. (China).

Italy's Novamont, the leading European company and pioneer in the field of starch plastics, continues to expand its Mater-Bi® brand of starch-based materials with a reported 2009 capacity of 80 000 tonnes per year, which is forecast to double in the next three years. Novamont has recently announced plans to join forces with Thantawan Industry, a Thai packaging manufacturer, to distribute Mater-Bi® in Thailand. Mater-Bi® film is presently sold into markets for agricultural films, packaging and kitchen films. Cereplast Inc, based in California, is the largest starch plastic producer in the United States with two product lines, Cereplast Compostable™ and Cereplast Hybrid™. Other thermoplastic starch producers include Rodenburg (Netherlands), Biotec (Germany), Limagrain (France), BIOP (Germany), PaperFoam (Netherlands), Harbin Livan (China), Plantic (Australia) and Biograde (Australia).

The US company Metabolix Inc. has the highest profile in terms of manufacturing bacterial polyesters (i.e., PHAs). A joint venture has been established with the Archer Daniels Midland Company (ADM) and PHAs are now being produced and marketed through Telles, a joint sales company formed between Metabolix and ADM, under the Mirel™ trade name. The plans are to produce Mirel™ at 50 000 tonnes per year at a plant in Clinton, Iowa in the US (http://www.mirelplastics.com). A range of Mirel™ concentrates or masterbatches are available which meet standards for compostability and biodegradability. Furthermore, a black Mirel™ film is presently being tested for use in agricultural mulch films. Other PHA producers include Kaneka (Japan), Meredian Inc. (USA), Biomer (Germany), Tianan Biologic Material Co. Ltd (China) and Tianan Green BioSciences Ltd (China), which is a joint venture with DSM.

There are numerous other commercial developments in the biopolymer field, especially in the area of bioplastics. For example, Biome Bioplastics recently announced a strong, translucent, low-noise, flexible and biodegradable film with high renewable content. JC Hagen of Austria has also added biodegradable resins to its portfolio, which are suitable for blown film extrusion and thermoforming. FKuR's Bio-Flex is now being used as the world's first compostable soap wrapping for application in Umbria Olii's EcoLive brand of laundry soap. A glance at the web site http://www.biopolymers.net provides an idea of the many different companies involved in this field as well as the biodegradable, but not necessarily 100% renewable material-based, products from major companies such as BASF (Ecoflex®), Bayer (BAK®) and DuPont (Biomax®).

In 2007, bioplastics represented 0.1% of total world commodity plastics. The predicted capacity for 2012 is in the 500 000–1000 000 tonne range, which is still a small niche in the total plastics market, which was forecast to reach 220 million tonnes in 2010. The market focus varies from region to region. In the European Union, the main areas of use are fresh food packaging and carrier bags, while in Japan it is mainly durable goods and textiles. According to Pira (http://www.pira.com), global bioplastic packaging consumption was projected to reach 125 000 tonnes in 2010 with a market value of $454 million; however, a report from BCC Research (http://www.bccresearch.com) entitled "Bioplastics: Technologies and Global Markets" issued in September 2010 gives a figure of 571 712 metric tonnes for usage of bioplastics in 2010. A product overview and market projection for emerging bio-based plastics was commissioned by the European Polysaccharide Network of Excellence (EPNOE) (http://www.epnoe.eu) and European Bioplastics (http://www.european-bioplastics.org) and completed in June 2009 [10].

The successful commercial introduction of bio-based plastics depends upon various factors. Economically, the most important of these factors are the price of oil, the price of bio-feedstocks, investment risks, fiscal policy initiatives, and the availability of capital at competitive interest rates. Technologically, success typically depends upon the patent situation, the reliability of new technologies and the pace at which technology can be developed. Other issues determining success are the availability of trained personnel with knowledge of the sector, collaboration with companies in the agro-industry chain, the availability of raw materials and the usefulness of co-products. Finally, from a market pull perspective, the most important issue will be demand from retailers and producers of consumer goods, as influenced by the attitudes of consumers, policy makers and other stakeholders. Bio-based polymers represent a new era for the polymer manufacturers within the chemical industry and, although volumes are still low, the speed of development in this sector is relatively fast.

If we look at the most important current market for bioplastics (i.e., packaging), industry growth is likely to be motivated by issues such as consumers' positive interest in environmentally friendly materials, the introduction of sustainability programmes by retailers and brand owners to assist in product differentiation, growth in the availability of non-biodegradable plant-derived plastics, developments in terms of new suppliers and production capability, and industry or government initiatives aimed at certification, regulation and standards. Even the existence of bio-based, biodegradable plastics may be unknown to a broad swathe of the wider public, pointing to the need for more educational initiatives as well as continued product- and process-orientated R & D.

Seen from the present perspective, it would be a mistake to assume that all bio-derived plastics necessarily have a lower carbon footprint than conventional polymers or that bio-based production can in due course entirely replace the oil-based industry. Biopolymers are the basis of an industry which is still in an emerging phase and will only be successful if equal product functionality can be delivered at lower cost, new product functionality can be delivered at the right value for money or the footprint in the value chain can be improved. Although the bio-based polymer business was only a few tenths of a percent of the total polymer market in volume at the end of 2009, some predictions suggest that in the post-financial crisis world, the annual growth rate based on existing technologies could reach 20% by 2020. The introduction of new technological developments and new products could further increase this predicted growth rate.

1.5   OTHER TOPICS

It is important to note that there are a number of important topics which are necessarily outside the scope of this book on sustainable films and coatings. One of the most significant of these is the use of biomass, instead of petroleum derivatives, to manufacture conventional polymers. The prime example is the pioneering activity of the Brazilian company Braskem, which opened a commercial plant in Triunfo, Brazil in September 2010 to produce PE from sugarcane ethanol with an initial annual capacity of 200 000 tonnes. The Braskem process involves dehydration of ethanol to ethylene and then conversion to PE by traditional means. Interestingly, the first commercial plant to manufacture ethylene from ethanol was built and operated at Elektrochemische Werke GmbH in Germany as long ago as 1913 and from 1930 onwards ethanol dehydration plants were the sole source of ethylene in Germany, Great Britain and the United States. The seeds of the Braskem technology therefore go back many decades. Brazilian capacity for plant-based ethanol production is esti-mated at 25 billion litres in 2009 and is thought likely to have reached 28 billion litres in 2010. From a sustainability perspective, the World Bank and the FAO have confirmed that Brazilian ethanol has not raised sugar prices significantly and may be the only bio-fuel which is both competitive with petroleum-based diesel or gasoline and which saves on greenhouse gas emissions [12, 13].

The start-up of PE production by Braskem has attracted global interest because the PE produced by this method is for all practical purposes identical to that manufactured from petroleum. Braskem has already established agreements with Procter & Gamble, Toyota, Shiseido and Johnson and Johnson, and Tetra Pak has also recently announced trials using PE from Braskem in its packaging. While bio-based PE production is now starting, bio-based PP is still at the laboratory bench stage. However, at the end of 2009, Braskem established a partnership with the Danish company Novozymes to develop bio-based PP using Novozymes' core fermentation technology and Braskem's chemical technology and thermoplastics expertise. In another development, Dow and Crystalsev, a major Brazilian ethanol producer, announced a joint venture in 2007 with plans for bio-based PE production in 2011.

Industrial coatings and architectural paints for buildings are also generally outside the scope of this book. Manufacturers have for decades realized the need to use waterborne coatings where practically feasible and a transition to bio-derived resins in such coatings might eventually occur if the sort of technology being developed by Braskem and other companies was extended to allow manufacture of bio-alternatives to a wide range of other commonly used petroleum-derived resins (e.g., polyurethanes). In fact, bio-materials are already used to some extent in the coatings sector, for example in the case of hydroxyethyl celluloses. These non-ionic cellulose derivatives are universal thickeners for internal or external paints and coatings, preventing pigment settling, adding structural viscosity and water retention to paints and enhancing spreadability and water resistance. Medium-viscosity methyl hydroxyethyl cellulose and hydroxyethyl cellulose have been used in aqueous paints, whereas ethyl hydroxyethyl cellulose has been preferred for solvent-borne systems. In recent years, associative thickeners based on acrylate chemistry have been developed as an alternative to these cellulose ethers [14].

Life cycle assessment (LCA), although important in helping to clarify sustainability and cradle-to-cradle issues, is also not specifically addressed as a topic. The reader is therefore referred to a number of recent studies and books which have advanced knowledge in this area as it applies to biopolymers. LCA will no doubt continue to be very significant in terms of how industry and governments view bio-sourced materials and their overall contribution to future societies [10, 15–19].

Finally, the use of biopolymer films and coatings in medical-related applications, examples of which would include antimicrobial coatings for medical devices, treatment of wounds and burns, and films for dermal drug delivery, is not included as a specific topic in this book.

REFERENCES

[1] F. Pardos, Plastic Films – Situation and Outlook, RAPRA market report, RAPRA Technology Ltd, Shrewsbury, UK, 2004.

[2] A.P. Ambekar, P. Kukade, and V. Mahajan, *Popular Plastics and Packaging*, September 2010, 30–32.

[3] Accenture report *Trends in manufacturing polymers: Achieving high performance in a multi-polar world*, 2008.

[4] T.A. Osswald, Polymer processing fundamentals, Carl Hanser Verlag, Munich, Germany, 1998.

[5] A.J. Peacock, A.R. Calhoun, *Polymer chemistry: Properties and applications*, Carl Hanser Verlag, Munich, Germany, 2006, pp 195–253.

[6] E. Linak, A. Kishi, Paint and coatings industry overview, *Chemical Economics Handbook*, SRI Consulting, 2009.

[7] W. McDonough, M. Braungart, *Cradle to Cradle*, North Point Press, New York, New York, 2002.

[8] ISO 14040:2006. Environmental management – Life cycle assessment – Principles and framework, International Standards Organization, Geneva, Switzerland.

[9] H. Pilz, B. Brandt, and R. Fehringer, *The impact of plastics on life cycle energy consumption and greenhouse gas emissions in Europe – Summary report*, June 2010, Plastics Europe - Association of Plastics Manufacturers, Brussels, Belgium.

[10] L. Shen, J. Haufe, and M.K. Patel, *Product overview and market projection of emerging bio-based plastics* (PRO-BIP 2009), commissioned by the European Polysaccharide Network of Excellence (EPNOE) and European Bioplastics, Utrecht, The Netherlands, June 2009.

[11] R. Harrington, Breakthrough may help PLA take the heat from hot-filled apps, *Food Production Daily*, 2 September 2010.

[12] A. Morschbacker, *Bioplastics Magazine*, 2010, 5, 52–55.

[13] A. Morschbacker, *J. Macromol. Sci. Part C: Polym. Rev.*, 2009, **49**, 79–84.

[14] J. Plank, *Appl. Microbiol. Biotech.*, 2004, **66**, 1–9.

[15] E.T. Vink, K.R. Rabago, D.A. Glassner, *et al.*, *Polym. Degr. Stab.*, 2003, 80, 403–419.

[16] S. Kim, B.E. Dale, *Int. J. LCA*, 2005, **10**, 200–210.

[17] M. Patel, C. Bastioli, L. Marini, *et al.*, Life cycle assessment of bio-based polymers and natural fiber composites in Biopolymers Online, John Wiley & Sons, 2005.

[18] R. Narayan, Drivers and rationale for use of bio-based materials based on life cycle assessment (LCA), in *Proceedings of the Global Plastics Environmental Conference*, Atlanta, 2004.

[19] J.H. Song, R.J. Murphy, R. Narayan, *et al.*, *Phil. Trans. R. Soc. B*, 2009, **364**, 2127–2139.

2

Production, Chemistry and Degradation of Starch-Based Polymers

Analía Vázquez and María Laura Foresti
Institute of Engineering in Technology and Science, Faculty of Engineering, University of Buenos Aires, Argentina

Viviana Cyras
Research Institute of Material Science and Technology, Faculty of Engineering, National University of Mar del Plata, Argentina

2.1 INTRODUCTION

Starch is the most abundant reserve polysaccharide in plants and as such is a renewable resource with many practical uses. Starch is biodegradable, produced in abundance at low cost and can exhibit thermoplastic behavior. The botanical sources of starch are seeds, roots and tubers, in which starch occurs as an organized structure called granules. Starch granules are insoluble in cold water. The main commercial sources of starch are maize, potato and tapioca; however, there are a significant number of species that have

Biopolymers – New Materials for Sustainable Films and Coatings, First Edition.
Edited by David Plackett.
© 2011 John Wiley & Sons, Ltd. Published 2011 by John Wiley & Sons, Ltd.

high starch contents, including legumes, grain (e.g., amaranth) and various nuts. Starch properties depend on the source, granule size distribution and morphology, genotype, amylose/amylopectin ratio and other factors such as composition, pH, and any chemical modifications. In starch-based products, gelatinization and reorganization behavior control the texture and stability of the final product [1–6].

Starch granules have two components, amylose and amylopectin, consisting of $\alpha(1 \rightarrow 4)$ linked D-glucose units. Amylose is linear, whereas amylopectin is a highly branched polymer with 5–6% of $\alpha(1 \rightarrow 6)$ linkages. A schematic representation of structural levels of the starch granule and the involvement of amylose and amylopectin is shown in Figures 2.1a and b. The ratio between amylose and amylopectin varies depending on the starch source. In many starches, amylose constitutes about 15–30% by weight [3, 7–8] (Table 2.1).

Starch granules exhibit a crystalline/amorphous structure, in which amylopectin is the major crystalline constituent, while amylose and the amylopectin branches form the amorphous part. The native structure of starch is made of helices that are more or less organized radially forming a granule, with concentric rings representing semi-crystalline shells separated by essentially amorphous regions. The x-ray scattering study of crystallites shows that helices are organized according to two main crystalline lattices: an A-type allomorph mainly found in cereal starches and a B-type allomorph found in tubers and amylose-rich starches. A C-type structure consisting of a mixed organization of A and B crystalline forms also occurs naturally, such as in pea and bean starches [3, 9–11]. There is also a V-type conformation which is a result of amylose being complexed with substances such as fatty acids or emulsifiers. The main difference between A and B types is that the former adopt a close-packed arrangement with water molecules between each double helical structure while the B-type is more open with more water molecules. [12–13]. The various starch types can be distinguished by their different x-ray diffraction (XRD) patterns (Figure 2.2).

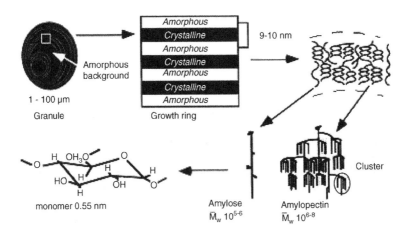

Figure 2.1 (a) Schematic representation of structural levels in the starch granule [3]. Reprinted from A. Buléon, P. Colonna, V. Planchot, *et al.*, Starch granules: structure and biosynthesis, *Int. J. Biol. Macromol.*, 23, 85–112. Copyright (1998) with permission from Elsevier

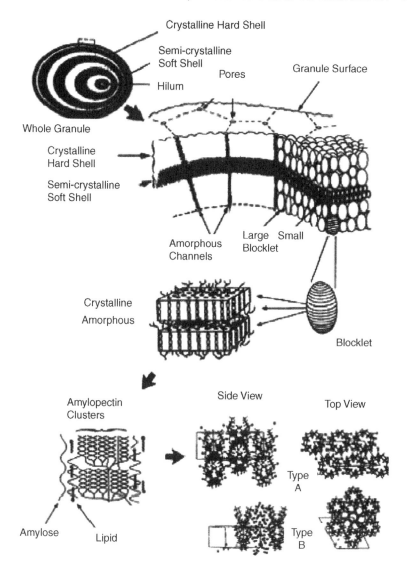

Figure 2.1 (b) Starch granule structure showing organization in semi-crystalline and crystalline shells, blocklet structure in association with amorphous channels, internal blocklet structure and the crystal structures of starch [7]. Reprinted from D. Gallant, B. Bouchet and P. Baldwin, Microscopy of starch: evidence of a new level of granule organization, *Carbohydr. Polym.* 32, 177–191. Copyright (1997) with permission from Elsevier

Starch granules contain minor components, of which lipids are the most important fraction. The main surface constituents are proteins, enzymes, amino acids and nucleic acids. Some components can be extracted without granule disruption, for example approximately 10% of proteins and 10–15% of lipids. Triglycerides

Table 2.1   Amylose and amylopectin content, granule size and crystallinity for different starches [14]. Reprinted from L. Avérous, Biodegradable multiphase systems based on plasticized starch: A review, *J. Macromol. Sci. Polym. Rev.* C44, 231–274. Copyright (2004) with permission from Taylor & Francis

Starch	Amylose content (%)	Amylopectin content (%)	Granule size (μm)	Crystallinity (%)
Wheat	26–27	72–73	25	36
Maize	26–28	71–73	15	39
Waxy starch	<1	99	15	39
Amylomaize	50–80	20–50	10	19
Potato	20–25	70–74	40–100	25

represent a major fraction of the surface lipids of maize and wheat. The presence of internal lipids is a characteristic of cereal starches. In contrast to tubers and legume starches, cereal starches are characterized by the presence of monoacyllipids and lysophospholipids in amounts positively correlated to amylose content [12].

2.2   GELATINIZATION

Gelatinization is an irreversible order–disorder transition, which occurs when starch is heated in the presence of water over a temperature range characteristic of the particular starch source. The gelatinization process in excess water involves primary hydration of amorphous regions. This in turn facilitates molecular mobility in the amorphous regions (swelling) which then provokes an irreversible molecular transition. This irreversible step involves dissociation of double helices and a radial expansion of granules, loss of birefringence, uptake of heat, loss of crystalline order,

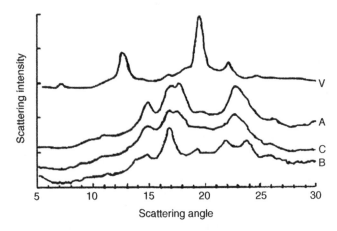

Figure 2.2   X-ray diffractograms for starches of different types [12]. Reprinted from J.J.G. van Soest and J.F.G. Viegenthart, Crystallinity in starch plastics: consequences for material properties, *Trends Biotechnol.*, 15, 208–213. Copyright (1997) with permission from Elsevier

uncoiling and dissociation of double helices and amylose leaching. Heat is taken up, according to the characteristic gelatinization endotherm and can be measured using differential scanning calorimetry (DSC). The gelatinization transition temperatures, To (onset), T_p (mid-point), T_c (conclusion) and the enthalpy of gelatinization (ΔH) are influenced by the molecular architecture of the crystalline region. The temperature at which the gelatinization process starts depends on the starch concentration.

Various techniques are used to study starch gelatinization. Birefringence is employed to follow the size of the granule during the gelatinization process. XRD can be used to study the crystallinity of starch and Fourier transform infrared spectroscopy (FTIR) has been used to describe the organization and structure of starch at various moisture contents. Amorphous starch can be characterized by an IR absorbance band at $\sim 1022\,cm^{-1}$ and the crystalline state identified by the development of a band at $1047\,cm^{-1}$ [15]. DSC measurements can also be used to study the loss of order that takes place during gelatinization. Amylose chains are released during gelatinization and these chains can be determined colorimetrically through formation of a blue complex with iodine [16]. Gelatinization affects the dielectric properties of the starch–water system and therefore conductance measurements can also be used to monitor the process [17]. Starch gelatinization can be followed by viscosity measurements because the viscosity of the starch–water mixture changes during the process due to swelling of the granules.

2.3 EFFECT OF GELATINIZATION PROCESS AND PLASTICIZER ON STARCH PROPERTIES

Thermoplastic starch can be obtained from native starch by disruption of starch granules and plasticization. This process occurs through the transformation of granules into a homogeneous material with the destruction of hydrogen bonds between the starch molecules and with the formation of hydrogen bonds between added plasticizer and starch molecules. Disruption can be achieved in the presence of an appropriate plasticizer by applying heat and shearing in a continuous process such as extrusion to obtain a homogeneous molten phase. The degree of gelatinization depends on the content and type of plasticizer and on processing parameters such as shear stress, melt viscosity, time and temperature [12].

Plasticizers such as glycerol, glycerol monostearate, glycol, xylitol, sorbitol, poly-ethylene glycol, sugars or oligosaccharides, fatty acids, lipids and derivatives are used to overcome film brittleness and to improve flexibility and extensibility. Small molecules containing the –CO–NH– functional group such as urea and formamide or formamide/urea mixtures can also plasticize native starch. The water resistance of formamide-plasticized thermoplastic starch is slightly improved relative to traditional glycerol-plasticized starch [1, 18–22].

The glass transition temperature (T_g) of starch–plasticizer systems is a function of plasticizer content. As an example, Figure 2.3 shows the effect of glycerol and water on starch T_g. Glycerol addition increases the toughness and strength of the materials due to strong hydrogen bonding with starch molecules [12].

Famá and co-workers (2006) observed that introducing antimicrobial potassium sorbate in edible films of cassava starch and glycerol resulted in a decrease in the degree of crystallinity and an increase in moisture content (Table 2.2) [23]. The increase in sorbate content displaced the T_g of a glycerol-rich phase toward lower

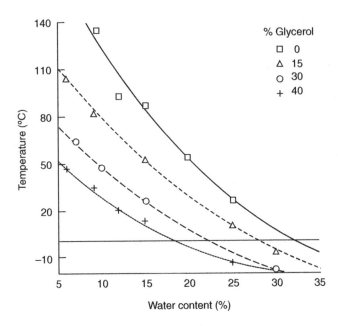

Figure 2.3 Starch glass transition temperatures as a function of glycerol and water content [12]. Reprinted from J.J.G. van Soest and J.F.G. Viegenthart, Crystallinity in starch plastics: consequences for material properties, *Trends Biotechnol.* 15, 208–213. Copyright (1997) with permission from Elsevier

temperatures. The decrease in crystalline fraction with antimicrobial increase is associated with a corresponding increase in moisture content.

Da Róz *et al* (2006) studied the effect of different additives on corn starch containing 28% amylose [24]. Ethylene glycol, propylene glycol, 1,4-butanediol, diethylene oxide glycol and sorbitol (shorter glycols) were effective in destructuring and plasticizing starch. Ethylene glycol was the most effective of these compounds. On the other hand, addition of 1-hexanol, 1-octanol, 1-dodecanol, 1-octadecanol, 1,6-hexanediol, 2,5-hexanediol, ethylene glycol monomethyl ether, polyethylene

Table 2.2 Moisture content and crystallinity of glycerol (33%)/cassava starch films two weeks after gelatinization when equilibrated at 57.5% humidity and 25 °C [23]. Reprinted from L. Famá, S. Flores, L. Gerschenson, *et al.*, Physical characterization of cassava starch biofilms with special reference to dynamic mechanical properties at low temperatures, *Carbohydr. Polym.* 66, 8–15. Copyright (2006) with permission from Elsevier

Sorbate content (g/100 g starch)	Crystallinity (%)	Moisture content (g/100 g starch)
0	33.4	25
1.28	24.2	22.3
2.19	22.2	38
2.69	17.9	44.9

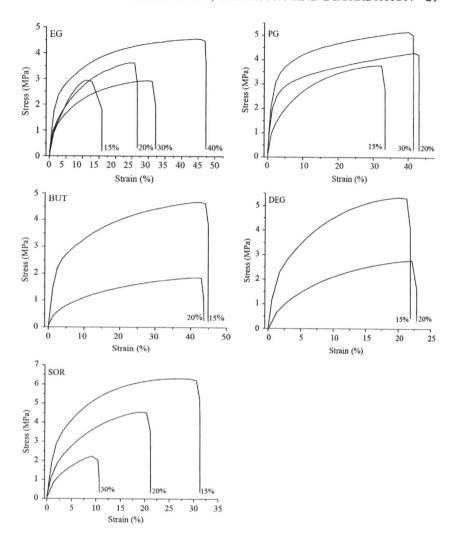

Figure 2.4 Stress–strain curves of plasticized starch [24]. Reprinted from A.L. Da Róz, A.J.F. Carvalho, A. Gandini, *et al.*, The effect of plasticizers on thermoplastic starch compositions obtained by melt processing, *Carbohydr. Polym.* 63, 417–424. Copyright (2006) with permission from Elsevier

oxide glycol, polypropylene oxide glycol (high-molecular-weight glycols) did not plasticize starch. The stress–strain curves obtained by Da Róz *et al.* (2006) are shown in Figure 2.4. The plots are linear at low strains, showing high modulus values, thus displaying the typical behavior of semi-crystalline materials. With ethylene glycol (EG) and propylene glycol (PG) as plasticizers, the modulus increased with increasing plasticizer content, except for ethylene glycol contents above 30 wt %, for which the modulus was affected in two opposite ways by the plasticizer, involving a decrease induced by plasticization and an increase caused by a corresponding increase in crystallinity. With 1,4-butanediol (BUT), diethylene oxide glycol (DEG) and D-sorbitol (SOR), the modulus decreased as the plasticizer content increased, which suggests that for these additives the first effect prevailed.

2.4 RETROGRADATION

Gelatinized starches suffer ageing during storage and cooling and there is a tendency for amylose and amylopectin to interact forming a more ordered structure. These molecular interactions are called retrogradation [25]. The retrogradation properties of starches are indirectly influenced by the structural arrangement of starch chains in the amorphous and crystalline regions of the non-gelatinized granule, which influences the degree of granule breakdown during gelatinization and the interactions that occur between the starch chains during gel storage [26]. In the work of Karim et al. (2000), the transition temperatures of retrogradation were found to be lower than the gelatinization temperatures. This could be because recrystallization causes a lower order in amylopectin chains than is present in native starch [27].

Starch retrogradation is accompanied by increases in the degree of crystallinity and gel firmness, exudation of water and the appearance of a 'B-type' x-ray pattern [25, 28]. Since starch retrogradation is a kinetically controlled process, the alteration of time, temperature, and water content during processing can produce a variety of end-products [29]. After processing, the properties of the metastable starch–water system can also be influenced by moisture content, the botanical source of starch, storage time, and storage temperature [30–31]. At lower concentrations, water acting as a plasticizer is well known to affect the T_g of semi-crystalline polymers. Water acts as a plasticizer of amorphous and partially crystalline starch systems and, as a result, water content influences the T_g and therefore also the properties, processing and stability of many starch-based foods [32]. When a stored starch gel is reheated in a DSC experiment, an endothermic transition occurs that is not present in the DSC scan of a recently gelatinized sample. This transition is generally attributed to the melting of recrystallized amylopectin.

Many research groups have studied the use of polysaccharides to control the retrogradation rate of starch-based products, including mixtures of amylose and dextran [33], amylose and galactomannan [34], potato starch and maltodextrin, and potato starch and xanthan gum [35]. These formulations enhance the interaction between starch granules, promoting retrogradation. Lii et al. (1998) reported that rice starch (33% wt/wt) in the presence of maltodextrins (5–20%) with a high average degree of polymerization strongly promoted retrogradation [36]. The retrogradation of wheat starch (30% wt/wt) was advanced by higher molecular weight chains of amylopectin [37]. The retrogradation rate of tapioca starch increased in the presence of xyloglucan [38].

Many methods including rheological techniques, sensory evaluation of texture, DSC, turbidimetry, XRD, nuclear magnetic resonance spectroscopy (NMR), Raman spectroscopy and FTIR spectroscopy have been used to study retrogradation [27]. Syneresis data and turbidity measurements yield information on both amylose and amylopectin crystallization, whereas DSC and NMR can provide information on amylopectin crystallization and changes in water mobility during retrogradation respectively. Sometimes, annealing-type processes are confused with retrogradation. However, annealing of starch granules is a process that retains granular structure and original order. Retrogradation occurs as amorphous α-glucan chains form double helices and, perhaps eventually, align themselves in crystallites.

2.5   PRODUCTION OF STARCH–POLYMER BLENDS

Starch-based plastics have some negative aspects, including limited stability caused by water absorption, ageing-induced retrogradation, inferior mechanical properties and poor processability. To overcome these limitations, the use of plasticized starch with other biodegradable polymers has been explored as a way to obtain low-cost, compostable materials.

Blending thermoplastic starch and other biodegradable polymers permits: (i) reduction in production costs; (ii) adjustable rates of degradation; (iii) a combination of properties derived from those of the individual polymers; and (iv) improved mixing during processing. Mixing is the process by which a combination of materials is made homogeneous. The quality of mixing can be defined by the intensity of segregation and the scale of segregation of the components. The scale of segregation can be related to the interfacial area and the intensity of segregation to the local concentration gradients as illustrated schematically in Figure 2.5.

A homogeneous mixture is obtained when the physical or chemical properties do not vary within the mixture. The rate of change of the intensity of segregation as depicted in Figure 2.5 will depend on diffusion coefficients. It is also necessary to know the characteristic length scales in the mixing field (i.e., the scale of segregation). Some authors have defined other variables including exposure or the potential to reduce the segregation, a nonlinear function of the scale of segregation [39]. The decrease in the scale of segregation occurs when a break point occurs, which is dependent on the viscous forces acting on a drop surface overcoming the stabilizing surface tension [40].

Homogeneity also depends on the level at which it is considered. When the dimensions of any inhomogeneity increase, the blend approaches a composite material in which the phases have clearly different properties [41]. Starch can be modified by mixing with other polymers, by adding reinforcements or by forming multilayer structures (Figure 2.6). However, in this chapter we will only discuss blends of starch with different polymers.

When two polymers are dissolved in a suitable solvent to produce a low viscosity solution, mixing can be done by means of an agitator, turbine, or mixer with blades or helix, which promote convective mixing. Films can then be obtained from solution by casting or solvent evaporation. However, when two polymers are mixed with zero or low solvent content, viscosity increases and an intensive mixer (laboratory scale), twin-screw extruder or co-kneader (industrial scale) must be used. These types of equipment can develop the high shear forces which are necessary for full mixing. As indicated, a break point occurs when shear force reaches a particular value and the final material shape will depend on interfacial tension values. Figure 2.7 shows the possible outcome of shear mixing on aggregates. As discussed later in this chapter, a potential way to change the quality of mixing and to obtain better dispersion in polymer blends is to introduce a compatibilizer. Melt processing of thermoplastic

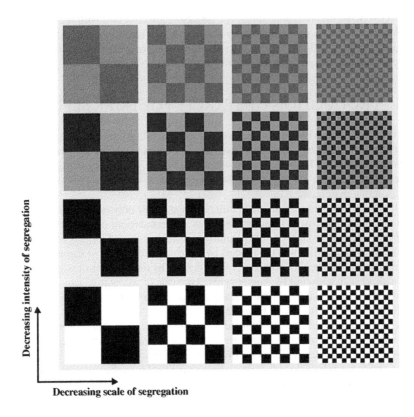

Figure 2.5 Changes in the quality of mixing of two components by decreasing the intensity and scale of the segregation

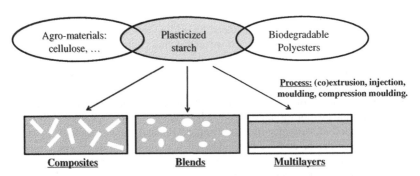

Figure 2.6 Scheme showing various starch-based materials [14]. Reprinted from L. Avérous, Biodegradable multiphase systems based on plasticized starch: A review, *J. Macromol. Sci. Polym. Rev.* C44, 231–274. Copyright (2006) with permission from Taylor & Francis

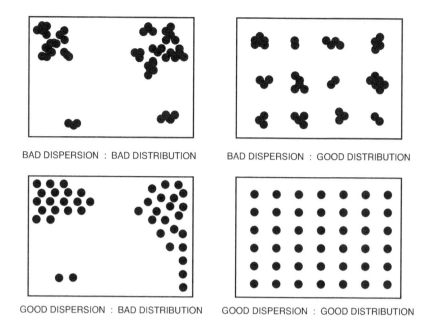

BAD DISPERSION : BAD DISTRIBUTION   BAD DISPERSION : GOOD DISTRIBUTION

GOOD DISPERSION : BAD DISTRIBUTION   GOOD DISPERSION : GOOD DISTRIBUTION

Figure 2.7   Schematic illustration of dispersion and distribution quality in polymer blends

starch on a commercial scale needs two or three steps and generally involves use of an extruder. After extracting thermoplastic starch from native starch, the material can be processed in a similar way to other thermoplastics. Excess water should be avoided in the processing and the steam pressure should not exceed 1 bar in order to avoid production of foams [42].

The important extrusion processing parameters are temperatures, screw speed and residence time. During extrusion, native starch granules are subjected to shear stress and as a consequence the crystalline granules are deformed and disordered, transforming them into a homogeneous and amorphous material [42].

The process to obtain a blend through extrusion may take place in one or two steps (Figure 2.8). In the case of a one-step process, native starch, plasticizer and a second polymer are fed into a twin-screw extruder in different zones. For example, native starch granules are fed in the first zone, plasticizer and water are fed in a low-pressure second zone, and a native starch/plasticizer mix is prepared in a third zone. This third zone has kneading blocks which produces high shear stresses. The second polymer is then added in a low-pressure zone with kneading blocks. Water is vented from the extruder as steam.

In a two-step process, a dry blend is first produced. A plasticizer and water are then added to the starch in certain amounts. Swelling occurs as a result of plasticizer diffusion into the starch granules. This blend is located in the extruder and mixing with a second biodegradable thermoplastic is performed [44–46].

One of the key parameters in starch processing is starch type because this determines the amylose content. Lower energy input is required when the starch has a high content

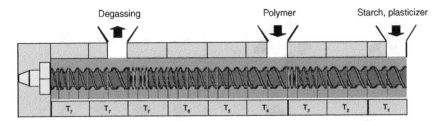

Figure 2.8   Twin-screw extruder profile (Adapted from [43]). Reprinted from E. Landreau, L. Tighzert, C. Bliard, *et al.*, Morphologies and properties of plasticized starch/polyamide compatibilized blends, *Eur. Polym. J.* 45, 2609–2618. Copyright (2009) with permission from Elsevier

of amylose because: (i) amylopectin forms the crystalline part of the starch granule and has to be broken down; (ii) the viscosity of the melt is higher in the case of high amylopectin content starch, since despite some degradation, the molecular weight of amylopectin is higher; and (iii) amylopectin has a branched structure [47].

There are several factors which influence the morphology of starch blends, such as the viscosity ratio of the components, their surface tension, the relative elasticity of the phases, and the distribution and ratio of molecular weights [42]. Subsequent extrusion of pellets into a mould involves high shear and discontinuous melt flow. Injection moulding conditions which affect final material properties include ram speed, barrel temperature profile, injection pressure, holding time and pressure of holding, mould temperature and cooling conditions. Avérous (2004) analyzed the effect of compatibility when two polymers are injected [14]. When two incompatible polymers are mixed there is a migration of the low-viscosity polymer to the wall in the injection machine. If two polymers with different viscosities are injected together (e.g., polyester and starch), the low-viscosity polymer migrates in the direction of the mould surface. After cooling of a polyester-starch blend, a polyester-rich skin and a starch core are obtained. This was demonstrated by use of imaging NMR to show that a pseudo-multilayer structure was produced. The structure obtained has higher water resistance than thermoplastic starch alone due to the polyester surface.

In the literature there are many publications on starch-based polymer blends. Thermoplastic starch has been mixed with agro-polymers such as proteins or pectins, and also with polycaprolactone (PCL), polyesteramide (PEA), poly(hydroxybutyrate-*co*-hydroxyvalerate) (PHBV), poly(butylene succinate adipate) (PBSA), poly(butylene adipate-*co*-terephthalate) (PBAT), and poly(lactic acid) (PLA). The mechanical properties of the blends depend on the property of each individual polymer (e.g, PLA at ambient temperature can be rigid while PCL, PBSA, and PBAT can be soft). A number of commercial starch-based blends such as Mater-Bi® from Novamont (Italy) or Bioplast® from Biotec (Germany) are available.

Starch and hydrophobic polymers are generally immiscible and mixing produces separated phases with poor interface. Interfacial bonding can be improved by additives or by chemical grafting of the polymers. To improve compatibility between the phases, starch can be chemically modified (e.g., using maleic anhydride, pyromellitic anhydride, polyacrylic acid, polyglycidyl methacrylate or a urethane). Another possibility is to obtain starch-grafted polyesters. For example, PCL-grafted

Table 2.3 Starch phase sizes in blends with PLA-PCL [50]. Reprinted from H.T. Liao and C.S. Wu, Preparation and characterization of ternary blends composed of polylactide, poly (ε-caprolactone) and starch, *Mater. Sci. Eng. A*, 515, 207–214. Copyright (2009) with permission from Elsevier

Starch (wt%)	Phase size (μm) $PLA_{70}PCL_{30}$/starch	Phase size (μm) $PLA_{70}PCL_{30}$-g-AA/starch
10	7.6 ± 0.8	2.8 ± 0.5
20	12.8 ± 1.0	3.3 ± 0.6
30	16.3 ± 1.1	3.9 ± 0.7
40	19.2 ± 1.3	4.2 ± 0.8
50	22.8 ± 1.5	4.5 ± 0.8

starch or dextran and PLA-grafted polysaccharides have been studied. The length of the grafts can be controlled to obtain specific structures [14].

The compatibility of polymer blends can be investigated through shifts in T_g values after blending, changes in phase size and distribution identified by electron microscopy, changes in tensile properties, peel tests, and by calculation of the theoretical work of adhesion from the contact angles of probe liquids. Table 2.3 shows the decrease in the size of the separated starch when compatibilization was performed by chemical grafting with acrylic acid. Shi and co-workers (2008) added citric acid to a polyvinyl alcohol blend, producing a small increase in T_g value as a consequence of cross-linking [48]. Sin *et al.* (2010) studied the synergistic interaction between PVOH and cassava starch using DSC [49].

Table 2.4 shows the mechanical properties of starch-biodegradable polymer blends reported in recent publications as a function of composition and method of processing. Final mechanical properties arise from the properties of the individual polymers in the blends, the processing methods and the type of starch. As a consequence it is very difficult to compare all the data in the literature. General tendencies are described in this chapter taking into account the various processing methods.

2.6 BIODEGRADATION OF STARCH-BASED POLYMERS

According to the International Union of Pure and Applied Chemistry (1993), biodegradation can be defined as the breakdown of a substance catalyzed by enzymes *in vitro* or *in vivo*. Biodegradation of polymers is generally carried out by bacteria, fungi, and yeasts that consume organic material and return it to the environment via the carbon cycle. However, due to the lack of water solubility and molecular size, microorganisms typically cannot transport polymers directly into the cells. Instead, extracellular enzymes are secreted which cleave the polymer chains outside the cell until fragments of sufficiently small size are produced. The fragments are then transported into the cell in which they undergo mineralization, a metabolic process in which the substrate is completely converted into carbon dioxide, water and new biomass. Methane can also be produced if degradation is performed under anaerobic conditions.

Table 2.4 Mechanical properties and processing methods as reported for different starch-based blends

Polymer/starch composition	Additive	Processing method	Tensile strength (MPa)	Elongation at break (%)	Modulus (MPa)	Reference
PVA/TPS 25:75	Citric acid 20wt%	Solution-casting technique	39.0	81.2		[48]
PLA/PBAT/TPS 50:30:20	Anhydride-functionalized polyester	Extrusion	14.0	6.0	1750	[51]
PLA$_{70}$PCL$_{30}$/TPS		Mixer and compression				[50]
100:0			41.0	350		
90:10			37.5	310		
80:20			30.0	260		
70:30			28.0	215		
60:40			22.0	190		
50:50			18.0	150		
PHB/starch granules		Solution-casting technique				[52]
100:0			18.29	3.32	1708	
90:10			17.20	9.8	1716	
80:20			19.7	6.0	1085	
70:30			19.23	9.4	949	
60:40			7.7	8.5	856	
50:50			10.06	5.27	694	
40:60			5.24	3.45	686	
30:70			4.99	4.3	579	
50:50 TS			11.13	2.65	1689	
70:30 TS			31.45	11.7	3334	

		(at break)			
PHB-18%HV/maize starch	Solution-casting technique				[53]
100:0		18	1200		
80:20		6	720		
70:30		3	300		
60:40		3	300		
50:50		1.8	150		
Ethylene-vinyl alcohol/ corn starch	Compression moulding	(at break)			[54]
without crosslinking		36.4	2160		
using $Na_3P_3O_9$ as a crosslinking reagent and Na_2CO_3		35.8	2770		
using $Na_3P_3O_9$ as a crosslinking reagent and NaOH		38.0	2310		
cellulose acetate/corn starch	Compression moulding	(at break)			[54]
without crosslinking		37.7	2390		
using $Na_3P_3O_9$ as a crosslinking reagent and Na_2CO_3		57.2	4590		
using $Na_3P_3O_9$ as a crosslinking reagent and NaOH		46.9	3400		
PCL/corn starch	Compression moulding	At break			[54]
without crosslinking		23.7	570		

(continued)

Table 2.4 (Continued)

Polymer/starch composition	Additive	Processing method	Tensile strength (MPa)	Elongation at break (%)	Modulus (MPa)	Reference
using $Na_3P_3O_9$ as a crosslinking reagent and Na_2CO_3			22.9		2440	[55]
using $Na_3P_3O_9$ as a crosslinking reagent and NaOH			23.4		1660	
PHBV (11.6% HV)/corn starch granular form		Compression moulding				
100:0			24 (yield)	13	378	
75:25 (high-amylose corn starch)			11	1.1	1180	
75:25 (waxy corn starch)			123.2	3.2	1200	[55]
PCL (40000 M_w)/corn starch granular form		Compression moulding				
100:0			13(yield)	850	318	
75:25 (high-amylose corn starch)			5	880	474	
75:25 (waxy corn starch)			3.2	19	495	[55]
PCL (80000 M_w)/corn starch granular form		Compression moulding				
100:0			15.3 (yield)	1100	258	
75:25 (high-amylose corn starch)			13.4	1020	500	
PCL/starch with 5 wt.% poly(ethylene glycol) Mn=600		Extrusion				[56]

Material	Plasticizer	Ratio	Plasticizer content				Ref
		100:0		11	100	120	
		95:5		11	135	250	
		90:10		8	225	225	
		80:20		8.5	125	150	
		70:30		3.5		155	
Polyamide/sodium carboxymethylcellulose/ wheat starch	Glycerol						[43]
		100:0		50	300		
		60:40		20	160		
		50:50		16	180		
		30:70		20	150		
PCL/modified corn starch/ soy protein isolate	Sorbitol						[57]
		100:0:0	0	24.9	5.5	231	
		40:40:0	20	4.6	1.8	142	
		40:36.5:3.5	20	7.6	2.4	109	
		40:29:11	20	5.7	1.9	159	

The biodegradability of polymers may be evaluated through field tests, simulation tests or laboratory tests. Field assays involving burial of samples in soil, placing them in real aqueous environments such as lakes or rivers, or performing a full-scale composting process, provide the closest approximation to real biodegradation conditions. These assays have the advantage of showing the effects of other nonbiotic processes such as photodegradation, chemical hydrolysis and thermal degradation which can take place in parallel with biodegradation processes. On the other hand, in field assays the proper control of temperature, pH and humidity and the monitoring of degradation become more challenging. Simulation tests performed in controlled laboratory reactors filled with seawater, soil, compost or landfill material, allow a good representation of environmental conditions and also provide for better control of conditions and for monitoring the course of biodegradation. Laboratory tests, in which a selected microbial population or specific extracellular enzymes able to depolymerize a targeted polymer are used, ensure controlled and reproducible biodegradation assays. In this case, degradation rates can appear to be higher than those observed under natural biodegradation conditions and should be analyzed with care [58]. Moreover, mineralization rates cannot be determined using enzyme assays [59].

Simulation tests based on soil burial, composting and exposure to water have been performed to determine the biodegradation of starch and starch-based polymer blends [60–65]. In addition, much laboratory-scale research has been devoted to enzyme-mediated degradation of starch and starch-based blends [66–71]. In particular, α-amylase-catalyzed degradation of starch has received much attention due to its industrial value for production of glucose and fructose syrups and other starch hydrolyzates. A summary of starch-converting enzymes and their enzymatic action is shown in Table 2.5. In reference to their pattern of action on starch granules, enzymes generally first adsorb onto the granule surface and then either erode it or digest channels from selected points leading towards the core of the granule [72].

Techniques available for determining the extent of the enzymatic degradation of starch include weight loss measurements and UV/visible spectroscopy, which can be used to quantify the release of reducing sugars. Moreover, different methods have been applied to identify and study the distribution of the degradation products, as well as to assess the properties of degraded starch. Among them, interesting results for the morphology, chemical properties, thermal properties, and crystallinity of degraded granules, as well as for the distribution of hydrolysis products have been obtained through FTIR spectroscopy, scanning electron microscopy, thin layer chromatography, x-ray diffraction, differential scanning calorimetry, tensile mechanical testing and high-performance anion-exchange chromatography with pulsed amperometric detection [70, 73].

Enzymatic degradation of starch is influenced by a number of factors which greatly affect hydrolysis rate and yield. These factors include the botanical source of the starch, the physical form in which starch is subjected to enzyme attack (e.g., granules, gelatinized starch, films) and the concentration of the starch solution/suspension. In addition, the type of enzyme, its source, concentration, whether it needs bound non-protein molecules (co-factors) to express activity, its thermostability, whether it is used alone or in combination with other enzymes (blends), and whether it is used in solution or in immobilized form, play a significant role in starch hydrolysis yield. Parameters such as the pH of the reaction medium and temperature also influence the

Table 2.5 Starch-converting enzymes and their mode of action

Starch-converting enzymes	Enzymatic action	Examples	End products
Endoamylases	Cleave α,1-4 glycosidic bonds in the inner part of amylose and amylopectin chains	α-amylase (EC 3.2.1.1)	Oligosaccharides of varying length and short chain branched amylopectin residues (α-limit dextrins)
Exoamylases	Cleave α,1-4 or α,1-4 + α, 1-6 glycosidic bonds of the external glucose residues of amylose and amylopectin	β-amylase (EC 3.2.1.2) Amyloglucosidase= glucoamylase (EC 3.2.1.3) α-glucosidase (EC 3.2.1.20)	Glucose, maltose and amylopectin residues of β-amylase action (β-limit dextrins)
Debranching enzymes	Cleave α,1-6 glycosidic bonds of amylopectin and pullulan	Isoamylase (EC 3.2.1.68) Pullulanase type I (EC 3.2.1.41) Isomaltase (EC 3.2.1.10)	Linear polysaccharides
Transferases	Cleave α,1-4 glycosidic bonds of the donor molecule and transfer part of the donor to a glycosidic acceptor	Amylomaltase (EC 2.4.1.25) Cyclodextrin glycosyltransferase (EC 2.4.1.19) Branching enzyme (EC 2.4.1.18)	New α,1-4 and α,1-6 glycosidic bonds: linear and cyclic oligosaccharides, highly branched dextrins, α,1-6 glycosidic bonds in side chains of glycogen

Table 2.6 Factors that influence the enzyme digestibility of starch*

Factor	Effect	Reference
Granule size	Negative	[67, 69, 74, 75]
Granule shape	Variable	[74]
Surface porosity and presence of cavities, fissures or indentations	Positive	[76, 77]
Extent of granule damage	Positive	[68]
Amylose content	Negative	[78–80]
Amylopectin unit-chain length distribution	Positive for shorter chains (DP~8–12), negative for longer chains (DP~16–26)	[67]
Strong interaction between starch chains	Negative	[68]
Type and proportion of polymorphic forms	A-type>C-type>B-type	[67, 68, 74, 81, 82]
High quality crystalline structures, fraction of crystalline structures	Negative	[68, 70]
Presence of amylose-lipid complexes	Negative	[66, 83–85]
Phosphorus content	Negative	[69, 86, 87]
Content of proteins and hydrolysis products	Negative	[66, 88]
Pasting properties (peak viscosity and breakdown)	Positive/Negative	[69, 89]
Growing conditions, processing history, storage conditions	Variable	[69, 72, 90]

*A 'positive' effect is given to factors when their value is positively correlated with starch digestibility. 'Negative' effects imply lower starch digestibility at higher values of the parameter.

course of starch degradation. Among these factors, the botanical source of starch plays a prominent role, since it determines granule size, size distribution and shape, as well as the structure and amylose/amylopectin ratio. In turn, these characteristics condition susceptibility to enzyme attack. Table 2.6 summarizes the main factors which influence enzyme digestibility of starch and starch-based polymers. As has been seen, granular structure plays a prominent role in defining the rate and extent of enzymatic hydrolysis.

Starch is rapidly metabolized and is an excellent base material for polymer blends, including those with more environmentally inert polymers, where it is metabolized to leave less residual polymer material on biodegradation [91]. The final cost of blends can also frequently be reduced by the addition of relatively cheap renewable materials based on starch. On the other hand, blending of starch with other polymers affects its biodegradation. This is due to interactions in which the less-degradable phase may inhibit the access of enzymes, moisture, oxygen or microorganisms. As a result, the rate and extent of degradation of a blend may not be predictable from the degradation patterns of its components.

Biodegradation of starch-based blends and composites is often studied through simulation tests carried out by placing polymer samples in reactors filled with seawater or landfill material, by burying samples in soil or by using composting methods. In these assays biodegradation is most often quantified through measurement of weight loss and by respirometric assays which measure the consumption/production of gases during sample mineralization (i.e., O_2, CO_2, CH_4). Changes in mechanical properties of test specimens (e.g., tensile properties due to the interest in the use of biodegradable plastics in packaging applications) are also frequently used as an indicator of biodegradation.

Table 2.7 contains examples of recent biodegradation studies on biodegradable polymer/starch blends. Due to its high hydrophilicity, preferential removal of the starch component is generally observed [50, 61, 62]. Starch addition enhances the biodegradation rate of the blend since starch particles act as weak links in the matrix and are sites for biological attack. The microbial consumption of starch in blends leads to increased porosity, void formation, and loss of matrix integrity [92]. Even polyolefins such as polyethylene and polypropylene have been reported to undergo defragmentation and biodegradation by addition of starch when used in combination with pro-oxidant additives or photo-sensitive components [92]. References included in Table 2.7 also show that blend degradation generally increases with increasing starch content. Concerning the effect of compatibilization techniques on blend degradation, results depend among other factors on the technique used and the quality of starch dispersion in the matrix. Dubois and Narayan (2003) found that compatibilization led to finer and more homogeneous starch dispersion within a PCL matrix [93]. These authors then explained the higher degradability of PCL/PCL-grafted dextran starch and PCL/PCL-grafted starch in terms of a larger starch surface content available for microorganisms. On the other hand, Liao and Wu (2009) observed a lower extent of biodegradation for $PLA_{70}PCL_{30}$-g-AA/starch when compared with $PLA_{70}PCL_{30}$/starch, which was explained in terms of a reduced water absorption in $PLA_{70}PCL_{30}$-g-AA/starch caused by formation of ester carbonyl functional groups in the grafted blend [50]. In comparison with PLA/starch, Wu (2005) observed a lower degradability of PLA-g-AA/starch, which was attributed to increased starch hydrophobicity in the blend [62].

2.7   CONCLUDING REMARKS

As reflected in the number of publications, research on starch and starch-based plastics is widespread and continuing to grow. The characteristics of starch plastics are highly influenced by the amylose/amylopectin ratio, humidity, type and content of plasticizer, processing method and final crystallinity. In this respect, it is very important to monitor processing and storage conditions because these will also influence the properties of starch-based plastics.

Thermoplastic starch is gelatinized starch and is usually based on starch with amylose content greater than 70%. The addition of specific plasticizers produces thermoplastic materials with good performance properties and inherent biodegradability. Starch is typically plasticized, destructured, and/or blended with other materials to obtain products with useful mechanical properties. Importantly,

Table 2.7 Biodegradation of biodegradable polymer/starch blends

Polymer/starch composition	Environment	Measurement technique	Extent of biodegradation		Reference
			Time (days)	Extent (%)	
Cellulose derivatives/ starch/additives (MaterBi-Y) ~ 38:38:22	Indoor soil burial	Weight loss, drop in mechanical properties	400	33	[60, 94]
PHB/soluble starch 7: 93 (25% glycerol)	Indoor soil burial	Weight loss	0.21 (5 h)	30	[61]
PHB/corn starch 7: 93 (25% glycerol)			0.21	15	
PHB/potato starch (25% glycerol)					
100:0			0.21	0	
7:93			0.21	15	
5:95			0.21	16	
3:97			0.21	17	
1:99			0.21	17	
PLA/starch	Indoor soil burial	Weight loss			[62]
100:0			90	5	
60:40			90	42	
40:60			90	58	
PLA-g-AA/starch	Indoor soil burial	Weight loss			[62]
100:0			90	6	
60:40			90	36	
40:60			90	52	
PCL/starch	Compost	Weight loss			[93]
100:0			140	21	

90:10			140	40	
80:20			140	42	
70:30			140	47	
60:40			140	53	
50:50			140	67	
PCL/PCL-grafted dextran starch 60:40	Compost	Weight loss	140	67	[93]
PCL/PCL-grafted starch 60:40	Compost	Weight loss	140	73	[93]
PCL/adipate modified starch/alkyl epoxy stearate plasticizer 50:35:15	Indoor soil burial	Respirometric test	120 / 120	Clay soil: max. 72 / Sandy soil: max. 60	[95]
$PLA_{70}PCL_{30}$/starch 100:0 70:30 50:50	Indoor soil burial	Weight loss	112 112 112	15 62 86	[50]
$PLA_{70}PCL_{30}$-g-AA/starch 100:0 70:30 50:50	Indoor soil burial	Weight loss	112 112 112	19 53 76	[50]
PVA/starch 70:30 (cross-linked with formaldehyde, 10 wt%)	Outdoor soil burial. Ambient conditions	Weight loss	50	> 50	[96]
PVA/starch	Anaerobic digestion in aqueous medium	Biogas production, volatile fatty acid analysis	4.2 (100h)		[65]
100:0				15	

(continued)

38 BIOPOLYMERS

Table 2.7 (*Continued*)

Polymer/starch composition	Environment	Measurement technique	Extent of biodegradation		Reference
			Time (days)	Extent (%)	
50:50			4.2	30	
25:75			4.2	40	
10:90			4.2	60	
PVA/starch/arabic gum and magnesium stearate (trays) 80:16:4	Placement in soil surface	Respirometric test. CO_2 evolution	60	36	[64]
	Soil burial	Respirometric test. CO_2 evolution	60	45	
	Compost	Respirometric test. CO_2 evolution	60	35	

thermoplastic starch can be processed on existing plastic processing equipment. High-starch-content plastics are highly hydrophilic and readily disintegrate on contact with water. The use of starch–polymer blends provides a route to materials with adjustable degradation rates. Chemical modification of starch to form new materials or for the purpose of compatibilization in blends is facilitated by the availability of free hydroxyl groups in starch, which can undergo reactions such as acetylation, esterification and etherification. The fine phase structure is an important parameter in terms of obtaining films with useful properties and is determined by the interface and interphase, the weight ratio of the components and features of the processing method such as shear, residence time and temperature.

Biodegradation of starch and starch-based polymers has been studied by means of laboratory tests using specific extracellular enzymes and also by means of simulation tests, mainly in soil burial and compost environments. Even if simulation tests do not entirely reflect the real biodegradation conditions in a natural environment, these assays have proved suitable for assaying the extent of polymer degradation in different environments whilst under controlled conditions. In the case of starch granules, enzymatic hydrolysis has been widely studied. The rate and extent of enzymatic degradation has proved to be dependent on a number of factors, among which the botanical origin plays the most important role by defining the granule size and shape, the amylose/amylopectin ratio and the starch morphology. In blends, starch generally enhances the degradation rate acting as the initial point of biological attack and starch concentration determines the extent of degradation of the blend. Although many research groups have been interested in studying the degradation of the starch-based polymers they have synthesized, the literature shows a wide diversity of assays and conditions used for such tests (e.g., time of exposure to degrading environment, temperature, humidity, method of measurement), sometimes making it difficult to compare results for the degradation of starch-based blends.

2.8   ACKNOWLEDGEMENT

The authors acknowledge support provided by the National Research Council of Argentina (CONICET).

REFERENCES

[1] S.J. Wang, W.Y. Gao, H.X. Chen, *et al.*, *J. Food Eng.* 2006, 76, 420–426.

[2] J. Singh, L. Kaur, and O.J. McCarthy, *Food Hydrocolloids* 2007, 21, 1–22.

[3] A. Buléon, P. Colonna, V. Planchot, *et al.*, *Int. J. Biol. Macromol.* 1998, 23, 85–112.

[4] A. Buléon, C. Gérard, C. Riekel, *et al.*, *Macromol.* 1998, 31, 6605–6610.

[5] S. Srichuwong, T. Candra Sunarti, T. Mishima, *et al.*, *Carbohydr. Polym.* 2005, 62, 25–34.

[6] Yu, L., Chen, L., Polymeric materials from renewable resources, in L. Yu (ed.), *Biodegradable Polymer Blends and Composites from Renewable Resources*, John Wiley & Sons Inc, Hoboken, NJ, USA, pp 1–15 (2009).

[7] D. Gallant, B. Bouchet, and P. Baldwin, *Carbohydr. Polym.* 1997, 32, 177–191.

[8] V.P. Cyras, M.C. Tolosa Zenklusen, and A. Vázquez, *J. Appl. Polym. Sci.* 2006, **101**, 4313–4319.

[9] Donald, A.M., Waigh, T.A., Jenkins, P.J. *et al.*, Internal structure of starch granules revealed by scattering studies, in P.J. Frazier, A.M. Donald and P. Richmond (eds.), *Starch: structure and function*, The Royal Society of Chemistry, London, United Kingdom, pp 172–179 (1997).

[10] J.P.J. Jenkins, R.E. Cameron, and A.M. Donald, *Starch/Starke* 1993, **45**, 417–420.

[11] D.R. Daniels, A.M. Donald, *Macromol.* 2004, **37**, 1312–1318.

[12] J.J.G. van Soest, J.F.G. Viegenthart, *Trends Biotechnol.* 1997, **15**, 208–213.

[13] H. Kim, K. Huber, *J. Cereal Sci.* 2008, **48**, 159–172.

[14] L. Avérous, *J. Macromol. Sci. Polym. Rev.* 2004, **C44**, 231–274.

[15] J. J. G. van Soest, H. Tournois, D. de Wit, *et al.*, *Carbohydr. Res.* 1995, **279**, 201–214.

[16] T. Baks, I.S. Ngene Ikenna, J.J.G. van Soest, *et al.*, *Carbohydr. Polym.* 2007, **67**, 481–490.

[17] T.D. Karapantsios, E.P. Sakonidou, and S.N. Raphaelides, *J. Food. Sci.* 2000, **65**, 144–150.

[18] A. Barrett, G. Kaletunc, S. Rosenburg, *et al.*, *Carbohydr. Polym.* 1995, **26**, 261–269.

[19] M.L. Fishman, D.R. Coffin, R.P. Konstance, *et al.*, *Carbohydr. Polym.* 2000, **41**, 317–325.

[20] X.F. Ma, J.G. Yu, *Polym. Int.* 2004, **53**, 1780–1785.

[21] L. Kaur, J. Singh, and N. Singh, *Food Hydrocolloids* 2005, **19**, 839–849.

[22] P.A. Perry, A.M. Donald, *Carbohydr. Polym.* 2002, **49**, 155–165.

[23] L. Famá, S. Flores, L. Gerschenson, *et al.*, *Carbohydr. Polym.* 2006, **66**, 8–15.

[24] A.L. Da Róz, A.J.F. Carvalho, A. Gandini, *et al.*, *Carbohydr. Polym.* 2006, **63**, 417–424.

[25] T. Hughes, R. Hoover, Q. Liu, *et al.*, *Food Res. Int.* 2009, **42**, 627–635.

[26] C. Perera, R. Hoover, *Food Chem.* 1999, **64**, 361–375.

[27] A. Karim, M.H. Norziah, and C.C. Seow, *Food Chem.*, 2000, **71**, 9–36.

[28] M.J. Miles, V.J. Morris, P.D. Orford, *et al.*, *Carbohydr. Res.* 1985, **135**, 271–281.

[29] D.K. Fisher, D.B. Thompson, *Cereal Chem.* 1997, **74**, 344–351.

[30] J. Longton, G.A. Legrys, *Starch/Stärke*, 1981, **33**, 410–414.

[31] M. Gudmundsson, A.-C. Eliasson, *Carbohydr. Polym.* 1990, **13**, 295–315.

[32] H. Bizot, P. Le Bail, B. Leroux, *et al.*, *Carbohydr. Polym.* 1997, **32**, 33–50.

[33] M.T. Kalichevsky, P.D. Orford, and S.G. Ring, *Carbohydr. Polym.* 1986, **6**, 145–154.

[34] M. Alloncle, J. Lefebvre, G. Llamas, *et al.*, *Cereal Chem.* 1989, **60**, 90–93.

[35] B. Conde-Petit, A. Pfirter, and F. Escher, *Food Hydrocolloids* 1997, **11**, 393–399.

[36] C-Y. Lii, M-F. Lai, and K-F. Liu, *J. Cereal Sci.* 1998, **28**, 175–185.

[37] K. Kohyama, J. Matsuki, T. Yasui, *et al.*, *Carbohydr. Polym.* 2004, **58**, 71–77.

[38] T. Temsiripong, R. Pongsawatmanit, S. Ikeda, *et al.*, *Food Hydrocolloids* 2005, **19**, 1054–1063.

[39] A. Kukukova, J. Aubin, and S.M. Kresta, *Chem. Eng. Res. Des.* 2009, **87**, 633–647.

[40] Curry, J.E., Practical aspects of processing of blends, in M.J. Folkes, P.S. Hope (eds.), *Polymer Blends and Alloys*, Chapman and Hall, London, United Kingdom, pp 46–73 (1993).

[41] Arridge, R.G.C., Theoretical aspects of polymer blends and alloys, in M.J. Folkes, P.S. Hope (eds.), *Polymer Blends and Alloys*, Chapman and Hall, London, United Kingdom, pp 126–162 (1993).

[42] Czigány, T., Romhány, G., and Kovács, J.G., Starch for injection molding purposes, in S. Fakirov, D. Bhattacharyya (eds.), *Handbook of engineering biopolymers: homopolymers, blends and composites*, Hanser Gardner Publications, Munich, Germany, pp 81–108 (2007).

[43] E. Landreau, L. Tighzert, C. Bliard, et al., *Eur. Polym. J.* 2009, 45, 2609–2618.

[44] L. Avérous, C. Frigant, and L. Moro, *Polymer* 2001, 42, 6565–6572.

[45] L. Avérous, L. Moro, P. Dole, et al., *Polymer* 2000, 41, 4157–4167.

[46] L. Avérous, C. Frigant, *Polym. Eng. Sci.* 2001, 41, 727–734.

[47] U. Funke, W. Bergthaller, and M.G. Lindhauer, *Polym. Degrad. Stab.* 1998, 59, 293–296.

[48] R. Shi, J. Bi, Z. Zhang, et al., *Carbohydr. Polym.* 2008, 74, 763–770.

[49] L.T. Sin, W.A.W.A. Rahman, A.R. Rahmat, et al., *Carbohydr. Polym.* 2010, 79, 224–226.

[50] H.T. Liao, C.S. Wu, *Mater. Sci. Eng. A* 2009, 515, 207–214.

[51] J. Ren, H. Fu, T. Ren, et al., *Carbohydr. Polym.* 2009, 77, 576–582.

[52] S. Godbole, S. Gote, M. Latkar, et al., *Biores. Technol.* 2003, 86, 33–37.

[53] K.C. Reis, J. Pereira, A.C. Smith, et al., *J. Food Eng.* 2008, 89, 361–369.

[54] D. Demirgöz, C. Elvira, J.F. Mano, et al., *Polym. Degrad. Stab.* 2000, 70, 161–170.

[55] M.F. Koenig, S.J. Huang, *Polymer* 1995, 36, 1877–1882.

[56] V. Chiono, G. Vozzi, M. D'Acunto, et al., *Mater. Sci. Eng. C* 2009, 29, 2174–2187.

[57] P.D.S.C. Mariani, K. Allganer, F.B. Oliveira, et al., *Polym. Test.* 2009, 28, 824–829.

[58] Müller, R.-J., Biodegradability of polymers: Regulations and methods for testing, in A. Steinbüchel (ed.), *General aspects and special applications*, Biopolymers 10, Wiley-VCH, Weinheim, Germany, pp 365–392 (2003).

[59] Innocenti, F.D., Biodegradation behaviour of polymers in the soil, in C. Bastioli (ed.), *Handbook of Biodegradable Polymers*, Rapra Technology Limited, Shrewsbury, United Kingdom, pp 57–102 (2005).

[60] V.A. Alvarez, R.A. Ruscekaite, and A. Vázquez, *Polym. Degrad. Stab.* 2006, 91, 3156–3162.

[61] S.-M. Lai, T.-M. Don, and Y.-C. Huang, *J. Appl. Polym. Sci.* 2006, 100, 2371–2379.

[62] C.-S. Wu, *Macromol. Biosci.* 2005, 5, 352–361.

[63] M.O. Rutiaga, L.J. Galan, L.H. Morales, et al., *J. Polym. Environ.*, 2005, 13, 185–191.

[64] E. Chiellini, P. Cinelli, V.I. Ilieva, et al., *J. Cell. Plast.*, 2009, 45, 17–32.

[65] M.A.L. Russo, C. O'Sullivan, B. Rounsefell, et al., *Biores. Technol.* 2009, 100, 1705–1710.

[66] Z. Konsula, M. Liakopoulou-Kyriakides, *Process Biochem.* 2004, 39, 1745–1749.

[67] S. Srichuwong, T. Candra Sunarti, T. Mishima, et al., *Carbohydr. Polym.* 2005, 60, 529–538.

[68] Y. Zhou, R. Hoover, and Q. Liu, *Carbohydr. Polym.* 2004, 57, 299–317.

[69] T. Noda, S. Takigawa, C. Matsuura-Endo, et al., *Food Chem.* 2008, 110, 465–470.

[70] S.H. Imam, S.H. Gordon, A. Mohamed, et al., *Polym. Degrad. Stab.* 2006, 91, 2894–2900.

[71] M.A.L. Russo, R. Truss, and P.J. Halley, *Carbohydr. Polym.* 2009, 77, 442–448.

[72] W.J. Wang, A.D. Powell, and C.G. Oates, *Carbohydr. Polym.* 1995, 26, 91–97.

[73] H.S. Azevedo, F.M. Gama, and R.L. Reis, *Biomacromol.* 2003, 4, 1703–1712.

[74] J.-C. Valetudie, P. Colonna, B. Bouchet, et al., *Starch/Stärke*, 1993, 45, 270–276.

[75] N. Goyal, J.K. Gupta and S.K. Soni, *Enzyme Microb. Technol.*, 2005, 37, 723–734.

[76] M. Sujka, J. Jamroz, *J. Int. Agrophys.* 2007, **21**, 107–113.
[77] J.E. Fannon, J.M. Shull, and J.N. BeMiller, *Cereal Chem.* 1993, **70**, 611–613.
[78] H. Fuwa, M. Nakajima, and A. Hamada, *Cereal Chem.* 1977, **54**, 230–237.
[79] M.E. Karlsson, A.M. Leeman, I.M.E. Björck, *et al., Food Chem.,* 2008, **100**, 136–146.
[80] T. Noda, T. Kimura, M. Otani, *et al., Carbohydr. Polym.* 2002, **49**, 253–260.
[81] J.-L. Jane, K.-S. Wong, and A.E. McPherson, *Carbohydr. Res.* 1997, **300**, 219–227.
[82] C. Gerard, P. Colonna, A. Buléon, *et al., J. Sci. Food Agric.* 2001, **81**, 1281–1287.
[83] R. Cui, C.G. Oates, *Food Chem.* 1999, **65**, 417–425.
[84] E. Nebesny, J. Rosicka, and M. Tkaczyk, *Starch/Stärke,* 2002, **54**, 603–608.
[85] F. Tufvesson, V. Skrabanja, I. Björck, *et al., LWT–Food Sci. Technol.* 2001, **34**, 131–139.
[86] T. Noda, S. Tsuda, M. Mori, *et al., Food Chem.* 2006, **95**, 632–637.
[87] J.-I. Abe, Y. Takeda, and S. Hizukuri, *Biochim. Biophys. Acta* 1982, **703**, 26–33.
[88] B. Ozbek, S. Yuceer, *Process Biochem.* 2001, **37**, 87–95.
[89] P. Hu, H. Zao, Z. Duan, *et al., J. Cereal Sci.* 2004, **40**, 231–237.
[90] T. Heitmann, E. Wenzig, and A. Mersmann, *Enzyme Microb. Technol.,* 1997, **20**, 259–267.
[91] R. Jayasekara, I. Harding, I. Bowater, *et al., J. Polym. Environ.* 2005, **13**, 231–251.
[92] Bastioli, C., Starch-Based Technology, in C. Bastioli (ed.), *Handbook of Biodegradable Polymers,* Rapra Technology Limited, Shrewsbury, United Kingdom, pp 257–286 (2005).
[93] P. Dubois, R. Narayan, *Macromol. Symp.* 2003, **198**, 233–243.
[94] A. Vázquez, V.A. Alvarez, Starch-cellulose fibers composites, in L. Yu (ed.), *Biodegradable Polymer Blends and Composites from Renewable Resources,* John Wiley & Sons Inc, Hoboken, NJ, USA, pp 241–286 (2009).
[95] M.E.F. César, P.D.S.C. Mariani, L.H. Innocentini-Mei, *et al., Polym. Test.,* 2009, **28**, 680–687.
[96] X. Han, S. Chen, and X. Hub, *Desalination* 2009, **240**, 21–26.

3

Production, Chemistry and Properties of Polylactides

Anders Södergård and Saara Inkinen

Laboratory of Polymer Technology, Åbo Akademi University, Turku, Finland

3.1 INTRODUCTION

Polylactide is a versatile material with a well-established position in coatings, films, fibers and medical applications. The starting chemical for all types of polylactide (PLA) is lactic acid (LA), and in most cases this is lactic acid of natural origin. From a chemistry viewpoint, lactic acid is a highly interesting monomer because it is chiral and has two stereoisomers, $L(+)$-LA and $D(-)$-LA (Figure 3.1). In addition, the LA molecule possesses both carboxylic and hydroxyl functional groups, which are able to undergo inter- and intramolecular esterification reactions in polymerization processes.

Features contributing to the extensive use of LA in a variety of applications and products are its preserving effect, ability to form mineral salts and its occurrence in mammalian metabolic systems [1]. The history of LA goes back to its discovery in the late 18th century [2]. Later on it was noticed that lactic acid polymerizes at elevated temperatures with formation of a six-membered ring lactide if the heating persists. The polymer resulting from the self-condensation reaction (step-growth polymerization) is generally named poly(lactic acid). The ring-formed lactide can undergo ring-opening polymerization (ROP) under certain conditions, whereby a high-molecular weight

Biopolymers – New Materials for Sustainable Films and Coatings, First Edition.
Edited by David Plackett.

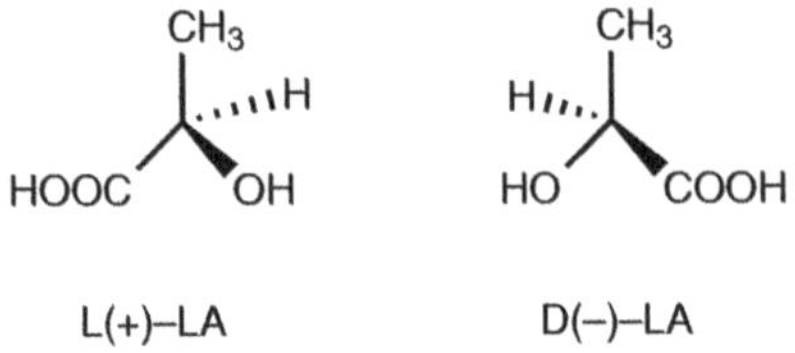

Figure 3.1 Stereoisomers of lactic acid

polymer, generally termed polylactide, is formed [3]. Polylactide can be of L-(poly (L-lactide)) or D-forms (poly(D-lactide)), or a combination thereof (poly(D,L-lactide)). Both types of LA-based polymers are often referred to as PLA, independent of their manufacturing route. The equilibrium of both the ROP and polycondensation (PC) reactions can be shifted in desired directions by manipulating reaction conditions such as temperature, as well as the type and amount of catalytic species present [4].

The manufacturing of PLA is rather complex and involves several different and competing reactions including chain growth by condensation or ring-opening, ring-formation resulting from intramolecular transesterification, as well as different degradation reactions and racemization. The molecular weight of the polycondensed reaction product can range from a few hundred Da in an uncatalyzed polycondensation to several hundreds of thousands of Da by polycondensation in solvents [5] or by melt/solid state polycondensation [6]. In addition to the dehydrated dimer of LA (i.e., lactide), polycondensation products have over the last decade been introduced as potential commercial products for use in adhesives [7], as precursors [8] and as modifiers [9]. Lactide is an important reaction product since it is the starting monomer for PLA manufactured by ring-opening polymerization. This is the most frequently used method for preparing a high-molecular weight lactic acid polymer due to the readily controllable manufacturing process [10]. Ring-opening polymerization has been extensively described in the literature over the years, and several companies have been developing industrial manufacturing processes. In comparison, polycondensation of lactic acid has drawn much less industrial attention [11]. The manufacturing and use of polylactide are closely connected to sustainability issues such as carbon dioxide neutrality, life-cycle assessment and waste management. Quite recently, the end-of-life options of bio-based polymers have been brought into the discussions about sustainability. This can be seen in a number of suggested approaches on how to deal with PLA waste materials from the manufacturing process or recycling options for end-products after use. The conversion of PLA into lower-molecular weight polymers has been described, as well as the option of completely hydrolyzing the polymer into lactic acid as a source of new starting chemicals for biosolvents or as building blocks for lactic acid-based polymers [12, 13].

3.2 PRODUCTION OF POLYLACTIDES

3.2.1 Lactic Acid and its Production

Lactic acid (2-hydroxypropanoic acid, $CH_3CHOHCOOH$) is a naturally occurring α-hydroxy acid. Outside the polymer field, LA is used in a variety of applications in the food, pharmaceuticals, textile, leather, and chemical industries [14, 15]. As a chiral molecule, LA exists in two stereoisomeric forms: L-lactic acid (L-LA) and D-lactic acid (D-LA). D-lactic acid is not common in nature but the L-form is found in almost all types of living organism. LA can be produced chemically by, for example, hydrolysis of lactonitrile; however, its present industrial production is based on microbial carbohydrate fermentation due to the technical and economical limitations of chemical synthesis routes [14]. Chemically produced LA is always a racemic mixture of D-LA and L-LA but, depending on the microorganism used, optically pure and racemic LA can be produced by means of fermentation. The optical purity of LA is essential in PLA production, since small amounts of enantiomeric impurities can cause drastic changes in properties such as crystallinity and rate of biodegradation.

In microbial LA production, the fermentation stage is typically followed by cell mass and protein removal, after which the LA is recovered and purified [16]. Polymerization grade LA is generally supplied as an 88–90% aqueous solution of L-LA, D-LA, or racemic LA. Commercial LA also contains small amounts of several impurities including arsenic (<1 ppm), iron (<5 ppm), heavy metals (<5 ppm), chloride (<10 ppm), sulphates (<10 ppm), sulphated ash (maximum 0.05%), reducing sugars, methanol and methyl ester [6]. In addition, carbohydrates and amino acids remaining in the LA solution after purification cause colouring of the polymer, while racemization caused by cations (e.g., Na^+) can cause significant deterioration in polymer properties when compared to those of an optically pure product [16]. Lactic acid fermentation can be both homofermentative and heterofermentative, but homofermentative species such as *lactobacillus* are typically used. A significant advantage of the microbial production route over chemical synthesis is the possibility to use cheap, abundant, and annually renewable raw materials such as whey, molasses, starch waste, beet, or cane sugar [17, 18]. Refined carbohydrates result in the purest product and they are therefore currently used in the industrial production of LA. However, pure sugars are economically unfavourable due to their cost and the need for additional nutrient supplements. Since raw material expenses constitute a major part of the overall cost of LA production, alternative starch-based or cellulosic feed sources obtained from agricultural wastes or side-products are continuously being explored [19]. The yield of LA, as well as the choice of microorganism, depend on the raw material used, because the ability and efficiency of the microorganisms to utilize different nutrition sources varies greatly [20]. Despite the notable progress made in recent decades, there is still need for improvements in fermentative LA production. However, recent developments related to more efficient fermentation and purification techniques [14], genetic modifications of lactic acid bacteria (LAB) [21], and, for example, amylolytic bacterial fermentation methods [22] seem to offer promise in terms of more economic and efficient LA production processes.

3.2.2   Production Methods for Polylactide

Polylactide can be manufactured either in a polycondensation-based process or by first converting the polycondensate into a lactide which is then polymerized in a separate ROP process. The polycondensation-based process can be further categorized into bulk polycondensation or solvent-assisted polycondensation. If there is a need for increasing the molecular weight, solid-state polycondensation and various chain-extension technologies can be further applied in a subsequent process step.

The ring-opening polymerization route can be divided into different process types depending on the conditions applied. In bulk polymerization, the temperature is kept below the melting point of PLA (typically below 160 °C), whereas in melt polymerization the temperature is kept above the PLA melting point. Solution polymerization is a third method, in which the lactide is dissolved in an organic solvent together with initiator(s) and catalyst(s) and polymerization takes place in the solvent. If the polymer is not soluble in the solvent it will precipitate and can thus be readily separated. However, in case the PLA is soluble in the polymerization medium it needs to be precipitated in a non-solvent in a separate process step. The fourth available technique is dispersion polymerizsation. In this case the lactide is dissolved in a polymerization medium (organic solvent), which is then dispersed in another non-miscible solvent under agitation. The polymerization takes place in the lactide solvent phase and can be thought of as multiple mini-bulk polymerizations [23].

3.2.2.1   Bulk-Polycondensation of Lactic Acid

Lactic acid undergoes self-condensation reactions in which water is released and the reaction equilibrium proceeds towards poly(lactic acid) as the product (Figure 3.2). Water removal becomes more difficult with increasing poly(lactic acid) molecular weight due to the increased viscosity of the reaction mixture. Transesterification and side

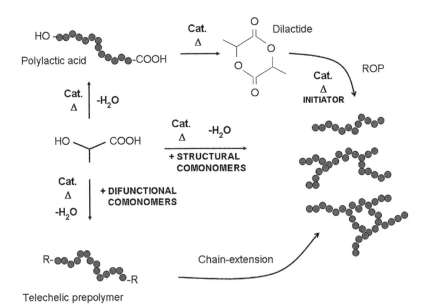

Figure 3.2   Production routes of PLA and ways to affect its structure

reactions occurring during polycondensation may result in colouring of the polymer and formation of ring structures of different sizes [24]. The formation of ring structures eventually results in production of lactide, which is the most stable ring form. The formation of lactide cannot be avoided, but it can be suppressed under optimized reaction conditions, for instance by lowering the reaction temperature or by using a partial condenser (reflux condenser) for recycling lactide to the reaction mixture. Addition of stabilizers, such as anti-oxidants or phosphorus compounds can reduce discolouration, which generally increases as a function of polycondensation time [25]. The general rule for optimizing polycondensation is that the rate of water removal should be as high as possible without allowing the reaction mixture to undergo transesterification reactions or chain-terminating reactions (by minimizing the amount of monofunctional alcohols or carboxylic acids). Traditional polycondensation catalysts are strong acids (e.g., p-toluenesulphonic acid, H_2SO_4), metallic or organometallic compounds. Catalysts based on Ca, Na, La, Ge, Sb, Zn, Fe, Al, Ti, and Sn have been reported as useful polycondensation catalysts [26–28]. A frequent disadvantage in producing poly(lactic acid) by direct polycondensation is a limited molecular weight in combination with a low yield. One way of increasing the molecular weight of PLA is by applying sequential melt/solid-state polycondensation [29, 30]. In solid-state polycondensation (SSPC), the melt-polycondensed PLA is cooled below its melting temperature and the solid PLA is subjected to crystallization, whereby both a crystalline phase and an amorphous phase can be identified. It is believed that the reactive end groups, as well as the catalyst, are concentrated in the amorphous phase, in between the crystals, and that polycondensation continues, even though the reaction is performed in the solid state at a low temperature [31]. An increase in the molecular weight of polycondensation PLA can also be achieved by the use of comonomers with functionality higher than two [32]. The resulting poly(lactic acid) has a higher molecular weight than a polymer prepared without the use of a comonomer; however, if the comonomer amount is large enough, the polymer eventually becomes hydroxyl- or carboxyl-terminated, which limits the molecular weight that can be achieved. On the other hand, if the multifunctional compound is used in extremely small amounts, the polycondensation reaction mixture becomes a blend of star-shaped PLA and linear PLA.

3.2.2.2   Solvent-Assisted Polycondensation

In solvent-assisted polycondensation, dehydration follows the same principal stages as direct melt condensation but the viscosity increase is eliminated because polycondensation is performed in a solvent (e.g., toluene, xylene). The removal of the reaction water from the medium is hence easier and a poly(lactic acid) of higher molecular weight can be obtained. The racemic purity is also easier to retain because of the lower reaction temperature. The solvent has to be recycled and dried (e.g., using molecular sieve) or fresh, dry organic solvent has to be added during the reaction. The prepared polymer has to be collected from the solvent, typically by using a non-solvent for the polymer, and dried [5, 33], These steps are time- and resource-demanding and environmentally unsound, which make this polycondensation method less attractive than melt polycondensation.

3.2.2.3   Chain Extension of Lactic Acid-Based Polymers

The molecular weight is a crucial parameter as far the technical feasibility of PLA production is concerned. A certain molecular weight is required in order to achieve sufficient mechanical and thermal properties. Chain extension is one way of reaching a higher molecular weight and the use of diisocyanates is the most commonly applied route for aliphatic polyesters. A large number of different diisocyanates have been used as linking molecules in the preparation of aliphatic poly(ester-urethane)s, but the most frequently used is 1,6-hexamethylene diisocyanate [34–36]. The LA-based polymer is generally prepared by polycondensation in the presence of a diol in order to make the prepolymer dominantly hydroxyl-terminated. The use of equimolar amounts of diisocyanate and prepolymer results in a rapid chain-extension reaction within a few minutes [37], after which the molecular weight starts to decrease due to thermal degradation. Side reactions occur more frequently if the amount of diisocyanate is increased, which can be seen from an increase in the molecular weight distribution. Important parameters for the chain extension with diisocyanate are the acid number of the prepolymer [26, 38], reaction conditions, and the catalyst used. The choice of catalyst will affect not only the reaction rate, but also the racemization, which in turn will affect the crystallinity and thermo-mechanical properties of the polymer.

Carboxylic acid end groups can also be utilized for chain-extension reactions. In the 1960s, bis-2-oxazoline was described as being useful for chain extension of aromatic polyesters [39, 40]. Bis-2-oxazolines have more recently also been applied in the linking of telechelic LA-based prepolymers (e.g., by using 2,2′-bis(2-oxazoline)) [41]. 2-Oxazolines are inert towards aliphatic alcohols [42] and therefore react selectively with carboxyl end groups of an LA-based prepolyester forming a poly(ester-amide). The molecular weight of the poly(ester-amide) is highly dependent on the polymerization temperature and the molar ratio of oxazoline and carboxylic acid end groups [30]. High-molecular weight polymers can be produced, but only within a narrow range of reactant ratio and temperatures. Thermal degradation takes place at reaction temperatures above 200 °C, but the linking reaction rate is insufficient at lower temperatures [30]. The selective reactivity of oxazolines also provides the possibility to perform dual linking processes of polycondensated lactic acid with both diisocyanates and oxazolines [37]. Bis-epoxies have also been reported as useful for chain extension of polylactide. The epoxy groups can react predominantly with carboxylic acid groups, whereby the ring opening of the epoxy group yields a secondary hydroxyl, which generally does not react further with remaining epoxy groups [43].

3.2.2.4   Lactic Acid-Based Polymers by Ring-Opening Polymerization

The ring-opening polymerization process involves three separate steps: lactic acid polycondensation, manufacturing of lactide, and the ring-opening step. The different process steps deal with a number of issues, which are partly related to the chemical properties of lactic acid and partly to the required process conditions. The three most crucial parameters, which need to be controlled are: (1) racemic purity; (2) lactide purity; and (3) residual monomer content. The optical purity of PLA is determined by both the racemic purity of LA and the racemization occurring during the three process steps. An increased amount of the antipodal form of the repeating units results in a decreased ability to crystallize, which in turn affects properties such as hydrolytic stability, barrier properties and heat distortion of the end-products [44, 45]. The presence of impurities in the lactide, such as carboxylic acids or oligomers, will affect the process by retarding polymerization (acids) or decreasing the molecular weight through the presence of excess initiating species (hydroxyls) [46]. The conversion of lactide is thermodynamically predetermined and a given amount of lactide will be present in the polymer. This will result in undesired properties for the end-products and the residual monomer needs to be removed in a post-polymerization process step [47].

Lactide is manufactured by depolymerization of poly(lactic acid), which preferably has a molecular weight ranging from 400 to 2500 Da [48]. In addition to lactide, other compounds are also formed (e.g., lactic acid, water, lactoyllactic acid, lactoyllactoyllactic acid and higher oligomers) and it is therefore necessary to purify the crude lactide in order to make it useful for polymerization purposes. Lactide can be manufactured batch-wise or in a continuous process and the choice of catalyst affects both throughput rate and the quality of the lactide [49, 50]. A typical industrial manufacturing process is conducted by heating poly(lactic acid) to 130–230 °C at reduced pressure in the presence of 0.05–1.0 wt % of tin dust [49]. The crude lactide generally contains different impurities, which means that the monomer mixture is unsuitable for direct ring-opening polymerization without purification. The racemic purity, acid number and yield of the lactide will accordingly determine the efficiency of the manufacturing process. Three optional purification approaches have been practiced: crystallization from a solvent [51], melt-crystallization [52], and purification in the gas phase [53].

The ring-opening polymerization of lactide in the melt is generally the most commonly applied method for preparing high-molecular weight polylactide because of the better possibility to control and tailor the properties of the PLA [1]. The reactor system can be of different designs ranging from a simple reaction vessel equipped with an agitator to combinations of reactors, static mixers and extruders [54, 55]. The catalyst is crucially important in the ROP of lactide. Tin and aluminum catalysts have been frequently applied [56] and stannous 2-ethylhexanoate (tin octanoate) is the most intensively studied and used. It is suggested that PLA polymerization mechanisms involve a pre-initiation step, in which the catalyst is converted to an alkoxide in a reaction with a hydroxyl compound (e.g., water or alcohol). The polymerization proceeds on the tin–oxygen bond of the alkoxide ligand, and the carboxylate itself is inactive in the process [57]. The highly active catalysts based on tin compounds for example are toxic [58] and efficient catalysts showing less toxicity based on Ca, Fe, Mg, Zn and other metals as well as metal-free catalysts have therefore been developed and studied for the polymerization of lactide [10, 59, 60]. However, many of these alternative catalysts tend to cause racemization of PLA, especially when polymerizing at high temperatures, and cannot replace tin octanoate in industrial processes.

3.3   POLYLACTIDE CHEMISTRY

3.3.1   Tacticity

Since the LA molecule is chiral and its polymerization involves acyl cleavage with preservation of chirality, PLA with different ratios of D- and L-LA stereoisomers can readily be prepared. When optical purity exceeds 72%, PLA can crystallize under suitable conditions [61] and 93% pure PLA is a semi-crystalline polymer, while polymers having a higher content of the opposite stereoisomer are amorphous. The need to control the tacticity of PLA is one reason for purification of the lactide monomer, while the possibility to use different enantiomeric mixtures of LA stereoisomers allows tuning of physical properties such as crystallinity and thermal transitions as well as the preparation of PLA stereocomplexes. As racemization occurs during polymerization, industrially produced PLLA typically contains 3.6–5.0% of the D-isomer, depending on its grade [62].

The polymerization reactivities of LL- and DD-lactide are the same, but the hydrolysis rate of *meso*-lactide is higher than that of the enantiomerically pure forms due to its differing conformational and configurational structure [63]. Since lactide is a dimer, *meso*-lactide imparts more regularity into the polymer structure in the form of diads than when polycondensing D-LA and L-LA in different feed ratios. Poly(L-lactide) and poly(D-lactide) are isotactic polymers while poly(*meso*-lactide) is syndiotactic.

3.3.2 Molecular Weight and its Distribution

Poly(lactic acid) produced by step-growth polymerization in the melt typically has a low molecular weight due to the high viscosity of the reaction mixture, the presence of impurities, and the decreasing amount of reactive end groups as the polymer synthesis proceeds [62]. Also, the thermodynamic equilibrium related to the hydration and dehydration balance (i.e., polymer formation by the elimination of water) and the equilibrium related to the formation of cyclic low-molecular weight compounds (involving depolymerization by back-biting) prohibit the formation of high-molecular weight polymers. In step-growth polymerization both the monomers and the oligomers and polymer chain ends can interact with each other in both intra- and intermolecular reactions, resulting in a broad molecular weight distribution. The growth of the molecular weight is thus only accelerated at a later stage of the polycondensation, and the polydispersity is dependent on statistics and the degree of conversion.

In contrast to polycondensation of lactic acid (M_w/M_n typically ~ 2), ROP yields a narrow molecular weight distribution (M_w/M_n ranging from 1.1 to 1.4) and a linear relationship between monomer conversion and molecular weight [62, 64]. The addition of esterification-promoting agents (e.g., bis(trichloromethyl) carbonate, dicyclohexylcarbodiimide, and carbonyl diimidazole) is reported to be effective in achieving higher PLA molecular weights [62]. The use of branched initiators or melt modification can also affect the rheological properties of polylactide.

3.3.3 Conversion and Yield

LA polycondensations are typically measured in terms of gravimetric yield. Living ROP is characterized by first-order kinetics in both monomer and catalyst concentration [65]. Catalysts enabling rapid, living ring-opening polymerization of lactide with narrow molecular weight distributions were already developed in the

1990s [62]. The monomer concentration and conversion can be calculated from the equations developed by Witzke as a function of reaction time [66]. Through the use of efficient initiators such as rare earth compounds based on lanthanum and yttrium initiators in the ROP of lactide, it is possible to achieve full conversion and low polydispersities in 15 minutes at room temperature [62]. However, in typical polymerization conditions (180–210 °C and 2–5 h) using 100–1000 ppm tin octanoate, 95% conversions are reached [65]. The conversion of the ROP process depends on various factors including processing conditions (e.g., polymerization time and temperature) as well as the types and amounts of reagents and catalyst used. A higher amount of catalyst typically increases the ROP reaction rate [16]. The catalyst concentration and the presence of carboxylic acid impurities are reported not to decrease the molecular weight of PLA produced by ROP. This is in contrast to hydroxylic compounds [16], the amount of which is inversely proportional to the PLA molecular weight that can be achieved. Deviations from the linear dependence between monomer concentration and molecular weight are typically attributed to slow initiation of the reaction and side reactions resulting in chain transfer and termination [64].

3.3.4 Copolymerization

In ROP of lactide, telechelic PLA can be prepared using bi- or multifunctional monomers as initiators. The most common comonomers used in polylactide are glycolide and ε-caprolactone, but δ-valerolactone, 1,5-dioxepan-2-one and trimethylene carbonate have also been used [67].

Poly(lactic acid) having different structures and types of end group can be obtained by step-growth copolymerization using monomers that can react with the hydroxyl or the carboxyl moiety in the LA molecule. Comonomers having a degree of functionality of more than two impart branching and cross-linking into the polymer structure and their amount needs to be adjusted in order to have a suitable cross-linking density. The molecular architecture of PLA affects, for example, its melt rheology and consequently the processing and properties of the polymer. However, it is important to remember that the chain termination of PLA produced by polycondensation is not absolute, as lactide and significant amounts of free chains not attached to the comonomers typically exist in the reaction mixture.

Predominantly hydroxyl-terminated PLA can be obtained by the incorporation of diols or multifunctional hydroxyl compounds such as 2-butene1,4-diol, glycerol, or 1,4-butanediol into the polymer structure [62]. Carboxyl-terminated PLA can, on the other hand, be obtained using diacids, anhydrides, or multifunctional carboxylic acids such as maleic, succinic, adipic, or itaconic acids. Hydroxyl- or carboxyl-functional polyesters can also be obtained by the implementation of post-polycondensation techniques, for example by linking monofunctional epoxides to the carboxylic chain end [68] or by converting the hydroxylic chain ends to carboxyl groups through reaction with anhydrides. The structure and the related properties of the polymer can further be conveniently tailored by the type, amount, and functionality of the comonomer (Figure 3.3). Due to environmental and sustainability issues, using biodegradable and biocompatible comonomers in combination with lactic acid would be ideal. Examples of potentially renewable building blocks used in PLA (block-)copolymers

	Tensile strength	ε	Impact strength	O_2 barrier	HDT	Hydrolytic stability
Polymer typology						
High molar mass	+	+/-	+	+	+	+
High crystallinity	+	-	-	+	+	+
Polymerization related issues						
High conversion	+	-	-	+	+	+
Block structure	+	-	+	+	+	+
Random order	-	+	-	-	-	-
Impact of structure						
Star initiated	+	+	+	+/-	+/-	+/-
Random long-branch	+/-	+	+	+/-	-	-
Crosslinked	+	-	+	+	+	+

Figure 3.3   Effect of polymer structure and polymerization-related issues on some of the most important end-use related properties of the polymer

include 1,4-butanediol, ethylene glycol, 1,2,3,4-butanetetracarboxylic acid and iso-sorbide [69–71].

3.3.5   Characterization of Lactic Acid Derivatives and Polymers

3.3.5.1   Analysis of Lactic Acid and Lactide

LA is not present solely as a monomer in its aqueous solutions as LA dimers, trimers and oligomers typically coexist. The water content of a lactic acid solution or a reaction mixture containing it can be readily determined using a Karl Fischer titrator and the degree of polymerization (DP) of the LA components of the solutions can be determined using, for example, combined gas chromatography–mass spectroscopy (GCMS) [72]. The acid number [73] and the hydroxyl number [74] determined by titration can also be utilized in estimation of the DP of the LA components in a LA solution and utilized in the design of a copolymer composition (i.e., molar ratio of the monomers) in step-growth polymerization processes. The enantiomeric purity of LA can be estimated using carbon nuclear magnetic spectroscopy (^{13}C-NMR). The D- and L-enantiomers of LA can also be separated and their relative amounts estimated by chiral high-performance liquid chromatography (HPLC), using ultra-violet (UV) detection [75].

In step-growth polymerization processes, lactic acid will typically be heated to 150–190 °C. Impurities such as proteins will cause discolouring of the reaction mixture during heating. The suitability of LA solutions for polymerization can be evaluated by a simple heating test. Commercial LA solutions are typically quite pure, but evaluation of heat stability can be necessary when testing experimental grade LA produced using novel nutrients or downstream purification techniques for polymerization purposes. Standardized American Public Health Association (APHA) colour solutions can be utilized in the visual evaluation of the colour of lactic acid.

The optical purity of the lactide monomer is important for the production of enantiomerically pure high-molecular weight PLA by ROP. The fact that D,L-lactide (116–119 °C) has a higher melting point than L- and D-lactide (94–96 °C) or *meso*-lactide (43–47 °C) can be used for quality control purposes in the ROP route. Enantiomeric impurities can be formed both in the step-growth polymerization of low-molecular weight PLA and in the depolymerization process resulting in lactide formation. Lactide is in fact more susceptible to racemization than PLA, and the conversion and the extent of racemization of lactide further depend on the reaction conditions, the catalyst used, and the molecular weight of the PLA used as the raw material [76]. Other ways to determine the optical purity of lactide include studying the methine region of the ring-formed dimer in single-frequency decoupled proton nuclear magnetic resonance (^{1}H-NMR) spectra or by GC using pure substances as references [77, 78].

3.3.5.2   Analysis of Poly(Lactic Acid)

Acid number titrations can be directly used on linear PLA homopolymers or on predominantly carboxyl-terminated copolymers to yield an estimate of the molecular weight. When using titrimetric methods for molecular weight determination on other types of PLA, both the acid- and hydroxyl number should be determined in order to obtain reliable results. The thermal transitions of PLA are typically determined by differential scanning calorimetry (DSC) or dynamic mechanical analysis (DMA). DSC can also be used to obtain information about the crystallinity and crystallization rate of the polymer.

Gel permeation chromatography (GPC) is the most common technique used to measure the molecular weight of PLA. However, the molecular weights obtained using this method cannot be considered absolute due to several factors, including issues related to solvent miscibility or the use of standards consisting of other polymers (polystyrene standards are typically used). The most common solvent used in the GPC analysis of PLA is chloroform, but for example tetrahydrofuran (THF), benzene, or ethyl acetate can also be used [62]. When using standards consisting of another polymer, Mark–Houwink constants are commonly applied in the GPC analysis of PLA by universal calibration techniques. Since Mark–Houwink constants are based on experimental results, the numerical values of the constants vary depending on the literature source and as a function of the type of PLA used (e.g., enantiomeric purity, molecular structure, crystallinity), the solvent used, and the temperature of the analyzed solution. Summaries of the Mark–Houwink constants determined for PLA can be found in several review articles [62].

Analysis based on quantitative determination of the hydroxyl end groups of PLA by ^{1}H-NMR can be used to estimate the molecular weight of linear PLA homopolymers or well-defined, hydroxyl-terminated PLA copolymers. ^{1}H-NMR can also be utilized in the detection of free chains not attached to the comonomer in theoretically carboxyl-terminated PLA. Such free chains exist especially if the PLA has been produced by polycondensation, since the predominant chain termination is rarely implicit. Quantitative ^{1}H-NMR can also be used to determine the amount of lactide monomer in the polymer.

The enantiomeric purity of PLA polymer can be determined, for example by hydrolyzing the polymer into its monomer constituents and then separating the stereoisomers by HPLC. NMR techniques also exist for evaluating the relative ratio of the stereoisomers in PLA. The optical purity of PLA can in addition be determined by dissolving the polymer in a suitable solvent (e.g., chloroform) and measuring its specific optical rotation at 589 nm ($[\alpha]_{589}$) using a polarimeter [75, 79]. The fractions of L- and D-lactoyl units can be calculated based on the obtained $[\alpha]_{589}$. However, the reported optical rotations of pure PLLA and PDLA homopolymers vary slightly, typically being between |140| and |156| [79–82].

The processability of PLA is generally tested in manufacturing quality control departments by measuring the melt-flow index (MFI) at given temperatures. Other standard quality control measurements are molecular weight (by GPC), residual lactide (by GC) and moisture content (by Karl Fischer titration).

3.4 PROPERTIES OF POLYLACTIDES

Since packaging is currently the most commercially interesting bulk application of polylactides, their properties will be discussed in this context. Packaging materials have to meet many requirements, among which the most important ones are to protect the content from contamination, prolong the shelf-life of the product, contribute to a decreased need for preservatives and provide product information. The use of plastics in the packaging industry has increased rapidly during the last few decades due to the apparent advantages of polymer materials compared to metal and glass (e.g., low weight and relatively low production costs). Some of the most important properties of the plastics used in packaging are processability, mechanical properties, stability, barrier properties, optical properties and surface properties.

3.4.1 Processability

Plastics used in packaging materials have to be converted into useful products by different processing methods such as compounding, extrusion, injection moulding, injection stretch blow moulding, casting, blown film formation, thermoforming, foaming, blending, and fiber spinning, which places high demands on polymer processability. The most important processing characteristics are related to melt flow behavior and thermal properties (e.g., melt stability, moldability, sealability). The processing of PLA is more demanding than that of commodity plastics due to its hygroscopic nature and relatively poor stability (i.e., the limited melt stability of the polymer and its ability to undergo hydrolytic degradation). The various means for improving the processability of PLA can be divided into those performed in the melt as a finishing process and those carried out as a subsequent and independent processing step. The processes performed in the melt such as catalyst deactivation are mainly focused on improving the melt stability. Deactivators used for PLA are phosphorus-containing compounds [83], antioxidants [84], acrylic acid derivatives [85], and organic peroxides [86]. Catalyst deactivation is generally combined with removal of the residual monomer (lactide), which is usually done in a devolatilization process (i.e., distillation) [87]. Another way of reducing the lactide content of PLA is to apply solid-state polymerization of the residual lactide-containing PLA below its melting point. Besides reducing the residual lactide content, this also increases the molecular weight of the polymer [88]. A necessary separate post-polymerization treatment of polylactide is drying. Drying of the polymer is generally applied directly before processing in order to minimize thermo-hydrolysis and consequent molecular weight reduction during melt processing. Suggested drying conditions for PLA are 60 °C under vacuum or under hot dry air [89]. The suitability of a certain PLA grade for different processing methods is to a large extent dependent on rheological behavior. Melt viscosity is highly related to molecular weight, shear rate and temperature, but branching and racemic purity also affect the rheology. Typical molecular weights (M_w) for injection-moulding grades are ~ 100 kDa and for film cast extrusion grades ~ 300 kDa [90]. The heat sealing of PLA film depends on the racemic purity of the polymer. Reported optimal sealing conditions for PLA (L/D ratio 94/6) are at temperatures around 110 °C for 1–5 s. It has been reported that shrinkage increases if the sealing temperatures exceeds 110 °C and that at 120 °C shrinkage can be as high as 20% [63].

3.4.2   Thermal Stability

Polylactides can be processed using conventional polymer processing equipment. However, limited thermal stability still restricts the suitability for many applications. The complex degradation phenomena of PLA are reported to be of first order [91] and include various radical and non-radical reactions involving depolymerization, inter- and intramolecular transesterification, hydrolysis, oxidative degradation and pyrolysis reactions [92–94]. Racemization also occurs at elevated temperatures. Factors that can accelerate the thermal degradation of PLA are high amounts of reactive end groups (i.e., low molecular weight) and high polydispersity resulting from intermolecular transesterifications, as well as the presence of residual catalyst or monomer (lactide or lactic acid), moisture and different impurities.

The limited thermal stability of PLA can be mainly attributed to intramolecular transesterification reactions occurring at temperatures close to the melting point, leading to formation and subsequent evaporation of ring-formed low-molecular-weight compounds such as lactide. These reactions are reversible due to the related thermodynamic equilibrium which, on the other hand, enables the formation of high-molecular weight PLA in ROP. It is also known that the thermal degradation of PLA in the presence of residual catalysts such as Sn compounds differs from that of pure metal-free PLA [95]. Reactions that only have minor importance for PLA thermal stability include radical degradation and pyrolytic eliminations leading to double bond formation in the polymer backbone. Since radical reactions only need to be considered at temperatures exceeding 250 °C, this should not be a concern at typical PLA polymerization or processing temperatures.

Polylactide stereocomplexes have significantly higher thermal resistance than enantiomerically pure PLA [96]. The pyrolysis of PLA stereocomplexes leads to formation of a larger relative amount of *meso*-lactide, an indication of intermolecular interactions occurring in the degradation process [96].

3.4.3 Hydrolytic Stability

Hydrolytic degradation of PLA can be considered as the reverse reaction in step-growth polymerization of LA. Depending on conditions, poly(lactic acid) degrades through random hydrolysis of ester linkages in the polymer backbone as well as through chain-end scission in the presence of water [97, 98]. The factors that make PLA sensitive to moisture include permeability to water vapour, which is partly dependent on crystallinity. These characteristics in turn affect the suitability of PLA for specific packaging applications. Another feature of the hydrolytic degradation of amorphous PLA is that hydrolysis can be autocatalytic through the action of carboxylic acid end groups [99] if the sample specimen is thick enough. PLA easily undergoes hydrolysis both in its solid state as well as in the molten state at elevated temperatures. The rate of hydrolysis is further dependent on the molecular weight of the PLA polymer. Hydrolysis of PLA typically occurs much faster above its T_g than at lower temperatures and takes place via slightly different mechanisms depending on how the samples are exposed to conditions that promote hydrolysis, such as high humidity or immersion in water.

The main strategies for improving the resistance of PLA to hydrolysis involve minimizing the presence of residual monomers, impurities and water, as well as preventing autocatalysis. Since amorphous PLA has a higher water permeability and uptake, leading to faster and higher mass loss in aqueous environments [97], the equilibrium moisture content can be regulated by controlling polymer morphology. This, on the other hand, can be accomplished by tuning the D-LA content in PLLA (or L-LA in PDLA), since PDLA is amorphous and therefore degrades significantly faster than optically pure D- or PLLA. PLA stereocomplex blends are reported to have a higher resistance to hydrolysis than enantiomerically pure PLA [100].

3.4.4 Thermal Transitions and Crystallinity of PLA

Factors influencing the T_g of PLA include the molecular weight and molecular architecture, tacticity and crystallinity, as well as the presence of any plasticizers [101]. The T_g of PLA is typically between 50 and 80 °C [63, 102], although oligomers or low-molecular weight polymers can exhibit T_g values below this range. The T_g of PLA increases as a function of polymer molecular weight up to a certain threshold value. The T_g is also dependent on the isomeric purity of the polylactide backbone, and pure PLLA typically shows slightly higher values. Higher T_g values can be obtained by copolymerization, incorporation of stiff monomers or by cross-linking the polymer structure [101, 103]. The physical properties of amorphous PLA mostly depend on its T_g as well as on the proximity of storage and usage temperatures to the T_g [63].

The equilibrium crystalline melting point of PLA is 207 °C [62], but the T_m of high-molecular weight PLA is typically in the 170–180 °C range due to smaller and imperfect crystals and because of the presence of enantiomeric and other impurities [62, 63]. Similar to the T_g, the T_m and the melting enthalpy (ΔH_m) of PLA increase as a function of molecular weight up to a threshold value regardless of which stereoisomer is used [61, 102]. Above this threshold, a decrease in crystallinity occurs for PLA with a high polydispersity index, despite levelling off of the T_m [61]. A value of 93.1 J/g [104] is typically cited as the melting enthalpy of 100% crystalline PLA, although other values are mentioned in the literature (e.g., 111 J/g (100% crystalline) [105] or 55.0 J/g (37.0% crystallinity)) [106]. The T_m of PLA stereocomplexes is higher than those of enantiomeric PLA homopolymers (usually 230–240 °C) [107], although this is dependent on various factors including PDLA and PLLA mixing ratio, molecular weights of the individual components, and any enantiomeric impurities [100]. A comparison of various physical properties of PLA with those of other biopolymers or conventional plastics can be found in recent review articles [63, 100, 102].

Poly(lactic acid) can exist in both amorphous and semi-crystalline states depending on its stereochemistry, composition, and thermal history [90]. The polymorphism (i.e., the solid-state structure of a polymer) generally affects not only its thermal properties, but also characteristics such as mechanical, electrical and solvent resistance properties, as well as biodegradability [108], providing wide opportunities to tailor polymer properties. High-molecular weight PLA having a threshold optical purity of at least 72% can usually crystallize [61], although PLA with optical purity as low as 43% has been reported to show detectable crystallinity [109]. The addition of nucleating agents such as, for example, PDLA (in PLLA) or talc [110], plasticizers such as polyethylene glycol (PEG) or acetyl triethyl citrate (ATC) [111], or exposure to temperatures above the T_g (annealing), have also been reported to increase the rate of crystallization.

Poly(lactic acid) can crystallize in the α-, α'-, β-, or γ-forms, depending on the method and temperature of preparation [108]. The α-form is the most common polymorph of PLA and is typically characterized by two anti-parallel chains in a left-handed 10_3 helix conformation and packed in an orthorhombic unit cell. Excellent summaries of different crystal structures and the related unit cell parameters of PLA can be found in recent review articles [63, 108]. The α'-form (i.e., the limiting disordered α-form) has been recognized to account for the differing crystallization kinetics and crystalline structures at low temperatures, resulting in the commonly observed multiple melting behavior of PLA [108]. The β-form crystals are typically formed upon stretching the α-crystals at high draw rate and at a high temperature. The β-form has been reported to have an orthorhombic unit cell involving six chains in the 3_1 conformation. The main difference between the α- and the β-form crystal structures comes from the molecular packing within the unit cell, since these two conformations have comparable conformational angles and approximately the same energy [108]. The β-form has been found to be less stable than its α-counterpart and it can transform back into the α-form at high temperatures [108]. The γ-form, characterized by two anti-parallel helices with 3_1 conformation packed in an orthorhombic unit cell, can form as a result of epitaxial crystallization [108]. Only the 3_1 helix exists in PLA stereocomplex crystals.

3.4.5 Barrier and Other Properties

One of the most important properties of food packaging is the barrier against water, grease and gases (e.g., CO_2, O_2, aroma) [112]. The barrier properties of polymers can be related to a variety of factors, including chemical structure, tacticity, crystallinity, density, orientiation, molecular weight, degree of cross-linking, cohesive energy density, glass transition temperature and free volume [113].

3.4.5.1 Carbon Dioxide Barrier

The barrier properties of PLA against carbon dioxide, oxygen and water are reported to vary depending on PLA grade and processing history. Auras *et al.* have measured the permeability coefficients for PLA resins with two different L/D ratios (98/2 and 94/6). The CO_2 permeability coefficients (kg m/m^2 s Pa) for PLA films (2.77 and 1.99×10^{-17}) were lower than previously reported values for crystalline polystyrene (PS) at 25 °C and 0% RH (1.55×10^{-16}), but higher than that for polyethylene terephthalate (PET) (1.73×10^{-18}).

3.4.5.2 Oxygen Barrier

The O_2 permeability (kg m/m^2 s Pa)for 98/2 L/D ratio PLA film was 3.5×10^{-18} and 11×10^{-18} at 5 and 40 °C respectively (0% RH), which are similar to the corresponding values for PET [63].

3.4.5.3 Water Vapour Barrier

The water vapour permeability (WVP) coefficient for 98/2 L/D ratio PLA film ranged between 2.25 and 1.5×10^{-14} kg m/m^2 s Pa in the temperature interval 10–37.8 °C. The reported WVP coefficients for PS and PET at 25 °C were 6.7 and 1.1×10^{-15} respectively [63].

3.4.5.4 Mechanical and Thermo-Mechanical Properties

Packaging materials have to resist mechanical loads during filling, loading and transportation. This requires certain minimum mechanical properties (e.g., impact strength, flexural modulus). Polylactide is by its nature brittle and glassy, which is a disadvantage in many applications. As a result, copolymerization, blending and chemical modifications have been investigated as means for improving flexibility and impact resistance [11, 114]. A lot of effort has also been devoted to the development of plasticizers and compatibilizers [9, 115, 116]. The heat resistance of PLA at elevated temperatures (above the T_g) is crucial for many end-products where, for instance, the package has to retain its shape and strength during storage and transportation and also in use. Depending on the measuring method, typical heat distortion temperatures for PLA are in the range 65–75 °C. Efforts to increase the usable temperature range of PLA have been made by reinforcing [117], the use of additives [118], or by utilizing the ability of antipodal PLA stereopolymers to form stereo complexes [100]. This last approach was recently demonstrated to yield a material moulded into the shape of a cup which could withstand boiling oil [119].

3.4.5.5 Optical Properties

Optical properties are important as in many cases it is desirable that the contents of packaging should be visible to the consumer and, for example, food packaging often requires high transparency. However, in some applications it may be preferable to protect the contents from light, in which case an opaque or semi-opaque film material can be desirable. The transparency or opacity of polymer films to visible and ultraviolet (UV) light can affect the quality of foodstuffs such as juices, vitamin and sports drinks, dairy products and edible oils. Milk is mainly affected by light in the 400–550 nm wavelength range, but other wavelengths (i.e., UV-B, UV-C and visible) also damage the quality of milk, juices and edible oils. PLA shows negligible UV transmission between 190 and 220 nm; however, at 225 nm UV light transmission increases significantly. At 250 nm, 85% of UV light is transmitted and as much as 95% at 300 nm. Hence, UV-C is not transmitted, but nearly all the UV-B and UV-A light can pass through PLA films. In order to retain taste and appearance and extend the shelf-life, it might therefore be necessary to add UV-blocking agents to PLA when used for packaging of light-sensitive foods [63].

3.4.5.6 Printability

Packaging should provide the consumer with essential information about the product (ingredients, best before date, etc.). This requires either an additional label or a surface on the packaging material which can be printed. Poly(lactic acid) can generally be printed using existing commercial printing lines, even without any surface pre-treatment. Important parameters to take into account when printing are the limited temperature resistance of the polymer (e.g., during drying processes) and the need for uniform film thickness.

3.5 CONCLUDING REMARKS

Polylactide-based materials are likely to have significant future potential in a variety of applications due to specific advantages when compared with conventional polymers derived from oil. In particular:

- the use of PLA in packaging materials has appeal due to its attractive appearance and unique combination of properties (e.g., sealability, barrier);
- tunable hydrolytic stability creates a platform for applications such as temporary or short-term technical solutions in adhesives and binders;
- the introduction of polylactide stereocomplex technology on a broader base enables the use of PLA in applications where high heat resistance is crucial, such as in textiles, microwaveable containers and electronic equipment.

Lactic acid-based polymers prepared by polycondensation have gained more and more interest due to this rather uncomplicated and cost-efficient manufacturing route. Future challenges will be concerned with concepts for handling PLA at the end of product life-cycles. Products based on polylactide fit into current municipal waste management systems (energy recovery, composting, etc.), but it should also be noted that the conversion of PLA back into useful starting chemicals for bio-solvents or for re-use as a monomer is highly interesting and is likely to be further explored in the future.

REFERENCES

[1] Holten, C. H., A. Müller, and D. Rehbinder (1971) *Lactic Acid: Properties and Chemistry of Lactic Acid and Derivatives*, Verlag Chemie GmbH, Weinheim.

[2] Scheele, C. W., Om mjolk, och dess syra, *Kongl. Vetenskaps Academiens Nya Handlingar*, Johan Georg Lange, Stockholm, pp 116–124 (1780).

[3] W. H. Carothers, G. L. Dorough, and F. J. Natta, *J. Am. Chem. Soc.* 1932, 54, 761–772.

[4] D. R. Witzke, J. J. Kolstad, and R. Narayan, *Macromol.* 1997, 30, 7075–7085.

[5] M. Ajioka, K. Enomoto, K. Suzuki, *et al.*, *Bull. Chem. Soc. Jpn.* 1995, 68, 2125–2131.

[6] T. Maharana, B. Mohanty, and Y. S. Negi, *Prog. Polym.Sci.* 2009, 34, 99–124.

[7] M. Viljanmaa, A. Södergård, and P. Törmälä, *Int. J. Adhes. Adhes.* 2002, 22, 219–226.

[8] S. Y. Gu, M. Yang, T. Yu, *et al.*, *Polym. Int.* 2008, 57, 982–986.

[9] O. Martin, L. Averous, *Polymer* 2001, 42, 6209.

[10] S. Jacobsen, H. Fritz, P. Degée, *et al.*, *Ind.Crops Prod.* 2000, 11, 265–275.

[11] A. Södergård, M. Stolt, *Prog. Polym. Sci.* 2002, 27, 1123–1163.

[12] EP 244114, Assignees: Imperial Chemical Industries PLC, UK. Inventor: Hutchinson, F. G., 1987.

[13] WO 2002012369, Assignees: Wako Pure Chemical Industries, Ltd. and Takeda Chemical Industries, Japan. Inventors: Yamamoto, K., Tani, T., Aoki, T. *et al.*, 2002.

[14] R. Datta, M. Henry, *J. Chem. Technol. Biotechnol.* 2006, 81, 1119–1129.

[15] K. Hofvendahl, B. Hahn-Hagerdal, *Enzyme Microb. Technol.* 2000, 26, 87–107.

[16] Henton, D. E., P. Gruber, J. Lunt, *et al.*, Polylactic acid technology, in A. K. Mohanty, M. Misra and L. T. Drzal. (eds), *Natural Fibers, Biopolymers, and Biocomposites*, CRC Press, USA, pp 527–577 (2005).

[17] G. Reddy, M. Altaf, B. J. Naveena, *et al.*, *Biotechnol. Adv.* 2008, 26, 22–34.

[18] P. S. Panesar, J. F. Kennedy, D. N. Gandhi, *et al.*, *Food Chem.* 2007, 105, 1–14.

[19] H. Oh, Y. Wee, J. Yun, *et al.*, *Bioresour. Technol.* 2005, 96, 1492–1498.

[20] W. L. Tang, H. Zhao, *Biotechnol. J.* 2009, 4, 1725–1739.

[21] S. K. Singh, S. U. Ahmed, and A. Pandey, *Process Biochem.* 2006, 41, 991–1000.

[22] R. P. John, G. S. Anisha, K. M. Nampoothiri, *et al.*, *Biotechnol. Adv.* 2009, 27, 145–152.

[23] J. Nieuwenhuis, *Clin. Mater.* 1992, 10, 59–67.

[24] S. Keki, I. Bodnar, J. Borda, *et al.*, *J. Phys. Chem. B* 2001, 105, 2833–2836.

[25] EP 937743, Assignees: Kabushiki Kaisha Kobe Seiko Sho, Japan and Kobe Steel Europe Ltd. Inventors: Qureshi, N. M. and Woodfine, B., 1999.

[26] G. Chen, H. Kim, E. Kim, *et al.*, *Eur. Polym. J.* 2006, 42, 468–472.

[27] S. I. Moon, C. W. Lee, M. Miyamoto, *et al.*, *J. Polym. Sci., Part A: Polym. Chem.* 2000, 38, 1673–1679.

[28] EP 0848026, Assignees: Kyowa Yuka Co., Ltd., Japan. Inventors: Maruyama, H., Murayama, T., Yanagisawa, N. *et al.*, 1998.

[29] S. I. Moon, C. W. Lee, I. Taniguchi, *et al.*, *Polymer* 2001, 42, 5059–5062.

[30] S. Moon, I. Taniguchi, M. Miyamoto, *et al.*, *High Perform. Polym.* 2001, **13**, S189–S196.
[31] EP 0953589, Assignees: Inc Mitsui Chemicals Japan. Inventors: Terado, Y., Suizu, H., Takagi, M., *et al.*, 1999.
[32] US 5434241, Assignees: Korea Institute of Science and Technology. Inventors: Kim, Y. H., Ahn, K. D., Han, Y. K., *et al.*, 1995.
[33] WO 9312160, Assignees: Mitsui Toatsu Chemicals, Inc., Japan. Inventors: Enomoto, K., Ajioka, M. and Yamaguchi, A., 1993.
[34] E. Ranucci, Y. Liu, M. S. Lindblad, *et al.*, *Macromol. Rapid Commun.* 2000, **21**, 680–684.
[35] C. J. Spaans, V. W. Belgraver, O. Rienstra, *et al.*, *Biomaterials* 2000, **21**, 2453–2460.
[36] WO 9601863, Assignees: Finland Alko Group Ltd. and Neste Oy. Inventors: Seppälä, J., Härkönen, M., Hiltunen, K., *et al.*, 1996.
[37] J. Tuominen, J. Kylmä, and J. Seppälä, *Polymer* 2002, **43**, 3–10.
[38] A. Helminen, J. Kylmä, J. Tuominen, *et al.*, *Polym. Eng. Sci.* 2000, **40**, 1655–1662.
[39] GB 1117798, Assignees: Sumitomo Chemical Co. Inventors: Fukui, K., Kagiya, T., Narisawa, S., *et al.*, 1968.
[40] EP 0020944, Assignees: Japan Teijin Ltd. Inventors: Inata, H., Matsumura, S., and Ogasawara, M., 1981.
[41] J. Tuominen, J. V. Seppälä, *Macromol.* 2000, **33**, 3530–3535.
[42] R. Po, L. Abis, L. Fiocca, *et al.*, *Macromol.* 1995, **28**, 5699–5705.
[43] US 5470944, Assignees: USA ARCH Development Corp. Inventor: Bonsignore, P. V., 1995.
[44] D. W. Grijpma, *Macromol. Chem. Phys.* 1994, **195**, 1649–1663.
[45] J. J. Kolstad, *J. Appl. Polym. Sci.* 1996, **62**, 1079–1091.
[46] X. Zhang, D. A. MacDonald, M. F. A. Goosen, *et al.*, *J. Polym. Sci., Part A: Polym. Chem.* 1994, **32**, 2965–2970.
[47] S. Jacobsen, H. G. Fritz, P. Degee, *et al.*, *Polymer* 2000, **41**, 3395–3403.
[48] WO 9509879, Assignees: USA Cargill Inc. Inventors: Gruber, P. R., Hall, E. S., Kolstad, J. J., *et al.*, 1995.
[49] EP 261572, Assignees: Boehringer Ingelheim K.-G., Fed. Rep. Ger. and Boehringer Ingelheim International G.m.b.H. Inventor: Mueller, M., 1988.
[50] WO 9200292, Assignees: du Pont de Nemours, E. I., and Co., USA. Inventors: Bellis, H. E. and Bhatia, K. K., 1992.
[51] J. W. Leenslag, A. J. Pennings, *Makromol. Chem.* 1987, **188**, 1809–1814.
[52] Stepanski, M., Purification of lactide by melt-crystallization, in *1st PLA World Congress*, Munich, Germany, 9–10 September 2008.
[53] EP 0531462, Assignees: Du Pont. Inventor: K. K. Bhatia, 1993.
[54] EP 661325, Assignees: Dainippon Ink Chemical Industry Co., Japan. Inventors: Ebato, H. and Imamura, S., 1995.
[55] EP 0618250, Assignees: Dainippon Ink Chemical Industry Co., Japan. Inventors: Ebato, H., Oya, S., Kakizawa, Y., *et al.*, 1994.
[56] K. M. Stridsberg, M. Ryner, and A. Albertsson, *Adv. Polym. Sci.* 2002, **157**, 41–65.
[57] A. Kowalski, A. Duda, and S. Penczek, *Macromol.*, 2000, **33**, 7359–7370.
[58] M. C. Tanzi, P. Verderio, M. G. Lampugnani, *et al.*, *J. Mater. Sci.: Mater. Med.* 1994, **5**, 393–396.
[59] M. Okada, *Prog. Polym. Sci.* 2001, **27**, 87–133.
[60] D. Bourissou, S. Moebs-Sanchez, and B. Martin-Vaca, *C. R. Chim.* 2007, **10**, 775–794.

[61] J. Ahmed, J.-X. Zhang, Z. Song, et al., J. Therm. Anal. Calorim. 2009, 95, 957–964.
[62] D. Garlotta, J. Polym. Environ. 2002, 9, 63–84.
[63] R. Auras, B. Harte, and S. Selke, Macromol. Biosci. 2004, 4, 835–864.
[64] N. E. Kamber, W. Jeong, R. M. Waymouth, et al., Chem. Rev. 2007, 107, 5813–5840.
[65] R. E. Drumright, P. R. Gruber, and D. E. Henton, Adv. Mater. 2000, 12, 1841–1846.
[66] Witzke, D. R. (1997) Introduction to properties, engineering, and prospects of polylactide polymers, Diss. Abstr. Int., B 1998, 59(1), Michigan State University,. Ann Arbor, MI, 389.
[67] Stolt, E. M. Doctoral Thesis. Lactoyl (Co)Polymers Prepared by Iron Carboxylate Catalysis, 2008, Åbo Akademi University, Turku, Finland.
[68] H. Inata, S. Matsumura, J. Appl. Polym. Sci. 1985, 30, 3325–3337.
[69] Y. Lemmouchi, M. Murariu, A. M. Dos Santos, et al., Eur. Polym. J. 2009, 45, 2839–2848.
[70] S. Inkinen, M. Stolt, and A. Södergård, Biomacromol. 2010, 11, 1196–1201.
[71] K. Hiltunen, M. Härkönen, J. V. Seppälä, et al., Macromol. 1996, 29, 8677–8682.
[72] C. Torres, C. Otero, Enzyme Microb. Technol. 2001, 29, 3–12.
[73] Annual book of ASTM Standards Vol 6.03, 1998 (D1980-87).
[74] Annual book of ASTM Standards vol. 06.03 (E1899-02).
[75] H. Tsuji, H. Daimon, and K. Fujie, Biomacromol. 2003, 4, 835–840.
[76] D. K. Yoo, D. Kim, and D. S. Lee, Macromol. Res. 2006, 14, 510–516.
[77] J. L. Robert, K. B. Aubrecht, J. Chem. Educ. 2008, 85, 258–260.
[78] T. Tsukegi, T. Motoyama, Y. Shirai, et al., Polym. Degrad. Stab. 2007, 92, 552–559.
[79] H. Tsuji, Y. Ikada, Macromol. Chem. Phys. 1996, 197, 3483–3499.
[80] H. Tsuji, Y. Ikada, Macromol. 1992, 25, 5719–5723.
[81] R. Slivniak, A. J. Domb, Biomacromol. 2002, 3, 754–760.
[82] M. Spasova, L. Mespouille, O. Coulembier, et al., Biomacromol. 2009, 10, 1217–1223.
[83] US 2683136, Assignees: E. I. du Pont de Nemours & Co. Inventor: Higgins, N. A., 1954.
[84] EP 0615532, Assignees: Cargill Inc. Inventors: Gruber, P., Kolstad, J. J., Hall, E. S. et al., 2000.
[85] EP 1070097, Assignees: Cargill Inc. Inventors: Kolstad, J. J., Witzke, D. R., Hartmann, M. H. et al., 2001.
[86] EP 0737219, Assignees: Fortum Oyj. Inventors: Södergård, A., Selin, J., Niemi, M. et al.
[87] EP 0615529, Inventors: Gruber, P., Kolstad, J. J. and Ryan, C., 1994.
[88] EP 664309, Assignees: Japan Shimadzu Corporation. Inventors: Ohara, H., Sawa, S. and Kawamoto, T., 1995.
[89] Crystallizing and Drying of PLA, NatureWorks LLC. http://www.natureworksllc .com/product-and-applications/ingeo-biopolymer/technical-resources/~/media/ Product%20and%20Applications/Ingeo%20Biopolymer/Technical%20 Resources/Processing%20Guides/ProcessingGuides_CrystallizingandDrying-PLA_pdf.ashx. (accessed 2/2, 2010).
[90] L.-T. Lim, R. Auras, and M. Rubino, Prog. Polym. Sci. 2008, 33, 820–852.
[91] M. C. Gupta, V. G. Deshmukh, Colloid Polym. Sci. 1982, 260, 308–311.
[92] F.-D. Kopinke R., Polym. Degrad. Stab. 1996, 53, 329–342.
[93] I. C. McNeill, H. A. Leiper, Polym. Degrad. Stab. 1985, 11, 309–326.

PRODUCTION, CHEMISTRY AND PROPERTIES OF POLYLACTIDES 63

[94] I. C. McNeill, H. A. Leiper, *Polym. Degrad. Stab.* 1985, 11, 267–285.
[95] Y. Fan, H. Nishida, Y. Shirai, *et al.*, *Polym. Degrad. Stab.* 2004, 84, 143–149.
[96] Y. Fan, H. Nishida, Y. Shirai, *et al.*, *Polym. Degrad. Stab.* 2004, 86, 197–208.
[97] C. Shih, *J. Controlled Release* 1995, 34, 9–15.
[98] G. Schliecker, C. Schmidt, S. Fuchs, *et al.*, *Biomaterials* 2003, 24, 3835–3844.
[99] H. Tsuji, *Polymer* 2002, 43, 1789–1796.
[100] H. Tsuji, *Macromol. Biosci.* 2005, 5, 569–597.
[101] G. L. Baker, E. B. Vogel, and M. R. Smith III, *Polym. Rev.* 2008, 48, 64–84.
[102] Y. Ikada, H. Tsuji, *Macromol. Rapid Commun.* 2000, 21, 117–132.
[103] WO2008056136, Assignees: Finland. Inventors: Södergård, N. D. A., Stolt, E. M., and Inkinen, S., 2008.
[104] E. W. Fischer, H. J. Sterzel, and G. Wegner, *Kolloid-Z. Z. Polym.* 1973, 251, 980–990.
[105] R. Vasanthakumari, A. J. Pennings, *Polymer* 1983, 24, 175–178.
[106] D. Cohn, H. Younes, and G. Marom, *Polymer* 1987, 28, 2018–2022.
[107] K. Fukushima, Y. Kimura, *Polym. Int.* 2006, 55, 626–642.
[108] P. Pan, Y. Inoue, *Prog. Polym. Sci.* 2009, 34, 605–640.
[109] J. Sarasua, R. E. Prud'homme, M. Wisniewski, *et al.*, *Macromol.* 1998, 31, 3895–3905.
[110] H. Tsuji, H. Takai, N. Fukuda, *et al.*, *Macromol. Mater. Eng.* 2006, 291, 325–335.
[111] H. Li, M. A. Huneault, *Polymer* 2007, 48, 6855–6866.
[112] D. V. Plackett, V. K. Holm, P. Johansen, *et al.*, *Packag. Technol. Sci.* 2006, 19, 1–24.
[113] K. S. Miller, J. M. Krochta, *Trends Food Sci. Technol.* 1997, 8, 228–237.
[114] W. Amass, A. Amass, and B. Tighe, *Polym. Int.* 1998, 47, 89–144.
[115] N. Ljungberg, B. Wesslen, *J. Appl. Polym. Sci.* 2002, 86, 1227–1234.
[116] N. Choi, C. Kim, K. Y. Cho, *et al.*, *J. Appl. Polym. Sci.* 2002, 86, 1892–1898.
[117] Plackett, D., A. Södergård, Polylactide-based biocomposites, in A. K. Mohanty, M. Misra and L. T. Drzal. (Ed.), *Plastics Fabrication and Uses*, CRC Press, USA, pp 579–596 (2005).
[118] Dartee, M., Performance and appearance improvements of PLA, in *1st PLA World Congress*, Munich, Germany, 9–10 September, 2008.
[119] US 2008207840, Assignees: Tate & Lyle PLC. Inventors: Södergård, N. D. A., Stolt, E. M., Siistonen, H. K., *et al.*, 2008.

4

Production, Chemistry and Properties of Polyhydroxyalkanoates

Eric Pollet and Luc Avérous

LIPHT-ECPM, Université de Strasbourg, Strasbourg, France

4.1   INTRODUCTION

Polyhydroxyalkanoates (PHAs) are a family of biopolymers synthesized by many bacteria as intracellular carbon and energy storage granules. Since PHAs are mainly produced from renewable resources by fermentation, PHAs are compatible with the growing worldwide interest in environmentally friendly materials and sustainable development. The PHAs are biodegradable (i.e., potentially suitable for short-term packaging uses) and also exhibit biocompatibility when in contact with living tissues (e.g., for biomedical applications such as tissue engineering).

PHAs are generally classified into short-chain-length (sCL-PHA) and medium-chain-length types (mCL-PHA) based on the different number of carbons in the repeating units. For instance, sCL-PHA contains four- or five-carbon repeating units, while mCL-PHA contains six or more carbon atoms in the repeating units. PHA

Biopolymers – New Materials for Sustainable Films and Coatings, First Edition.
Edited by David Plackett.
© 2011 John Wiley & Sons, Ltd. Published 2011 by John Wiley & Sons, Ltd.

Table 4.1   Main PHA homopolymer structures based on Figure 4.1

Chemical Name	Abbreviation	x Value	R Group
poly(3-hydroxypropionate)	P(3HP)	1	Hydrogen
poly(3-hydroxybutyrate)	P(3HB)	1	Methyl
poly(3-hydroxyvalerate)	P(3HV)	1	Ethyl
poly(3-hydroxyhexanoate) or poly(3-hydroxycaproate)	P(3HHx) or P(3HC)	1	Propyl
poly(3-hydroxyheptanoate)	P(3HH)	1	Butyl
poly(3-hydroxyoctanoate)	P(3HO)	1	Pentyl
poly(3-hydroxynonanoate)	P(3HN)	1	Hexyl
poly(3-hydroxydecanoate)	P(3HD)	1	Heptyl
poly(3-hydroxyundecanoate)	P(3HUD)or P(3HUd)	1	Octyl
poly(3-hydroxydodecanoate)	P(3HDD) or P(3HDd)	1	Nonyl
poly(3-hydroxyoctadecanoate)	P(3HOD) or P(3HOd)	1	Pentadecanoyl
poly(4-hydroxybutyrate)	P(4HB)	2	Hydrogen
poly(5-hydroxybutyrate)	P(5HB)	2	Methyl
poly(5-hydroxyvalerate)	P(5HV)	3	Hydrogen

nomenclature may still be in a state of flux as new structures continue to be discovered. The main polymer of the PHA family is the polyhydroxybutyrate homopolymer (PHB), but different poly(hydroxybutyrate-co-hydroxyalkanoate) copolyesters exist, such as poly(hydroxybutyrate-co-hydroxyvalerate) (PHBV), poly(hydroxybutyrate-co-hydroxyhexanoate) (PHBHx), poly(hydroxybutyrate-co-hydroxyoctanoate) (PHBO) or poly(hydroxybutyrate-co-hydroxyoctadecanoate) (PHBOd) (Tables 4.1 and 4.2).

Figure 4.1 shows the generic formula for PHAs in which x is one or higher and R can be either hydrogen or hydrocarbon chains of up to around C16 in length.

A wide range of PHA homopolymers and copolymers has been produced, in most cases at the laboratory scale. A few of them have attracted industrial interest and have been commercialized in the past few decades.

Copolymers of PHA vary in the type and proportion of monomers, and are typically random in sequence. Poly(3-hydroxybutyrate-co-3-hydroxyvalerate) or

Table 4.2   Most common PHA homopolymers and copolymers with their usual abbreviations

Conventional abbreviations (short)	Full abbreviations	Structures
PHB	P(3HB)	Homopolymer
PHV	P(3HV)	Homopolymer
PHBV	P(3HB-co-3HV)	Copolymer
PHBHx	P(3HB-co-3HHx)	Copolymer
PHBO	P(3HB-co-3HO)	Copolymer
PHBD	P(3HB-co-3HD)	Copolymer
PHBOd	P(3HB-co-3HOd)	Copolymer

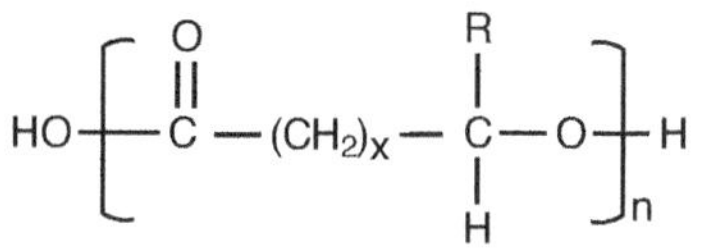

Figure 4.1   Chemical structure of the polyhydroxyalkanoates

P(3HB-*co*-3HV) is based on a random arrangement of two monomers with R = methyl and with R = ethyl. Poly(3-hydroxybutyrate-co-3-hydroxyhexanoate) consists of two monomers with R = methyl and propyl. The most common PHA homopolymers and copolymers are listed in Table 4.2.

4.2   POLYHYDROXYALKANOATE SYNTHESIS

4.2.1   Background

In 1923, French bacteriologist Maurice Lemoigne reported that the bacterium *Bacillus megaterium* generated 3-hydroxybutyric acid, and in 1927 he described the procedure for obtaining material from the bacterial cells. At the time this product was considered to be a lipid and the biochemistry and chemistry communities only fully realized the existence of these natural polyesters in the late 1950s.

For almost 30 years, Lemoigne and co-workers published numerous papers on PHB analysis, reporting in particular that a variety of bacteria could produce PHB as a 'reserve material'. It was only in the late 1950s that the important role of PHB in the overall metabolism of bacterial cells was discovered and understood by Stanier and Wilkinson and their co-workers, who showed that the PHB granules in bacteria serve as an intracellular food and energy reserve and that the polymer is produced by cells in response to nutrient limitations and in order to avoid starvation if an essential element becomes unavailable [1–3].

For a long time 3-hydroxybutyrate (3HB) was considered to be the only material produced by bacteria until, in 1974, Wallen and Rohwedder identified 3-hydroxy-valerate and 3-hydroxyhexanoate in chloroform extracts [4]. Then, De Smet *et al.* obtained a polymer formed principally of 3-hydroxyoctanoate units when cultivating *Pseudomonas oleovorans* in *n*-octane, [5] thus demonstrating the possibility of preparing various PHAs as a function of the substrate. Later on, it was found that many different microorganisms are able to produce PHAs [6]. In the 1970s, growing attention was paid to PHAs, resulting in a large number of published studies. Intensive work on biochemistry led to detailed knowledge about the physiological function of PHB and the enzymology of PHB synthesis [6] as well as PHB metabolism and its regulation [7]. All this research allowed optimization of the PHB elaboration process by defining the conditions that favour PHB accumulation [8]. In the 1980s, extensive work was done to identify all potential HA units, resulting not only in the description of a wide range of 3HA units, but also in the discovery of 4HA and 5HA units in the polymer chain [9]. With the development of molecular biology in the 1980s, genes coding for enzymes involved in PHA biosynthesis were cloned from *Ralstonia eutropha* (since then renamed *Wautersia eutropha* and more recently *Cupriavidus necator*) and shown to be active in *Escherichia coli* [10–12]. This development may lead in the near future to production of tailor-made PHAs through even more efficient processes.

Bacterial biosynthesis is logically considered nowadays to be the most important process for PHA production. This biosynthesis will be presented in detail in the following section. In addition to bacterial synthesis, chemical synthesis of PHA via ring-opening polymerization of lactones (e.g., butyrolactone) is also possible. Recently, increased attention has been paid to biosynthesis using genetically modified plants. These alternative production routes are also briefly discussed hereafter.

4.2.2   Bacterial Biosynthesis of Polyhydroxyalkanoates

Historically, bacterial biosynthesis has been the preferred way to produce PHAs and a lot of work has been conducted to fully understand the principles and mechanisms of the process, especially for PHB [13, 14].

PHB is accumulated by a wide variety of bacteria under unbalanced growth conditions, that is to say when the cells are subjected to limitation of an essential nutrient (e.g., phosphate, sulphate, magnesium, ammonium, iron) but are exposed to an excess of carbon.

The basic biochemical pathway of PHB synthesis is shown in Figure 4.2. In the first step, a selected carbon source is converted to acetate. Then, an enzyme co-factor is attached via the formation of a thioester bond. The enzyme, called coenzyme A (CoA), is a universal carrier of acyl groups in biosynthesis and acetyl-CoA is a basic metabolic molecule found in all PHA-producing organisms. A dimer of acetoacetyl-CoA is subsequently formed via reversible condensation and then reduced by nicotinamide adenine dinucleotide phosphate (NADPH) to the actual monomer unit (R)-3-hydroxybutyryl-CoA. In the next step, PHB is formed via the polymerization of this monomer unit, retaining the asymmetric centre of the molecule. The propagation step proceeds by the bonding of another monomer unit to the free thiol group of the active site, followed by another thioester–oxyester exchange reaction to form the trimer and so on. The polymerization occurs spontaneously in a thermodynamically favourable process since the oxyester shows higher bond energy compared with that of the thioester. Thus, the polymerization reaction proceeds by a continuous series of insertion reactions and the synthase acts as both initiator and catalyst for the polymerization reaction. It is worth pointing out that all natural PHAs are totally isotactic since the enzyme is specific for monomers with the [R] configuration and will not polymerize identical compounds having the [S] configuration.

Today, it is estimated that more than 250 different bacteria can produce PHAs, mainly PHB, under various conditions. These microorganisms generate PHAs with different levels of efficiency, yield and quality. Besides PHB, a broad range of homopolyesters, copolyesters such as PHB-co-HV, terpolyesters and so on can be obtained. Even more impressive is the variety of substrates which the bacteria are able to convert. Depending on the type of strain and substrate, PHAs can even bear functional groups, such as double bonds, halogens, and phenyl or epoxy moieties. However, the number of bacterial strains which can be used to synthesize biopolyesters at an industrially acceptable level of production remains limited. From this viewpoint, the Pseudomonas genus currently seems to provide the most versatile

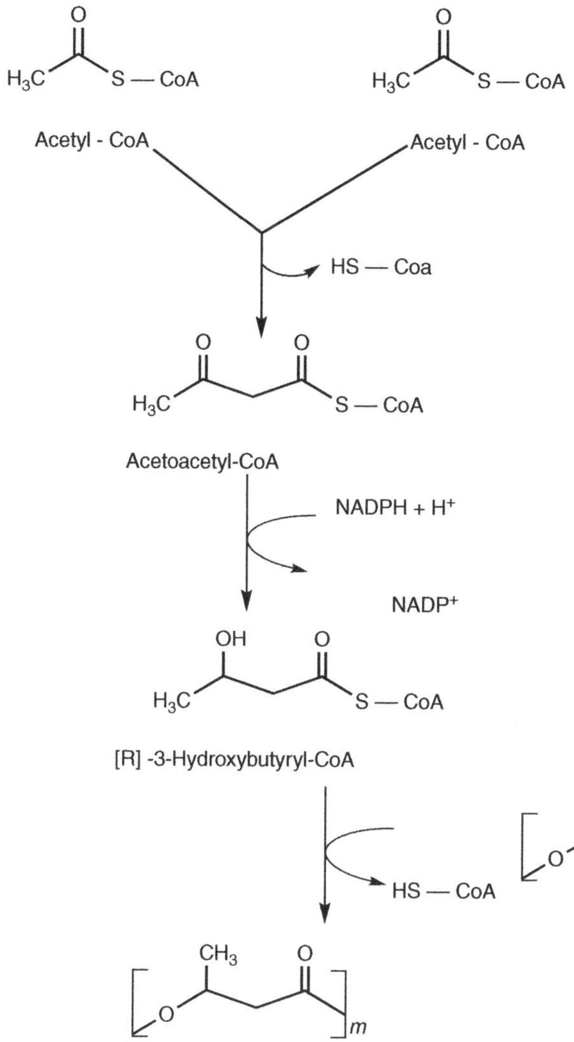

Figure 4.2   Simplified representation of the biochemical pathway of PHB synthesis

accumulators of PHAs. Among the efficient bacteria that have been tested industrially for this purpose, one can cite *Pseudomonas oleovorans*, *Azotobacter vinelandii*, some methylotrophs, recombinant *Escherichia coli*, *Azohydromonas lata* (ex *Alcaligenes latus*) and *Cupriavidus necator* (formerly known as *Wautersia eutropha*, *Ralstonia eutropha* or *Alcaligenes eutrophus*).

Apart from the P3HB homopolymer, growing attention has been paid to copolymers such as poly-3-hydroxybutyrate with either 3-hydroxyvalerate or 4-hydroxybutyrate. This is mainly because these copolymers present good properties and might be produced using economically viable technology. Initially, *A. eutrophus* (now named *Cupriavidus necator*) was grown on a variety of substrates, including sugars and ethanol, to synthesize PHB. Later, it was found that the addition of propionate or valerate to the medium could lead to incorporation of HV comonomer units [15]. In such a feeding medium, the HV content was limited to the range 0–47 mol % because of the fast metabolic pathway of propionyl-CoA to acetonyl-CoA in the cell [16, 17]. Later on, copolymers with a wider range of composition (0–95 mol % of HV units) were produced by using mixtures of butyric and valeric acids in the medium [18]. Then, the possibility to incorporate 4-hydroxybutyric acid [9, 19, 20] and 3-hydroxypropionic acid was also demonstrated. *Cupriavidus necator, Alcaligenes latus* (now *Azohydromonas lata*) or *Delftia acidovorans* were shown to produce random copolymers of 3-hydroxybutyric acid (3HB) and 4-hydroxybutyric acid (4HB), P (3HB-co-4HB), in a broad composition range using various carbon sources [9, 21–23]. For instance, when 4-hydroxybutyric acid was used as a unique carbon source for *Azohydromonas lata*, a P(3HB-*co*-4HB) random copolyester containing 37 per cent of 4-HB units was obtained [9]. When 4-hydroxybutyric acid or 1,4-butanediol was used as the sole carbon source for *Delftia acidovorans*, the homopolymer of 4-hydroxybutyric acid P(4HB) was obtained [23]. A short overview of the various PHAs synthesized by different bacteria is presented in Table 4.3.

PHA recovery remains a key process issue. Isolation of the polyester from biomass proceeds through the destruction of the cell membrane performed mechanically, chemically or enzymatically [24]. This step is followed by polymer dissolution in a suitable solvent, such as chloroform, methylene chloride, 1,2-dichloroethane or pyridine. Cell wall residues are removed by filtration and/or centrifugation. In a last step, extraction using mixed solvents (e.g., water/organic solvent) can be conducted

Table 4.3 Overview of the various PHAs produced by different bacteria

Bacteria	PHA produced	Specific substrates
Cupriavidus necator	PHB & PHBV, P(3HB-co-4HB) P(3HB-co-3HV-co-5HV) P(3HB-co-4HB-co-3HV) P(3HB-co-3HV-co-4HV)	
Azohydromonas lata	PHB & PHBV, P(3HB-co-3HP) P(3HB-co-4HB)	
Rhodospirillum rubrum	P(3HB-co-3HV-co-3H4-pentenoate)	4-Pentenoic acid or pentanoic acid
Chromobacterium violaceum	P(3HV) homopolymer	Sodium valerate
Delftia acidovorans	P(3HB-co-4HB)	1,4-Butanediol
Methylobacterium	P(3HB-co-4HB)	Methanol + 3-hydroxypropionate
Comomonas testosterone	P(3HB-co-3HC) P(3HB-co-3HC-co-3HO)	
Sphaerotilus natans	P(3HB-co-3HV)	Glucose + sodium propionate

for the final purification. The difficulty of solubilizing PHAs in a very limited number of mostly chlorinated solvents represents one of the major issues for the process. Future large-scale applications of PHA will be dependent on production cost. It is thus very important to develop low-cost PHA production routes and tremendous research efforts are currently being made to develop more efficient technologies, such as continuous and non-sterile processes based on mixed cultures and substrates (e.g., using wastes or other cheap carbon sources).

4.2.3 Production of Polyhydroxyalkanoates by Genetically Modified Organisms

Important genetic engineering research has been conducted with the aim to improve the production rate and/or widen the range of species able to produce PHAs. The first approach that was considered consisted of cloning the appropriate genes and transferring them into other organisms. In 1988, Dennis and co-workers cloned the entire set of genes in R. *eutropha* (now *Cupriavidus necator*) for the enzymes involved in PHB synthesis from acetyl-CoA [11]. The genes were clustered and successfully introduced into E. *coli* so the bacteria were then able to synthesize PHB in large quantities from a wide range of organic substrates. Some recombinant E. *coli* are also able to produce PHBV copolyesters [25]. In addition, some strains containing only the PHA synthase gene can express this protein in sufficiently large quantities for isolation and purification [26]. The purified enzyme can then be used for *in vitro* polymerization of various 3- and 4-hydroxyalkanoate-CoA monomers. Since these reactions can present a living character, polymers with very high molecular weights can be obtained [27].

A second approach relies on the genetic modification of plants to produce polymers through more or less standard agricultural processes, while introducing the same mechanisms as used by bacteria. In the early 1990s, Somerville and co-workers reported the successful application of genetic engineering to direct production of PHAs in plants [28]. The reductase and synthase genes of *Cupriavidus necator* (formerly A. *eutrophus*) were transferred into *Arabidopsis thaliana*, a plant which can also produce acetoacetyl-CoA. As a result, the transgenic plant accumulated up to 14% of its dry weight in PHB [29]. Several other studies dealing with the production of PHA (mainly PHB) in genetically modified plants were later reported [30–32]. Among these studies, a procedure allowing the achievement of genetically modified plants able to produce small quantities of PHB was developed by Monsanto. Steinbüchel and co-workers achieved the formation of transgenic PHB-producing potato (*Solanum tuberosum*) and tobacco (*Nicotiana tabacum*) plants using a variety of chimeric constructs but the amount of PHB formed was still rather low (less than 0.5 wt %, dry weight) [33].

The synthesis of PHAs in crop plants is considered to be an attractive route for large-scale and low-cost production of these bioplastics. PHA synthesis has been demonstrated in a number of plants, including monocots and dicots. Although an accumulation of as much as 40% PHA per gram dry weight has been reported in *Arabidopsis thaliana*, accumulation is commonly below 10%. Thus, except for some promising studies on transgenic *Arabidopsis*, only a few successful studies on PHA production in plants have been reported up to now and the yields obtained are still too low to consider high-volume PHA production using such technology. Moreover, PHA production in plants often leads to chlorosis, reduced growth or even premature death of the plants. The main challenge for the future is to succeed in synthesizing different PHAs of controlled composition at high levels that do not compromise plant productivity, and in developing economical methods for efficient extraction of the biopolymers. This will require significant advances in biorefinery concepts as well as a deeper understanding of plant biochemical pathways. However, it is believed that progress in genetic engineering as a route to PHA bioplastics could also offer new perspectives for agricultural development in the future.

4.2.4 Chemical Synthesis of Polyhydroxyalkanoates

It is worth recalling that the majority of PHAs can be chemically synthesized from the appropriate substituted lactones. A detailed and instructive review on the chemical synthesis of PHAs was published by Müller and Seebach [34]. The ring-opening polymerization of the lactones is generally carried out using zinc- or aluminum-based catalysts with water as a co-catalyst. Several alternative routes have also been investigated, for example PHBV can be produced through butyrolactone and valerolactone polymerization promoted by an oligomeric aluminoxane catalyst. Polyesters with partially stereoregular blocks can also be obtained. More recently, several studies have demonstrated the possibility to produce PHAs by enzymatically catalyzed ring-opening polymerization of lactones as an even greener route [35, 36]. Regarding stereoregularity, synthetic PHAs can be almost identical to the corresponding bacterial biopolymers [37, 38] and this results in materials retaining excellent biodegradation behavior [16].

Due to the specificity of PHA synthase, the biosynthesis route does not allow much control over composition and structure in PHA copolyesters. In contrast, the chemical synthesis can permit better control and fine tuning of the final structure and composition of the polyesters. However, biosynthesis of PHA leads to much higher molecular weights when compared with those achieved using chemical methods. Thus, although academically interesting, owing to the high costs of lactone monomers the synthetic homologues of bacterial PHAs are unlikely to be competitive with PHAs produced by fermentation. Moreover, since biosynthesis of PHA is conducted by microorganisms grown in an aqueous solution containing sustainable resources such as starch, glucose, sucrose, fatty acids, and even nutrients in waste water at 30–37 °C and atmosphere pressure, it is obviously a more environmentally friendly and sustainable production method.

4.3 PROPERTIES OF POLYHYDROXYALKANOATES

4.3.1 Polyhydroxyalkanoate Structure and Mechanical Properties

Polymers based on 3-hydroxyalkanoic acids are reported to be in the [R] configuration resulting from the stereospecificity of the enzyme involved in the polymerization step [14]. However, some structures presenting S units have also been reported [39]. The molecular weight of the simplest poly-3-hydroxybutyrates depends on the bacteria used and the conditions of synthesis. The means of separation and purification may also have a certain influence.

Biodegradability is a key property and a major advantage of PHAs which can to a certain extent counterbalance some drawbacks relative to conventional plastics, such as higher price. Nevertheless, a comprehensive evaluation of the ultimate properties of PHAs is a requirement before any applications as everyday commercial products can be considered. Thus, a comparison of the properties of both PHB and a few PHA copolymers with those of polypropylene and low-density polyethylene is given in Table 4.4.

The properties of PHB are rather similar to those of polypropylene, outperforming polyethylene in most parameters. The low water vapour permeability of PHAs is certainly the most interesting property, allowing possible applications in packaging. The weakest point is the low deformation at break, related to low film toughness and unacceptable rigidity and brittleness. The reason for the PHB brittleness arises mainly from its high crystallinity and the large size of the spherulites. Nevertheless, the high strength and modulus represent a good starting point for further polymer modifications to increase deformability and toughness.

The properties of PHAs can be tuned by varying the HV content, its increase resulting in an increase in the impact strength and a decrease in melting temperature and glass transition, tensile strength, crystallinity and water vapour permeability [40–42]. Regarding the mechanical properties of P3HB-co-3HV copolymers, studies suggest that the 3HB and 3HV units are isodimorphous because of their similarity in shape and size and that the 3HV units are incorporated into the P(3HB) crystal lattice [43, 44]. As a result, the properties of P(3HB-co-3HV) are not significantly improved in comparison with the P(3HB) homopolymer. However, copolymers with improved mechanical properties can be prepared via copolymerization of 3HB with longer-chain hydroxyalkanoic acids, which form a separate crystalline lattice or do not crystallize at all, such as is the case for copolymers of 3HB and 3-hydroxyhexanoate (3HH) [45].

PHA copolymers display interesting properties, much closer to those of LDPE, but their availability and price are still a hindrance to uptake of these materials as serious competitors for commodity polyolefins.

Table 4.4   Comparison of the physical properties of PHB and some PHA copolymers with those of polypropylene (PP) and low density polyethylene (LDPE). Adapted with permission from K. Sudesh, H. Abe, Y. Doi, *Prog. Polym. Sci.* 25, 1503–1555. Copyright (2000) with permission from Elsevier

	PHB	PHBV (20% HV)	P(3HB-co-3HA) (3%HD&3%HDD)	PP	LDPE
Melting temperature (°C)	175	145	133	170	110
Glass transition temperature (°C)	4	− 1	− 8	− 10	− 110
Crystallinity (%)	60	n.a.	n.a.	50	50
E modulus (GPa)	3.5	0.8	0.2	1.5	0.2
Tensile strength (MPa)	4	20	17	38	10
Elongation at break (%)	5	50	680	400	600

4.3.2 Polyhydroxyalkanoate Crystallinity and Characteristic Temperatures

PHB synthesized by bacteria and separated by standard procedures is a semi-crystalline polymer with a rather high crystallinity, which can reach 80%. Although the amorphous character of *in vivo* P3HB is fully accepted, questions remain regarding the mechanism triggering crystallization when the inclusions are isolated. The existence of *in vivo* plasticizers or nucleation inhibitors has been put forward as an explanation [14], but no proof has yet been found to validate this assumption. The crystalline structure of PHB is orthorhombic with basic crystalline cell dimensions of $a = 5.76$ Å, $b = 13.20$ Å, and $c = 5.96$ Å. PHB forms extremely thin lamellar crystals with a thickness ranging between 4 and 7 nm [46]. Due to low nucleation density, PHB shows a small number of very large and rather imperfect crystallites which are responsible for its poor mechanical properties, especially the low elongation at break and the brittleness of moulded products and films. This drawback can be eliminated, to a certain extent, by selecting appropriate crystallization conditions, especially the temperature [47]. The addition of nucleating agents and suitable post-processing treatments can also lead to improved properties.

Numerous studies have examined the glass transition and melting temperatures of P3HB and its copolymers with P3HV. The PHB melting temperature varies significantly with molecular weight. An almost linear increase is observed for low molecular weights (oligomers) after which the melting temperature levels off at about 180 °C for polymer molecular weights above 50 000. As far as P3HB-3HV copolymers are concerned, an almost linear decrease in Tm (from 180 to 137 °C) is observed for HV contents increasing from 0 to 25%. The melting enthalpy for 100% crystalline PHB is reported to be 146 J g^{-1} [47].

Regarding glass transition temperatures, PHAs show T_g values below, but still close to, room temperature so that the material is brittle at such temperatures. However, it is possible to lower the T_g and thus make the materials ductile, for instance by plasticization. PHB copolymers based on longer alkanoate monomer units usually exhibit lower T_g values than that of the corresponding homopolymer. For instance, P (3HB-co-HHx) with 5 and 12% of HHx units displays T_g values of $+4$ °C and -3 °C respectively with corresponding T_m at 144 and 112 °C [44]. Interestingly, the homopolymer of 4-hydroxybutyric acid, P(4HB), shows much lower values for both its glass transition and melting temperatures (respectively ~ -50 and $+60$ °C) when compared with other common PHAs [23, 48]. P(4HB) homopolymer and P(3HB-co-4HB) copolymers are therefore considered very promising materials for future applications.

4.4 POLYHYDROXYALKANOATE DEGRADATION

4.4.1 Hydrolytic Degradation of PHAs

Even if PHAs only take up minor quantities of water upon storage, the possibility of polymer degradation by hydrolysis must be taken into account. Doi *et al.* examined the *in vitro* degradation of PHB solvent-cast films in phosphate buffer at 55 °C and pH 7.4 [49]. The hydrolytic degradation of PHB films occurred throughout the whole polymer matrix (only at the surface for enzymatic degradation) and the molecular weights decreased with time as a result of random chain scission. The weight of the PHB film was unchanged for 48 days, whereas the M_n decreased from 768 000 to 245 000. In the same period, the film thickness increased from 65 to 75 µm, showing that water permeated the polymer matrix during hydrolytic degradation. Whereas hydrolytic degradation of PHB proceeds relatively slowly, the rate of mass loss from films of PHBV copolymers appears to be faster. The degradation behaviour of PHO has been examined during hydrolysis incubation of solution-cast films [50–52]. PHO cast films underwent a simple hydrolytic degradation process characterized by water absorption, gradual molecular weight decrease, and negligible mass loss after 24 months of incubation. The degradation first occurred in the amorphous zone, followed by attack in the crystalline domain. The process is very slow due to the presence of long hydrophobic pendant chains and after a two-year incubation period, weight loss remained lower than 1%. M_w and M_n both slowly decreased with incubation time, reaching approximately 30% of the initial molecular weight without any release of soluble low-molecular weight oligomers.

The generally slow process of hydrolytic degradation in PHAs, occurring as a result of high crystallinity and the presence of pendant hydrophobic alkyl chains, can be increased by addition of other polymers or plasticizers. Amorphous or hydrophilic additives lead to higher water absorption and acceleration in the rate of hydrolysis. In contrast, a reduced degradation rate is observed with the addition of hydrophobic species such as citrate plasticizers [53]. Hydrolytic degradation is also strongly accelerated at higher temperatures or in an alkaline medium [54].

Several studies have shown that the presence of a second component, whatever its chemical nature, could also perturb PHA crystallization and increase hydrolytic degradation. Nevertheless, water absorption in the polymer, directly linked to matrix hydrophobicity, is the most important factor influencing the hydrolytic process.

4.4.2   Biodegradation of PHAs

PHAs can degrade in natural environments such as soil, sludge, fresh water, seawater and compost in which microorganisms can use the degradation products as a carbon source. The ability to degrade PHAs is widely distributed among fungi and bacteria, and depends on PHA depolymerase enzymes, which are carboxyesterases, and on the physical state of the polymer (amorphous or crystalline). As previously mentioned, all polyesters are to some extent prone to degradation by simple hydrolysis and thus in a natural environment it is difficult to separate biodegradation and hydrolysis. PHAs can be rapidly hydrolyzed to their monomers by intracellular or extracellular depolymerase enzymes secreted by a wide variety of bacteria and fungi and many different types of these depolymerases have been identified. As far as biodegradation of PHAs is concerned, only extracellular biodegradation will be discussed here. A considerable amount of knowledge about the biochemical properties of extracellular depolymerases has been gathered in recent years [55, 56]. All the depolymerases contain an N-terminal catalytic domain, a C-terminal substrate-binding domain, and a linker region connecting the two domains. Similar catalytic and binding domains have been identified in other depolymerizing enzymes that hydrolyze water-insoluble polysaccharides such as cellulose, xylan or chitin [57–59]. Further detailed information on the structure and mechanisms of PHA depolymerases can be found elsewhere [55].

Since PHAs are high-molecular weight water-insoluble polymeric materials, which cannot be transported through the cell wall, bacteria and fungi excrete extracellular PHA depolymerases [60] which hydrolyze the material into the monomer and soluble oligomers. Low-molecular weight degradation products are then transported in the degrading microorganisms and subsequently metabolized as carbon and energy sources by intracellular depolymerases. Finally, H_2O and CO_2 are the ultimate products of biodegradation.

Various studies have been conducted on the environmental factors that influence biodegradation, such as the enzymology of the process and the polymer composition. Water content and temperature have been found to be important, along with the microbial activity in any given environment. For instance, in aquatic ecosystems, even under extreme conditions (such as low temperatures, high hydrostatic pressure and absence of sunlight) plastic articles made from PHA are degraded. Besides the environmental conditions, the microstructure and properties of the PHA materials themselves can significantly affect the degradation rates [61–63]. This includes factors such as polymer composition, crystallinity and surface area. Indeed, electron microscopy analysis of PHA films revealed that degradation occurs at the surface by enzymatic hydrolysis. Thus, the degradation rate depends on the surface area available for microbial colonization.

It is worth pointing out that the ability to degrade short-chain-length (scl) PHAs is widespread among bacteria and fungi and a large number of depolymerases have been identified and characterized [64]. However, few data concerning the degradation of medium-chain-length (mcl) PHAs have been reported so far [65–68]. More detailed information on PHA biodegradation can be found elsewhere [69].

4.4.3   Thermal Degradation of PHAs

The very high susceptibility of PHB and other PHAs to thermal degradation is a major issue, especially with respect to processing. The thermal degradation of P(3HB) proceeds mainly by a *cis*-elimination reaction, releasing crotonic acid and linear oligomers of 3HB containing crotonyl chain ends as volatile products [70]. Considering its melting point, the PHB processing temperature should be at least 190 °C. At this temperature, thermal degradation rapidly occurs, so it is impossible to avoid substantial decomposition resulting in a decrease in molecular weight and, consequently, a detrimental effect on the mechanical and other ultimate properties [71]. Transesterification reactions must also be considered when blending PHB with other polymers, especially if they present hydroxyl, carboxyl or other reactive moieties.

Stabilization of PHB by conventional methods is almost impossible. Antioxidants are ineffective, as are other common additives, and many inorganic species, such as aluminum compounds and fumed silica can act as prodegradants [72]. The same negative impact was observed for impurities present in technical PHB when compared with a carefully purified sample [73]. Given the high rate of thermal degradation of PHB, it is necessary to look for processing procedures operating at lower temperatures. Thus, the addition of plasticizers can be an option since this leads to a decrease in the melting temperature of ~20 °C. Nevertheless, this requires rather high amounts of plasticizer addition, up to 30 wt %, and it has also been shown that some plasticizers have a negative impact on PHB thermal stability [74]. Interestingly, the addition of a second monomer in PHA copolymers results in a slightly increased thermal stability, as revealed by thermogravimetric analyses on PHB and its copolymers with hydroxyvalerate and hydroxyhexanoate [75, 76].

In spite of some copolymers showing more acceptable properties, the poor thermal stability of PHAs, especially PHB and PHBV, is a serious issue which makes their processing by conventional thermoplastic technologies quite challenging, as will be discussed below.

4.5   PHA-BASED MULTIPHASE MATERIALS

4.5.1   Generalities

Extrusion of PHA-based materials is generally linked with other processing steps such as thermoforming, injection moulding, fiber drawing, film blowing, bottle blowing, or extrusion coating. The properties of the final material will therefore depend on the specific conditions used during the processing steps. The key parameters during the melt processing will be temperature, residence time, moisture content and atmosphere. However, since PHAs show a low degradation temperature compared to their melting temperature, the major problem in the manufacturing of PHA-based products is the high thermal sensitivity and narrow processing temperature window. PHB thermal [77–84] and thermo-mechanical [85, 86] stability has been well described in the literature. In extrusion processes, increasing the shear level, temperature, and/or residence time [87] leads to a rapid decrease in molecular weight due to macromolecular chain cleavage and as a result a concomitant decrease in melt viscosity. In comparison, PHBV copolymers are more adapted for melt processing because an increase in HV content results in lower melting and glass transition temperatures [42]. To overcome drawbacks such as thermal instability or to introduce new properties in PHAs, a great number of multiphase materials have been developed, mainly by mixing PHB or PHBV with other products.

4.5.2   PHA Plasticization

Many routes including plasticization have been investigated to ease PHB transformation [88] and a number of authors have noted the changes in PHA properties on plasticization (e.g., with addition of citrate ester or triacetin) [40, 89, 90].

Recently, Wang et al. [91] tested different plasticizers such as dioctyl phthalate, dioctyl sebacate, and acetyl tributyl citrate (ATBC) with poly(3-hydroxybutyrate). From the DSC measurements, these researchers concluded that only the addition of ATBC led to an obvious decline in T_g and improved other thermal characteristics. However, ATBC addition does little to improve the mechanical properties.

The effects of biodegradable plasticizers on the thermal and mechanical properties of poly(3-hydroxybutyrate-co-3-hydroxyvalerate) (PHBV) were studied by Choi and Park [92] using thermal and mechanical analyses. Soybean oil (SO), epoxidized soybean oil (ESO), dibutyl phthalate (DBP) and acetyl tributyl citrate (ATBC) were tested as plasticizing additives. PHBV/plasticizer blends were prepared by evaporating solvent from solutions. DPB and ATBC were more effective than soybean oils (SO and ESO) in depressing the glass transition temperature as well as in increasing elongation at break and impact strength of films. Based on the thermal and mechanical properties of the plasticized PHBV, it could be concluded that ATBC or DBP are better plasticizers than SO and ESO for PHBV. From various different studies, it seems that acetyl tributyl citrate (ATBC) may be the most efficient plasticizer for PHB and PHBV.

4.5.3   PHA Blends

Polymer blending offers interesting possibilities for preparing inexpensive materials with adequate mechanical properties and combining the properties of two or more polymers. In this context, PHAs are an important source of biodegradable additives. Several blending studies have been performed on PHB or PHBV copolymers with other compounds [93, 94]. The mechanical properties, morphology, biodegradability, and thermal or crystallization behavior of melt-blended or solvent-cast PHAs have been investigated using various PHBs blended with polymers such as poly(vinyl alcohol) or polysaccharides [95–97], poly(ε-caprolactone) [98, 99], poly(vinyl phenol), poly(vinyl acetate) [100], poly(lactic acid) [101], xylogen [102], and even dendritic polyester oligomers [103].

Reactive blending has also been developed as a method to increase the compatibility between components (e.g., by melt blending with peroxides). The peroxides used in reactive melt blending decompose during processing to form free radicals which crosslink the polymers in the blend [99]. The miscibility and compatibility of PHAs with functional polymers is well documented [94, 104].

4.5.4   PHA-Based Multilayers

Multilayer co-extrusion has been widely used in the past to combine the properties of two or more polymers into one single multilayered structure [105]. However, some problems inherent to the multiphasic nature of the flow are likely to occur during co-extrusion operations, including nonuniform layer distribution, encapsulation, and interfacial instabilities, which are critical since they directly affect the quality and functionality of the multilayer products. There have been extensive experimental and theoretical investigations on these phenomena [106]. The layer encapsulation phenomenon corresponds to the surrounding of the more viscous polymer by the less viscous one. Experimental investigations have shown that viscosity differences between the respective layers dominates over elasticity ratios in regards to the shape of the interface [105]. In experimental studies, Dooley et al. [107] investigated layer rearrangement during co-extrusion and the importance of channel geometry. Their findings indicated that co-extrusion of polymers with matched viscosities can still lead to layer rearrangement as a result of factors associated with die geometry.

Despite the number and diversity of studies on multilayer flows and stability, only a limited number of articles have reported the use of biopolyesters in co-extrusion processes. Different stratified structures have been processed by co-extrusion and studied; however, only a very few studies has been carried out with PHAs and most of these dealt with combinations of PHA (e.g., PHBV) and plasticized starch [105, 108].

Applications of multilayer products based on PHAs as commodities are primarily limited by cost and sometimes by PHA availability. Thus, attention is being focused on products with plastics constituting only a minor part of the material, such as plastic film for moisture barrier in food or drink cartons and in sanitary napkins.

4.5.5   PHA Biocomposites

Biocomposites are obtained by the association of macro-fillers (mainly ligno-cellulose fibers) into a bio-matrix. One of the main advantages of PHAs for biocomposite applications is their polar character since PHAs show better adhesion to ligno-cellulose fibers compared to conventional polyolefins [109]. Consequently, there are now many papers which discuss PHA biocomposites. The addition of cellulose fibers and different fillers has often been proposed as a solution for increasing the mechanical performance and particularly the toughness of PHB and PHBV [109–115].

In terms of crystallization and thermal behavior, no significant effect of cellulose on PHB crystallinity was reported. A slight increase of T_g and delay in the crystallization process has however been observed [111]. The presence of cellulose fibers as a filler increased the rate of PHBV crystallization due to a nucleating effect, while thermal parameters such as percent crystallinity remained unchanged. Studies on the crystallization behavior of PHB/kenaf fiber biocomposites showed that the nucleation of kenaf fibers affected the crystallization kinetics of the PHB matrix [113]. Differences in the effect of cellulose fibers on the crystallization process have been attributed to the lignin content at the surface/interface of the cellulose fiber. Increase in HV content, addition of compatibilizers and increased cellulose fiber content have all been shown to affect the mechanical performance of the corresponding biocomposites.

The addition of HV leads to a reduction in the matrix stiffness and also to increased elongation at break. In reinforced PHBV, a 50–150% enhancement in tensile strength, 30–50% in bending strength and 90% in impact strength have been reported [112]. The addition of increasing HV content to PHB polymers improved the toughness of the natural fiber composites and increased ductility; however crystallization rate was reduced. It has been suggested however that the combination of coupling agents and HV improved the storage modulus and led to a reduction in tan δ [109]. This was attributed to an improvement in the interfacial bonding between PHB and fibers and an increase in transcrystallinity near the fiber interfaces. It has been reported that addition of cellulose fibers led to some improvement in tensile strength and stiffness, but the composites remained brittle [110]. At low contents, the incorporation of cellulose fibers lowered the stiffness; however, higher amounts of cellulose fibers greatly improved the mechanical properties of PHB.

For biocomposites based on cellulose fibers and PHB, the effect of fiber length and surface modification on the tensile and flexural properties of the biocomposites has been investigated. For example, the fracture toughness values of PHB composite materials containing 10–20 wt % straw fibers were higher than those of pure PHB, while biocomposites containing 30–50 wt % straw fibers presented almost the same values as neat PHB [111]. The interfacial shear strength in PHB/flax fiber biocomposites has been improved through addition of interface modifiers [116].

4.5.6 PHA-Based Nano-Biocomposites

Nano-biocomposites are obtained by incorporation of nano-fillers in a bio-matrix. In the literature [117], nano-biocomposites based on PHAs, mainly PHB and PHBV, using different layered silicate clays (e.g., montmorillonite, sepiolite) and preparation methods (i.e., casting, melt processing) have been reported [118–120]. However, until now, full exfoliation of such nano-fillers has not been reported even if the beginning of clay exfoliation has been indicated in few studies [118]. Only intercalated or well-intercalated structures and microcomposites were obtained using either organo-modified or unmodified nanoclays. However, although fully exfoliated structures have not been obtained, improvements in mechanical and thermal properties as well as crystallization and biodegradation rates have been reported. The structure–property relationships in PHA/organically modified montmorillonite (OMMT) nano-biocomposites have been investigated and are generally in good agreement with conclusions drawn in previously reported studies on synthetic polymer-based nano-composites. As a result of heat sensitivity, attention has been paid to PHA degradation in nanocomposite systems [120–122]. Possible effects of the clay organomodifier and/or the mineral clay layer itself on the polymer during processing have been pointed out which may explain the limited nature of PHA improvements even with the addition of well-dispersed nanoclays.

When research findings to date are considered, the poor thermal stability of PHAs and the role of (organo-modified) clays in this respect could present significant obstacles to the production of technically competitive materials. Thus, scientists have been interested in other PHA-based nanocomposites (e.g., incorporating layered double hydroxides or LDHs) [123, 124], cellulose whiskers [125–127] and hydroxy-apatite (HA) [128], the latter being particularly used for biomedical and tissue engineering applications. Hsu et al. [123, 124] reported nanofiller exfoliation when using LDH organically modified by poly(ethylene glycol) phosphonates (PMLDH). The crystallization behavior of the PHB/PMLDH compounds was comparable to that of PHB/OMMT nanocomposites [123, 124]. Regarding the nano-hydroxyapatite filler, as in the case of polymer/clay nanocomposites, the good dispersion of this inorganic nano-filler in PHBV inevitably benefits the improvement of mechanical properties of the materials [128]. Furthermore, these authors pointed out the enhanced material bioactivity in PHBV/hydroxyapatite nanocomposites, a specific property which is desirable for the repair and replacement of bone.

PHA/cellulose whisker materials have been obtained by using a latex of poly (3-hydroxyoctanoate) (PHO) [129] as a matrix and a colloidal suspension of hydro-lyzed cellulose whiskers as a natural and biodegradable filler. Due to the geometry and aspect ratio of cellulose whiskers, the formation of a rigid network and percolation phenomena, higher mechanical properties in PHO/cellulose whisker composites were reported [125–127].

'Green nanocomposites' based on PHA represent a new generation of environmentally friendly materials and through enhancements in processing characteristics such as ductility, melt viscosity and thermal stability, could broaden the range of PHA applications. In addition, advances in macromolecular architecture and nanoparticle-based systems may allow the limitations of PHA-based materials such as high crystallinity, and brittleness to be gradually overcome.

4.6   PRODUCTION AND COMMERCIAL PRODUCTS

PHA production is currently shared between a number of companies (Table 4.5). Worldwide, about 24 companies are known to be presently engaged in PHA production and applications [69]. Compared with PLA, the world production of PHA is low and probably less than 50 000 metric tons per annum. However, it is difficult to be precise because there may be a wide gap between announcements in the media and true production figures. Usually, it is only the production capacities which are published.

The story of PHA industrial production is very long and starts in the 1950s. In the 1970s, Zeneca (formerly ICI) produced several metric tons of PHA copolymers under the trade name Biopol®. In the 1990s, Zeneca UK produced P(3HB-co-3HV) in a pilot plant by bacterial fermentation using a mixture of glucose and propionic acid. In 1996, Zeneca sold its Biopol® business to Monsanto, where investigations started by Zeneca into production of PHA in genetically-modified crops were continued. Monsanto produced Biopol® P(3HB-co-3HV) commercially with HV contents reaching 20% by means of fermentation. However, production ceased at the end of 1999. Metabolix bought the Biopol® assets in 2001. In 2007, Metabolix and Archer Daniels Midland (ADM) formed a joint venture, called Telles, to produce PHA under the trade name Mirel™. ADM began to build the first plant in Clinton, Iowa (USA) in

Table 4.5   Main PHA producers

Company	Country	Trade name	PHA	Pilot/ Industrial scale
Biomatera	Canada	Biomatera	PHBV	Pilot
Biomer	Germany	Biomer	PHB, PHBV	Pilot
Bio-On	Italy	Minerv PHA	PHB, PHBV	Pilot (?)
Kaneka	Japan	Kaneka	PHBHx	Pilot/Ind. (?)
Meredian	USA		Copolymers	Pilot/Ind. (?)
Metabolix and ADM (Telles)	USA	Mirel	Copolymer	Ind. (Start)
PHB Industrial/ Copersucar	Brazil	Biocycle	PHB, PHBV	Pilot/Ind. (?)
PolyFerm Canada	Canada	VersaMer PHA	PHBV and copolymers	Pilot
Tianan	China	Enmat	PHBV	Ind.
Tianjin & DSM	China	GreenBio	Copolymers based on 3HB and 4HB	Pilot (?)
Tianzhu	China	Tianzhu	PHBHx	Pilot

2009, which will be able to produce 50 000 tons of resin per year and start-up was scheduled for late 2010. Metabolix has also developed production of PHA in genetically modified crops. In 2009, the company announced having completed a field trial of tobacco, genetically engineered to express PHA bio-based polymers. This company has also announced that, in greenhouse trials, switchgrass plants engineered using multi-gene expression technology can produce significant amounts of PHA in leaf tissue.

A number of small companies currently produce bacterial PHA. For example, PHB Industrial (Brazil) produces Biocycle® PHB and PHBV (HV = 12%) from sugarcane molasses [130]. Biocycle® production is planned to be 4000 tons/year in 2010 and then to be extended to 14 000 tons/year [131]. In 2004, Procter & Gamble (USA) and Kaneka Corporation (Japan) announced a joint development agreement for the completion of R&D leading to the commercialization of Nodax^{TM}, a large range of polyhydroxybutyrate-co-hydroxyalkanoates (PHBHx, PHBO, PHBOd) [132]. Although large-scale industrial production was planned with a target price around 2€/kg, the Nodax^{TM} development was stopped in 2006 [133]. In 2007 Meredian Inc. purchased P&G's PHA technology. Meredian plans to produce over 270 kTons annually. Besides, Kaneka plans to produce 50 000 metric tons per year of PHBHx in 2010. Tianan, a Chinese company has also announced increases in capacity from the current 2000 tonnes to 10 000 tonnes per year in 2010. The Dutch chemical company DSM has invested in a PHA plant together with a Chinese bio-based plastics company – Tianjin Green Bio-Science Co. The company will start up production of PHAs with an annual capacity of 10 000 tonnes.

In summary, production of PHAs reflects a commercial interest in new 'green' materials as replacements for synthetic nondegradable polymers in a wide range of applications [133]. These applications include products for use in packaging, agriculture, leisure, fast-food and hygiene. The biocompatible nature of PHAs will also continue to stimulate research and developments in medicine [134, 135].

REFERENCES

[1] M. Doudoroff, R. Y. Stanier, *Nature* 1959, **183**, 1440–1442.
[2] D. H. Williamson, J. F. Wilkinson, *J. Gen. Microbiol.* 1958, **19**, 198–209.
[3] R. M. Macrae, J. F. Wilkinson, *J. Gen. Microbiol.* 1958, **19**, 210–222.
[4] L. L. Wallen, W. K. Rohwedder, *Environ. Sci. Technol.* 1974, **8**, 576–579.
[5] M. J. de Smet, J. Kingma, H. Wynberg, *et al.*, *Enzyme Microb. Technol.* 1983, **5**, 352–360.
[6] E. A. Dawes, P. J. Senior, *Adv. Microb. Physiol.* 1973, **10**, 135–266.
[7] L. A. Kominek, H. O. Halvorson, *J. Bacteriol.* 1965, **90**, 1251–1259.
[8] H. G. Schlegel, G. Gottschalk, and R. Von Bartha, *Nature* 1961, **191**, 463–465.
[9] M. Kunioka, Y. Nakamura, and Y. Doi, *Polym. Commun. (Guildford)* 1988, **29**, 174–176.
[10] P. Schubert, A. Steinbüchel, and H. G. Schlegel, *J. Bacteriol.* 1988, **170**, 5837–5847.
[11] S. C. Slater, W. H. Voige, and D. E. Dennis, *J. Bacteriol.* 1988, **170**, 4431–4436.
[12] O. P. Peoples, A. J. Sinskey, *J. Biol. Chem.* 1989, **264**, 15293–15297.
[13] G. Braunegg, G. Lefebvre, and K. F. Genser, *J. Biotechnol.* 1998, **65**, 127–161.
[14] K. Sudesh, H. Abe, and Y. Doi, *Prog. Polym. Sci.* 2000, **25**, 1503–1555.

[15] P. A. Holmes, *Phys. Technol.* 1985, **16**, 32–36.
[16] T. Araki, S. Hayase, *J. Polym. Sci., Part A: Polym. Chem.* 1979, **17**, 1877–1881.
[17] Y. Doi, M. Kunioka, Y. Nakamura, *et al., J. Chem. Soc., Chem. Commun.* 1986, 1696–1697.
[18] Y. Doi, A. Tamaki, M. Kunioka, *et al., Appl. Microbiol. Biotechnol.* 1988, **28**, 330–334.
[19] Y. Doi, M. Kunioka, Y. Nakamura, *et al., Macromolecules* 1988, **21**, 2722–2727.
[20] M. Kunioka, Y. Kawaguchi, and Y. Doi, *Appl. Microbiol. Biotechnol.* 1989, **30**, 569–573.
[21] S. Nakamura, Y. Doi, and M. Scandola, *Macromolecules* 1992, **25**, 4237–4241.
[22] M. Hiramitsu, N. Koyama, and Y. Doi, *Biotechnol. Lett* 1993, **15**, 461–464.
[23] Y. Saito, S. Nakamura, M. Hiramitsu, *et al., Polym. Int.* 1996, **39**, 169–174.
[24] A. J. Anderson, E. A. Dawes, *Microbiol. Rev.* 1990, **54**, 450–472.
[25] S. Slater, T. Gallaher, and D. Dennis, *Appl. Environ. Microbiol.* 1992, **58**, 1089–1094.
[26] T. U. Gerngross, K. D. Snell, O. P. Peoples, *et al., Biochemistry* 1994, **33**, 9311–9320.
[27] L. Su, R. W. Lenz, Y. Takagi, *et al., Macromolecules* 2000, **33**, 229–231.
[28] Y. Poirier, D. Dennis, K. Klomparens, *et al., FEMS Microbiol. Rev.* 1992, **103**, 237–246.
[29] Y. Poirier, D. E. Dennis, K. Klomparens, *et al., Science* 1992, **256**, 520–523.
[30] L. Allenbach, Y. Poirier, *Plant Physiol.* 2000, **124**, 1159–1168.
[31] H. E. Valentin, T. A. Mitsky, D. A. Mahadeo, *et al., Appl. Environ. Microbiol.* 2000, **66**, 5253–5258.
[32] P. Saruul, F. Srienc, D. A. Somers, *et al., Crop Sci.* 2002, **42**, 919–927.
[33] K. Bohmert, I. Balbo, A. Steinbüchel, *et al., Plant Physiol.* 2002, **128**, 1282–1290.
[34] H. M. Müller, D. Seebach, *Angew. Chem. Int. Ed.* 1993, **32**, 477–502.
[35] S. Matsumura, *Adv. Polym. Sci.* 2006, **194**, 95–132.
[36] A. C. Albertsson, R. K. Srivastava, *Adv. Drug Del. Rev.* 2008, **60**, 1077–1093.
[37] D. E. Agostini, J. B. Lando, and J. R. Shelton, *J. Polym. Sci., Part A: Polym. Chem.* 1971, **9**, 2775–2787.
[38] J. R. Shelton, D. E. Agostini, and J. B. Lando, *J. Polym. Sci., Part A: Polym. Chem.* 1971, **9**, 2789–2799.
[39] G. W. Haywood, A. J. Anderson, D. R. Williams, *et al., Int. J. Biol. Macromol.* 1991, **13**, 83–88.
[40] M. A. Kotnis, G. S. O'Brien, and J. L. Willett, *J. Environ. Polym. Degrad.* 1995, **3**, 97–105.
[41] R. Shogren, *J. Environ. Polym. Degrad.* 1997, **5**, 91–95.
[42] W. Amass, A. Amass, and B. Tighe, *Polym. Int.* 1998, **47**, 89–144.
[43] N. Yoshie, M. Saito, and Y. Inoue, *Macromolecules* 2001, **34**, 8953–8960.
[44] J. J. Fischer, Y. Aoyagi, M. Enoki, *et al., Polym. Degrad. Stab.* 2004, **83**, 453–460.
[45] Y. Doi, S. Kitamura, and H. Abe, *Macromolecules* 1995, **28**, 4822–4828.
[46] D. Seebach, M. G. Fritz, *Int. J. Biol. Macromol.* 1999, **25**, 217–236.
[47] P. J. Barham, A. Keller, E. L. Otun, *et al., J. Mater. Sci.* 1984, **19**, 2781–2794.
[48] Y. Saito, Y. Doi, *Int. J. Biol. Macromol.* 1994, **16**, 99–104.
[49] Y. Doi, Y. Kanesawa, M. Kunioka, *et al., Macromolecules* 1990, **23**, 26–31.
[50] Y. Marois, Z. Zhang, M. Vert, *et al., Tissue Eng.* 1999, **5**, 369–386.
[51] Y. Marois, Z. Zhang, M. Vert, *et al., J. Biomater. Sci., Polym. Ed.* 1999, **10**, 483–499.
[52] Y. Marois, Z. Zhang, M. Vert, *et al., J. Biomed. Mater. Res.* 2000, **49**, 216–224.
[53] T. Freier, C. Kunze, C. Nischan, *et al., Biomaterials* 2002, **23**, 2649–2657.

[54] A. C. Albertsson, S. Karlsson, *Acta Polym.* 1995, 46, 114–123.

[55] D. Jendrossek, A. Schirmer, and H. G. Schlegel, *Appl. Microbiol. Biotechnol.* 1996, 46, 451–463.

[56] D. Jendrossek, R. Handrick, *Annu. Rev. Microbiol.* 2002, 56, 403–432.

[57] L. E. Kellett, D. M. Poole, L. M. A. Ferreira, *et al.*, *Biochem. J* 1990, 272, 369–376.

[58] T. Watanabe, K. Suzuki, W. Oyanagi, *et al.*, *J. Biol. Chem.* 1990, 265, 15659–15665.

[59] N. R. Gilkes, B. Henrissat, D. G. Kilburn, *et al.*, *Microbiol. Rev.* 1991, 55, 303–315.

[60] D. Jendrossek, *Polym. Degrad. Stab.* 1998, 59, 317–325.

[61] H. Abe, Y. Doi, *Macromolecules* 1996, 29, 8683–8688.

[62] H. Abe, Y. Doi, H. Aoki, *et al.*, *Macromolecules* 1998, 31, 1791–1797.

[63] H. Abe, Y. Doi, *Int. J. Biol. Macromol.* 1999, 25, 185–192.

[64] D. Jendrossek, *Appl. Microbiol. Biotechnol.* 2007, 74, 1186–1196.

[65] H. Kim, H. S. Ju, and J. Kim, *Appl. Microbiol. Biotechnol.* 2000, 53, 323–327.

[66] H. M. Kim, K. E. Ryu, K. S. Bae, *et al.*, *J. Biosci. Bioeng.* 2000, 89, 196–198.

[67] D. Y. Kim, J. S. Nam, and Y. H. Rhee, *Biomacromolecules* 2002, 3, 291–296.

[68] H. J. Kim, D. Y. Kim, J. S. Nam, *et al.*, *Antonie Van Leeuwenhoek* 2003, 83, 183–189.

[69] G.-Q. Chen, (2010), *Plastics from Bacteria: Natural Functions and Applications*, Springer-Verlag, Berlin, Germany, 450 pp.

[70] K. J. Kim, Y. Doi, and H. Abe, *Polym. Degrad. Stab.* 2006, 91, 769–777.

[71] A. Hoffmann, S. Kreuzberger, and G. Hinrichsen, *Polym. Bull.* 1994, 33, 355–359.

[72] K. Csomorová, J. Rychlý, D. Bakos, *et al.*, *Polym. Degrad. Stab.* 1994, 43, 441–446.

[73] F. D. Kopinke, M. Remmler, and K. Mackenzie, *Polym. Degrad. Stab.* 1996, 52, 25–38.

[74] I. Janigová, I. Lacík, and I. Chodák, *Polym. Degrad. Stab.* 2002, 77, 35–41.

[75] J. D. He, M. K. Cheung, P. H. Yu, *et al.*, *J. Appl. Polym. Sci.* 2001, 82, 90–98.

[76] S. D. Li, P. H. Yu, and M. K. Cheung, *J. Appl. Polym. Sci.* 2001, 80, 2237–2244.

[77] N. Grassie, E. J. Murray, and P. A. Holmes, *Polym. Degrad. Stab.* 1984, 6, 47–61.

[78] N. Grassie, E. J. Murray, and P. A. Holmes, *Polym. Degrad. Stab.* 1984, 6, 95–103.

[79] N. Grassie, E. J. Murray, and P. A. Holmes, *Polym. Degrad. Stab.* 1984, 6, 127–134.

[80] M. Kunioka, Y. Doi, *Macromolecules* 1990, 23, 1933–1936.

[81] Y. Aoyagi, K. Yamashita, and Y. Doi, *Polym. Degrad. Stab.* 2002, 76, 53–59.

[82] S.-D. Li, J.-D. He, P. H. Yu, *et al.*, *J. Appl. Polym. Sci.* 2003, 89, 1530–1536.

[83] H. Abe, *Macromol. Biosci.* 2006, 6, 469–486.

[84] F. Carrasco, D. Dionisi, A. Martinelli, *et al.*, *J. Appl. Polym. Sci.* 2006, 100, 2111–2121.

[85] D. H. Melik, L. A. Schechtman, *Polym. Eng. Sci.* 1995, 35, 1795–1806.

[86] R. Renstad, S. Karlsson, and A.-C. Albertsson, *Polym. Degrad. Stab.* 1997, 57, 331–338.

[87] D. H. S. Ramkumar, M. Bhattacharya, *Polym. Eng. Sci.* 1998, 38, 1426–1435.

[88] N. C. Billingham, T. J. Henman, and P. A. Holmes, Degradation and stabilisation of polyesters of biological and synthestic origin, in N. Grassie (ed.), *Developments in Polymer Degradation - Vol 7*, Elsevier Applied Science, London, UK, pp 81–121, (1987).

[89] R. L. Shogren, *J. Environ. Polym. Degrad.* 1995, 3, 75–80.

[90] R. C. Baltieri, L. H. I. Mei, and J. Bartoli, *Macromol. Symp.* 2003, 197, 33–44.

[91] L. Wang, W. Zhu, X. Wang, *et al.*, *J. Appl. Polym. Sci.* 2008, 107, 166–173.

[92] J. S. Choi, W. H. Park, *Polym. Test.* 2004, 23, 455–460.

[93] J.-C. Huang, A. S. Shetty, and M.-S. Wang, *Adv. Polym. Tech.* 1990, **10**, 23–30.
[94] H. Verhoogt, B. A. Ramsay, and B. D. Favis, *Polymer* 1994, **35**, 5155–5169.
[95] L. Finelli, M. Scandola, and P. Sadocco, *Macromol. Chem. Phys.* 1998, **199**, 695–703.
[96] G. Cárdenas T, J. Sanzana L, and L. H. Inoccentini Mei, *Bol. Soc. Chil. Quim.* 2002, **47**, 529–535.
[97] K. Sriroth, K. Sangseethong, *Acta Horticulturae* 2006, **703**, 145–151.
[98] F. Gassner, A. J. Owen, *Polymer* 1994, **35**, 2233–2236.
[99] M. Grimaldi, B. Immirzi, M. Malinconico, *et al., J. Mater. Sci.* 1996, **31**, 6155–6162.
[100] Y. An, L. Li, L. Dong, *et al., J. Polym. Sci., Part B: Polym. Phys.* 1999, **37**, 443–450.
[101] L. M. W. K. Gunaratne, R. A. Shanks, *Polym. Eng. Sci.* 2008, **48**, 1683–1692.
[102] K. Weihua, Y. He, N. Asakawa, *et al., J. Appl. Polym. Sci.* 2004, **94**, 2466–2474.
[103] S. Xu, R. Luo, L. Wu, *et al., J. Appl. Polym. Sci.* 2006, **102**, 3782–3790.
[104] R. Sharma, A. R. Ray, *Polymer Reviews* 1995, **35**, 327–359.
[105] O. Martin, E. Schwach, L. Avérous, *et al., Starch - Starke* 2001, **53**, 372–380.
[106] P. D. Anderson, J. Dooley, and H. E. H. Meijer, *Appl. Rheol.* 2006, **16**, 198–205.
[107] J. Dooley, K. S. Hyun, and K. Hughes, *Polym. Eng. Sci.* 1998, **38**, 1060–1071.
[108] L. Avérous, C. Fringant, and O. Martin, Coextrusion of biodegradable starch-based materials, in P. Colonna and S. Guilbert (eds.), *Biopolymer Science: Food and non food applications*, INRA, Versailles, France, pp 207–212, (1999).
[109] R. A. Shanks, A. Hodzic, and S. Wong, *J. Appl. Polym. Sci.* 2004, **91**, 2114–2121.
[110] P. Gatenholm, J. Kubát, and A. Mathiasson, *J. Appl. Polym. Sci.* 1992, **45**, 1667–1677.
[111] M. Avella, E. Martuscelli, B. Pascucci, *et al., J. Appl. Polym. Sci.* 1993, **49**, 2091–2103.
[112] A. K. Mohanty, M. Misra, and G. Hinrichsen, *Macromol. Mater. Eng.* 2000, **276–277**, 1–24.
[113] M. Avella, G. Bogoeva-Gaceva, A. Buzõarovska, *et al., J. Appl. Polym. Sci.* 2007, **104**, 3192–3200.
[114] V. P. Cyras, M. S. Commisso, A. N. Mauri, *et al., J. Appl. Polym. Sci.* 2007, **106**, 749–756.
[115] N. M. Barkoula, S. K. Garkhail, and T. Peijs, *Ind. Crop. Prod.* 2010, **31**, 34–42.
[116] S. Wong, R. A. Shanks, and A. Hodzic, *Compos. Sci. Technol.* 2007, **67**, 2478–2484.
[117] P. Bordes, E. Pollet, and L. Averous, Potential Use of Polyhydroxyalkanoate (PHA) for Biocomposites Development, in A. K. T. Lau, F. Hussain and K. Ladfi (eds.), *Nano- and Biocomposites*, CRC Press, Boca Raton, Fl., USA, pp. 193–225, (2009).
[118] P. Bordes, E. Pollet, S. Bourbigot, *et al., Macromol. Chem. Phys.* 2008, **209**, 1474–1484.
[119] P. Bordes, E. Pollet, and L. Avérous, *Prog. Polym. Sci.* 2009, **34**, 125–155.
[120] L. Cabedo, D. Plackett, E. Giménez, *et al., J. Appl. Polym. Sci.* 2009, **112**, 3669–3676.
[121] E. Hablot, P. Bordes, E. Pollet, *et al., Polym. Degrad. Stab.* 2008, **93**, 413–421.
[122] P. Bordes, E. Hablot, E. Pollet, *et al., Polym. Degrad. Stab.* 2009, **94**, 789–796.
[123] S.-F. Hsu, T.-M. Wu, and C.-S. Liao, *J. Polym. Sci., Part B: Polym. Phys.* 2006, **44**, 3337–3347.
[124] S.-F. Hsu, T.-M. Wu, and C.-S. Liao, *J. Polym. Sci., Part B: Polym. Phys.* 2007, **45**, 995–1002.
[125] D. Dubief, E. Samain, and A. Dufresne, *Macromolecules* 1999, **32**, 5765–5771.
[126] A. Dufresne, *Compos. Interfaces* 2000, **7**, 53–67.
[127] A. Dufresne, M. B. Kellerhals, and B. Witholt, *Macromolecules* 1999, **32**, 7396–7401.

[128] D. Z. Chen, C. Y. Tang, K. C. Chan, *et al.*, *Compos. Sci. Technol.* 2007, **67**, 1617–1626.

[129] A. Dufresne, E. Samain, *Macromolecules* 1998, **31**, 6426–6433.

[130] A. El-Hadi, R. Schnabel, E. Straube, *et al.*, *Polym. Test.* 2002, **21**, 665–674.

[131] L. Velho, P. Velho, *The development of a sugar-based plastic in Brazil*, Employment-oriented Industry Studies Report, Human Science Research Council, Pretoria, South Africa, (2006).

[132] I. Noda, P. R. Green, M. M. Satkowski, *et al.*, *Biomacromolecules* 2005, **6**, 580–586.

[133] S. Philip, T. Keshavarz, and I. Roy, *J. Chem. Technol. Biotechnol.* 2007, **82**, 233–247.

[134] S. F. Williams, D. P. Martin, D. M. Horowitz, *et al.*, *Int. J. Biol. Macromol.* 1999, **25**, 111–121.

[135] M. Zinn, B. Witholt, and T. Egli, *Adv. Drug Del. Rev.* 2001, **53**, 5–21.

5

Chitosan for Film and Coating Applications

Patricia Fernández-Saiz and José M. Lagaron

Novel Materials and Nanotechnology Group, IATA-CSIC, Paterna, Spain

5.1 INTRODUCTION

This chapter reviews the properties of chitosan and chitosan derivatives and assesses the value of the biopolymer for coating and film applications, particularly in the food area. The chapter places special emphasis on describing advances that have enabled a greater understanding of the biocide capacity of the biopolymer and that have also allowed the biomaterial to adapt to particular functionalities and application requirements in film form.

Chitin, a polymer of N-acetylglucosamine (β-1,4 linked 2-acetamido-D-glucose), is a cellulose-like biopolymer present in the exoskeleton of crustaceans and in cell walls of fungi, insects, and yeast. In a similar role to that of cellulose in plants, the biopolymer acts as a supportive and protective material for biological systems. Chitin is said to be produced at approximately 1×10^9 metric tons annually and is the second most abundant natural biopolymer in the world. Chitosan is derived from chitin by deacetylation, to different degrees, in the presence of alkali. Therefore, chitosan is a copolymer consisting of β-$(1 \rightarrow 4)$-2-acetamido-D-glucose and β-$(1 \rightarrow 4)$-2-amino-D-glucose units with the latter usually exceeding 80% [1].

Biopolymers – New Materials for Sustainable Films and Coatings, First Edition.
Edited by David Plackett.
© 2011 John Wiley & Sons, Ltd. Published 2011 by John Wiley & Sons, Ltd.

Chitin is a commercial by-product or a waste from crab, shrimp, and crawfish processing industries. However, isolation and preparation of chitin from various crustacean shells has been carried out [2–7]. Chitin and chitosan offer a wide range of applications, including clarification and purification of water and beverages, uses in pharmaceuticals and cosmetics, as well as in the agricultural, food, and biotechnology sectors [4].

Recently, efficient utilization of marine biomass resources has become an environmental priority and therefore interest in chitin and chitosan has intensified. Early applications included the treatment of wastewater and use as heavy metal adsorption agents in the industry, immobilization of enzymes and cells, resins for chromatography, functional membranes in biotechnology, seed coatings and animal feed in agriculture, artificial skin, absorbable surgical sutures, and wound healing accelerators in the medical field. Chitin and chitosan have also been developed as new physiological materials lately since they possess antitumor activity by immuno-enhancing capacity, antibacterial activity, hypocholesterolemic activity, and antihypertensive action [8].

Chitosan (β-(1,4)-2amino-2-deoxy-D-glucose) is a biodegradable, biocompatible, nontoxic aminopolysaccharide. While it has been widely reported that this material has unique and wide potential interest as an inherent antimicrobial film-forming material, there is still insufficient awareness as to which mechanisms and processing conditions provide chitosan and chitosan derivatives with optimum biocidal performance [9, 10]. However, this understanding is of the outmost importance in devising methodologies, materials and processing technologies for inclusion of chitosan in applications involving active packaging, antimicrobial surfaces, antitumoral activity, hemostatic activity and accelerated wound healing. These applications are of interest in diverse industrial fields incuding food, biomedicine, biotechnology, environment, packaging and food coatings [9, 10].

The typical means of processing chitosan to obtain films has been by casting from organic acidic water solutions [11]. While there is no doubt that chitosan is already widely used internationally for various applications, the full potential for its use in film packaging and coating applications has yet to be realized. An opportunity for high-value chitosan-based materials in films and coatings now exists in Europe, particularly since the European Food Safety Authority (EFSA) has recently issued a new regulation dealing with active and intelligent packaging that finally legislates use of such technologies in the EU area (Commission Regulation (EC) No. 450/2009).

5.2 PHYSICAL AND CHEMICAL CHARACTERIZATION OF CHITOSAN

5.2.1 Degree of N-acetylation

Chitosan is the universally accepted nontoxic N-deacetylated derivative of chitin, where chitin is N-deacetylated to such an extent that it becomes soluble in dilute aqueous acetic and formic acids. In chitin, the acetylated units prevail (degree of acetylation typically = 0.90). Chitosan is the fully or partially N-deacetylated derivative of chitin with a typical degree of acetylation of less than 0.35. In order to define this ratio, attempts have been made with many analytical tools [12–21], which include IR spectroscopy, pyrolysis gas chromatography, gel permeation chromatography and UV spectrophotometry, first derivative UV spectrophotometry, ^1H-NMR spectroscopy, ^{13}C-solid-state NMR spectroscopy, thermal analysis, various titration schemes, acid hydrolysis and HPLC, separation spectrometry methods and, more recently, near-infrared spectroscopy [22].

5.2.2 Molecular Weight

Chitosan molecular weight distributions have been obtained using HPLC [23]. The weight-average molecular weight (Mw) of chitin and chitosan has been determined by light scattering [24]. Viscometry is also a simple and rapid method for the determination of molecular weight [25]. The weight-average molecular weight of chitin is 1.03×10^6 to 2.5×10^6, but the N-deacetylation reaction reduces this to 1×10^5 to 5×10^5 [26].

5.2.3 Solvent and Solution Properties

Both cellulose and chitin are highly crystalline, intractable materials and only a limited number of solvents are known which are applicable as reaction solvents. Chitin and chitosan degrade before melting, which is typical for polysaccharides with extensive hydrogen bonding. This makes it necessary to dissolve chitin and chitosan in an appropriate solvent to impart functionality. For each solvent system, polymer concentration, pH, counterion concentration and temperature effects on the solution viscosity must be known. Comparative data from solvent to solvent are not available. As a general rule, the maximum amount of polymer is dissolved in a given solvent that will provide a homogeneous solution. Subsequently, the polymer is regenerated in the required form [25].

Chitin/chitosan-based materials find application in the form of powders and flakes, but foremost as gels, beads, membranes, capsules, fibers, sponges and scaffolds. The methods for chitosan gel preparation described in the literature can be broadly divided into five groups: (1) solvent evaporation method (casting method), most commonly used; (2) neutralization method, by adding chitosan solution dropwise to a solution of NaOH; (3) cross-linking method, by the use of a cross-linking bath; (4) ionotropic gelation method, by adding an anionic polyelectrolyte solution (alginate, carrageenan, xanthan) dropwise into an acidic chitosan solution; (5) a freeze-drying method, which consists of freezing chitosan solutions or gels followed by lyophilization [27].

5.3 PROPERTIES AND APPLICATIONS OF CHITOSAN

As a result of of the diverse range of applications for chitosan, the potential and interest in this biomacromolecule has never been so high. In addition, by suitable chemical or enzymatic modification, polymer grafting, blending or selective depolymerization, new possibilities may be realized for adding value to this versatile bio-based material.

5.3.1   Waste/Effluent Water Purification

Treatment of waste waters is one of the major current applications of chitin/chitosan-based formulations due to their excellent coagulating, flocculating and metal-chelating properties. Awareness of the ecological and health problems associated with heavy metals and pesticides and their accumulation through the food chain has prompted the need for rigorous purification of industrial water [28, 29]. The ability of the free NH_2 groups of chitosan to form coordinate/covalent bonds with metal ions is of great interest in such applications [30]. Thus, chitosan powder and/or dried chitosan films are of considerable use in metal ion complexing because chitosan presents most of its free amino groups above the pKa value (≈ 6.5). Use of chitosan for potable water purification has been approved by the United States Environmental Protection Agency (EPA) up to a maximum level of 10 ppm [28]. Furthermore, chitosan, carboxymethyl chitosan, and cross-linked chitosan are effective in removing Cd^{2+}, Cu^{2+}, Hg^{2+}, Ni^{2+} and Zn^{2+} from waste water and industrial effluents [31, 32].

5.3.2   Cosmetics

Since chitosan becomes viscous upon being solubilized in acid solutions, it is often used in creams, lotions and nail lacquers. Depolymerized chitosan is being used as an active ingredient in products such as hair shampoo and conditioner since its aqueous solutions are viscous, moisture retaining and impart softness to hair and skin.

5.3.3   Fat Trapping Agent

Due to the ability of this biopolymer to form ionic bonds at low pH, chitosan can bind in vitro to different types of anions derived from bile acids or free fatty acids [33]. Large proportions of these bound lipids are thus excreted from the body. Inside the digestive tract, chitosan forms micelles with cholesterol in the alkaline fluids of the upper intestine, resulting in a depressed absorption of dietary cholesterol and circulation of cholic acid to the liver. The formation of cholic acid from the blood cholesterol in the liver tends to decrease blood cholesterol concentration. Deuchi et al. [34] hypothesized that chitosan is solubilized in the stomach to form an emulsion containing intragastric oil droplets and begins to precipitate in the small intestine at pH 6–6.5. With the aggregation of the polysaccharide chains, the oil droplets are entrapped in their matrices thereby passing through the lumen and being excreted through the faeces.

5.3.4   Pharmaceutical and Biomedical Applications: Controlled Drug Release, Tissue Engineering

Chitosan has been used in the pharmaceutical industry for a variety of applications. As a drug vehicle, chitosan gels are used in the form of beads, capsules, bioadhesive gels and films for various deliveries that include oral, parenteral, transdermal, ophthalmic and nasal routes [35]. A simple example is the reduced irritation effect of aspirin on the gastric mucosa due to both the swelling of chitosan in the acidic environment and the inherent antacid and antiulcer properties of chitosan [35]. Specific and advanced applications include the colon-targeted delivery of peptide drugs, which has been achieved by loading the drugs into PEG-grafted chitosan nanoparticles [36]. Through this system, the drugs are protected against degradation by peptide digestive enzymes and the low pH environment in both the stomach and small intestine. Other recent applications for chitosan-based controlled drug delivery systems are in implants [37, 38]. The nontoxic, nonthrombogenic and noncarcinogenic properties of chitosan make this polymer a viable biomolecule for this attractive application, allowing drug stability, biodegradability and sterility.

Tissue engineering makes use of polymer scaffolds with the capacity to promote adherence, proliferation and differentiation of cells [35]. During use, scaffolds are most commonly degraded or integrated with the tissue, ideally at a rate corresponding to the rate of new tissue formation. Other functions of polymer scaffolds in tissue engineering are space filling and controlled release of growth factors or nutrients. Chitosan is an appealing candidate for tissue engineering applications and as a result of its controllable biological, physical and chemical properties now belongs to the class of most frequently studied biopolymers, along with alginate, collagen and hyaluronic acid. A more recent concept in tissue engineering systems is so-called injectable technology. By making use of this approach, living cells or therapeutic agents are incorporated prior to injection and surgery is not required. For this last application, chitosan–calcium phosphate composites and temperature-sensitive neutral chitosan–polyol salt solutions have been studied [39, 40].

5.3.5   Antimicrobial Properties and Active Packaging Applications

Some chitosan formulations present inherent antimicrobial character against the growth of pathogen microorganisms as has already been widely demonstrated in a range of foods such as bread, strawberries, juices, mayonnaise, milk and rice cakes [41–46]. Moreover, the use of chitosan as an anti-infective biomaterial is also of relevance in other areas such as in the biomedical and pharmaceutical fields [27, 47–49].

Although the exact mechanism by which chitosan exerts its antimicrobial activity is currently unknown, it has been suggested that the polycationic nature of this biopolymer in acidic solutions below pH 6.5 is a crucial factor. Thus, it has been proposed that the positively charged amino groups of the glucosamine units interact with negatively charged components in microbial cell membranes, altering their barrier properties and thereby preventing the entry of nutrients and/or causing the leakage of intracellular contents [50–54]. Another reported mechanism involves the penetration of low-molecular weight chitosan into cells, binding to DNA and subsequent inhibition of RNA and protein synthesis [55]. Chitosan has also been shown to activate several defence processes in plant tissues and to inhibit the production of toxins and microbial growth due to its metal ion-chelating properties [25, 56].

Chitosan has shown tremendous potential as an antimicrobial component and therefore a clear understanding of the linkage between optimum biocidal performance and characteristics such as molecular structure, functional chemistry and physical state is of paramount importance from an applied viewpoint. The antimicrobial mechanism of chitosan was studied using FTIR spectroscopy by monitoring the intensity of the carboxylate biocide groups at 1546 and at 1405 cm^{-1} in various chitosan samples (Figure 5.1) [57, 58]. These authors demonstrated that high-molecular weight chitosan appears to act only in its acetate (salt) form and as carrier of activated protonated species which, by passing from chitosan film to the microbial solution leads to the inactivation of certain microorganisms. So, in essence, the biocidal properties of chitosan function through migration from the solid form. Furthermore, the higher the water dispersability the film presents, the higher the antimicrobial action that is exhibited. The same authors also developed a novel methodology based on the use of a normalized infrared band centered at 1405 cm^{-1} and its correlation with biocide properties [57].

Regarding its application in real food systems, the effectiveness of chitosan against food spoilage microorganisms has been tested in recent years by assessing its use as a natural preservative when applied as an edible coating. Specifically, Tsai et al. [59] studied application in fish preservation and obtained an increase in shelf-life of the product from five to nine days. Other research performed by Sagoo et al. [60] evaluated the effect of treatments with chitosan solutions and reached an increase in the shelf-life of raw sausages stored at chill temperatures from 7 to 15 days. More recent work on the effect of chitosan–gelatin blends as a coating for fish patties showed a difference of around 2 log cycles between the control and the coated

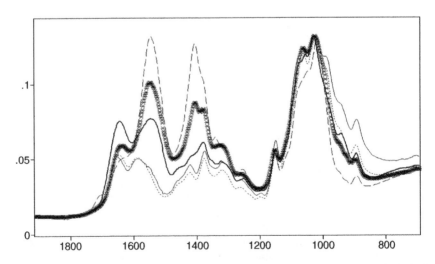

Figure 5.1 FTIR spectra of just-formed chitosan acetate film over an ATR crystal (dashed line), chitosan acetate film prior to microbial testing (line with crosses), chitosan acetate film after microbial testing (thicker line), neutralized chitosan acetate film (dotted line) and as-received chitosan powder (continuous line). Reprinted from *Biopolymers*, 83(6), 577–583. Copyright (2006) with permission from John Wiley & Sons

batches in terms of total bacterial counts of *Pseudomonas* and enterobacteria at 8–11 days of storage [61]. In this study, the addition of powdered chitosan to the mixture of patty ingredients was also investigated, but no antimicrobial effect was observed due to its poor solubility at neutral pH. The effect of the addition of low-molecular weight chitosan to raw rice before steam cooking was also analyzed by Tsai *et al.* [62] who found an effective inhibition of the total anaerobic bacteria and *B. cereus* in cooked rice stored at 37 °C and 18 °C for 48 and 72 h respectively. Other work performed by Juneja *et al.* [63] demonstrated that the incorporation of chitosan glutamate into beef or turkey can reduce the potential risk of *C. perfringens* spore germination and development during abusive cooling from 54.4 to 7.2 °C in 12, 15 or 18 h. In all these studies, chitosan was always applied as a food coating and, to our knowledge, very little research has been conducted to elucidate the biocidal properties of chitosan films in regard to application as an internal coating of packaging materials or as a product contact layer. In this respect, Ye *et al.* [64] studied the control of *L. monocytogenes* on ham steaks by making use of chitosan-coated plastic films. In this research the antimicrobial properties of chitosan became negligible when this polysaccharide was tested in the form of insoluble films, since diffusion through a solid medium such as agar was not feasible. In contrast, Ouattara *et al.* [65] obtained satisfactory results when studying the inhibition of surface spoilage bacteria in processed meats by application of chitosan-based films. More recent work with satisfactory results has been published by Fernández-Saiz *et al.* [66], who studied the antimicrobial properties of chitosan films in direct contact with fish soup against *S. aureus, Salmonella spp.* and *Listeria monocytogenes* at various incubation temperatures.

Focusing more in the context of active food packaging applications, several studies have already demonstrated the antibacterial and antifungal action of chitosan-based films for bioactive packaging applications [53, 67, 68]. Systematic research has also been reported on the effect of plasticizer concentration, storage time [69], acid types and concentrations [70], molecular weight [71] and degree of deacetylation [72] on the physical and active properties of chitosan films.

The biggest drawback in use of chitosan films is arguably their high hygroscopicity. In fact, depending on the molecular weight, chemistry, processing and storage conditions, this material may virtually dissolve in the presence of high moisture products. In food packaging, for example, the dissolution of the biopolymer could compromise packaging structure, physical integrity and organoleptic or microbiological food quality aspects and compromise its application. Chemical features such as the degree of deacetylation, the molecular weight and the functionalization of chitosan are crucial factors which determine the properties of the films obtained [73–78]. Interestingly, and in accordance with published reports, films obtained from high-molecular weight chitosan with lower degrees of deacetylation show better water resistance but, for instance, poorer biocidal properties [79–81]. Thus, as the solubility of chitosan films diminishes, its antimicrobial capacity is also reduced, in good agreement with the biocidal mechanisms described above. For example, as the number of protonated amine groups in chitosonium acetate film is reduced by alkaline neutralization, the film becomes water insoluble and no longer exhibits antimicrobial performance. Therefore, by means of physical or chemical treatment, film disintegration in water can be prevented, but, at the same time, biocidal properties are reduced. In this sense, the incorporation of cross-linking agents which form covalent bonds between chitosan chains, preventing polysaccharide-water interactions could be a means to retain film structure. Previous studies in this direction have involved incorporation of compounds such as glutaraldehyde, glyoxal or epichlorohydrin [82–84]. However, Tang *et al.* [85] confirmed that the antimicrobial capacity of chitosan films was strongly diminished with increase in cross-linker content. Other authors showed that cross-linking agents like geniposidic acid improved the mechanical properties of chitosan films and did not alter their antibacterial activity when the cross-linking degree was relatively low [86]. Of course, an additional concern is the inherent toxicity of the cross-linking agents which could restrict use in certain applications. Another strategy to overcome the film dissolution drawback is to blend chitosan with a more moisture-resistant polymer in order to obtain water-resistant chitosan-based antimicrobial films. An example of a recent study using this approach involved the formulation of novel blends of chitosan with EVOH copolymers (see later) [87].

From an application viewpoint, the effectiveness of chitosan films when applied in or coated over foods or packaging materials may deteriorate during film forming, distribution and storage. Therefore, all these parameters should be assessed and optimized for the application. The mechanical and barrier properties of chitosan films obtained under specific forming and storage conditions have been investigated to some extent. For example, it has been demonstrated that water vapour permeability and elongation at break of low-molecular weight chitosan films decreased with storage time at 23 °C and 50% RH [69]. It has also been shown that thermal treatment (120 °C, three hours) of chitosan films led to a significant strengthening of the films and reduced their solubility in aqueous media [88]. This effect was also observed by other researchers after storage of chitosan films under different conditions of temperature and relative humidity [89, 90]. With regard to film-forming conditions, it has been shown that infrared drying was faster and superior in preserving desirable physical characteristics such as water or oxygen barrier properties than oven drying or room temperature drying [91]. A number of studies have been reported in which the chemical changes in chitosan salts during film forming and even under different storage conditions have been followed using infrared spectroscopy [57, 58, 88, 90, 92, 93]. Furthermore, the optimum procedure to preserve the full biocidal activity of chitosan-based films with minimal loss of the activated antimicrobial species has been investigated by Fernández-Saiz *et al.*, [94]. These authors concluded that films maintained at high relative humidity (i.e., 75%) or at higher temperatures (i.e., 37 °C) showed a progressive yellowing and a gradual loss of antimicrobial capacity, due to water resistance induced by chemical and/or physical alterations (Figure 5.2).

5.3.6   Agriculture

Just as chitosan can be used for controlled drug release, this biopolymer can also be employed as a coating for controlled release of fertilizers, pesticides, herbicides, nematocides and insecticides in soil and as a way of reducing environmental damage caused by excessive use of agrochemicals. Coating of seeds and leaves with chitosan films can also prevent microbial infections and induce genes in plants that may help them resist diseases [95].

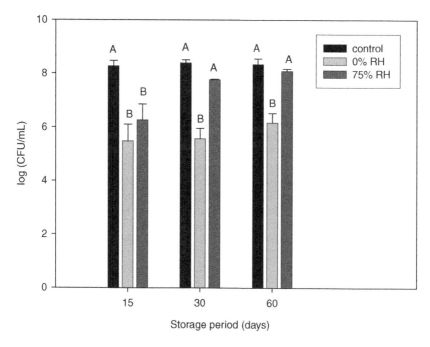

Figure 5.2   Antimicrobial properties of chitosonium acetate films stored at different humidity conditions against *S. aureus*. Different letters indicate significantly different groups determined by Tukey's test ($p < 0.05$). Reprinted from *Journal of Agricultural and Food Chemistry*, 2009, 57(8), 3298–3307. Copyright (2009) with permission from American Chemical Society

5.3.7   Biosensors – Industrial Membrane Bioreactors and Functional Food Processes

Enzyme immobilization is a method by which enzyme molecules are confined in a distinct phase separated from a bulk phase while allowing intercalation between these two phases. The immobilization of Penicillin G acylase on different physical forms of chitosan, namely beads, particles and powders was studied by Braun *et al.* [96], who observed activity retentions of 40, 93, and 100% respectively. Another study by Siso *et al.* [97], demonstrated that microencapsulation in chitosan beads is in fact an effective enzyme immobilization method for invertase and α-amylase. In general, chitosan gels as enzyme immobilization supports offer great potential for a multiplicity of applications including biosensors for both in situ measurement of environmental pollutants and metabolite control in artificial organs. The technique can also be useful for removing organic contaminants from wastewaters in the wine, sugar and fish industries.

5.3.8   Other Applications of Chitosan-Based Materials in the Food Industry

Coating fruit with semi-permeable films has generally been shown to retard ripening by modifying the endogenous CO_2, O_2 and ethylene levels of fruits.

Since chitosan films are more selectively permeable to O_2 than to CO_2, chitosan coatings are likely to modify the internal atmosphere in fruit without causing anaerobic respiration. Therefore, chitosan coatings with their ability to modify internal atmospheres in the tissue and with fungistatic properties have the potential to prolong storage life and control decay of fruit. Furthermore, edible coatings can be used as a vehicle for incorporating functional ingredients such as antioxidants, flavors, colors, antimicrobial agents and nutraceuticals. In this respect, several workers have endeavored to incorporate calcium, vitamin E or oleic acid into chitosan film formulations to prolong the shelf life and to improve the nutritional value of fruit [88–101].

Processing of clarified fruit juices commonly involves the use of clarifying aids, including gelatin, bentonite, silica sol, tannins, polyvinylpyrrolidone or combinations of these compounds [102]. Chitosan with a partial positive charge has been shown to possess acid-binding properties [103] and to be effective in aiding the separation of colloidal and dispersed particles from food-processing wastes [104, 105]. Clarification of fruit juices with chitosan has been attempted by Soto-Peralta et al. [102], Chatterjee et al. [106], and Rungsardthong et al. [107]. Chitosan (2% dissolved in water) was more effective in reducing the turbidity of juices than bentonite and gelatin. The appearance and acceptability of the juices after chitosan treatment could be significantly increased and the turbidity reduced. Chitosan may also be used to control acidity in fruit juices due to its acid-binding properties. The effect of chitosan treatment on the reduction of titratable acidity provides potential for acidity control in other food systems. Enzymatic browning in apple and pear juices can also be prevented by chitosan treatment, probably due to the ability of the positively charged polymer to coagulate suspended solids to which polyphenol oxidase (PPO) is bound.

Unlike most other polysaccharides, under controlled pH conditions chitosan possesses a positive ionic charge and also has both reactive amino and hydroxyl groups, providing the ability to chemically bond with negatively charged proteins. When the pH is less than 6.5, chitosan carries a positive charge along its backbone. Chitosan, with its polar groups, also provides additional stabilization due to hydration forces [108]. According to Filar and Wirick [109], chitosan only functions in acid systems as a thickener and stabilizer.

Meat and meat products are highly susceptible to lipid oxidation, which leads to rapid development of rancid or warmed-over flavor. Chitosan possesses antioxidant and antibacterial capacity [110], and may retard lipid oxidation and inhibit the growth of spoilage bacteria in meat during storage. Kanatt and others investigated the use of irradiated chitosan as a natural antioxidant to minimize lipid peroxidation of radiation-processed lamb meat [111]. Irradiation of chitosan using a 25 kGy dose of gamma radiation resulted in a six-fold increase in antioxidant activity as compared to nonirradiated chitosan, as measured by 1,1-diphenyl-2-picrylhydrazyl (DPPH) scavenging activity.

Seafood products are highly susceptible to quality deterioration due to lipid oxidation of unsaturated fatty acids, catalyzed by the presence of high concentrations of hematin compounds and metal ions in fish muscle [112]. Furthermore, seafood quality is highly influenced by autolysis, contamination by and growth of microorganisms, and loss of protein functionality [113]. Several studies indicate the antioxidant capacity of chitosan added to fish muscle, which depends on its molecular weight and concentration. Chitosan materials may retard lipid oxidation by chelating ferrous ions present in the fish model system, thus eliminating the pro-oxidant activity of ferrous ions or preventing their conversion into ferric ion. Coating with chitosan was very effective in reducing about 50% moisture loss and delaying lipid oxidation in salmon fillets compared with control non-coated fillets [59].

5.4 PROCESSING OF CHITOSAN

As mentioned before, chitin is easily obtained from crab or shrimp shells and fungal mycelia. For this reason, chitin production is associated with food industries such as shrimp canning. The production of chitosan–glucan complexes is associated with fermentation processes, similar to those for the production of citric acid from *Aspergillus niger*, *Mucor rouxii*, and *Streptomyces*, which involve alkali treatment yielding chitosan–glucan complexes. Alkali treatment simultaneously removes the protein and deacetylates chitin. The processing of crustacean shells mainly involves the removal of proteins and the dissolution of calcium carbonate which is present in crab shells in high concentrations. The resulting chitin is deacetylated in 40% sodium hydroxide at 120 °C for between one and three hours. This treatment produces 70% deacetylated chitosan [25].

As discussed, chitosan films and coatings are typically prepared by dissolving chitosan in dilute acids. Solutions are then spread on level surfaces and air-dried at room temperature (casting process). Films can also be prepared by spreading solutions on, for instance, polypropylene films and drying at 60 °C in an oven [69, 72]. These processes may clearly be considered time-consuming. Chitosan films have been prepared by wet casting followed by infrared (IR) drying [114]. This method is faster than the conventional lamination and drying method and no significant differences have been observed in the mechanical and barrier properties of the films obtained in comparison with those prepared using conventional drying methods [91]. However, in this case, biocidal properties were not evaluated. In general, the mechanical properties, permeability, thermal decomposition points, solvent stability, and biocidal capacity are important parameters in selecting the right chitosan processing conditions for specific applications [115].

Polymer surface modifications, such as plasma treatments, have been exploited to improve the adhesion between chitosan and other matrices such as PP [116–119].

Polymer blending can be an effective processing strategy for providing new polymeric materials for various applications. The addition of plasticizing agents can also be important as a way to overcome the excessive brittleness of some biopolymer films. Brittleness is an inherent quality attributed to the complex/branched primary structure and weak intermolecular forces of natural biopolymers including chitosan. Plasticizers soften the rigidity of the film structure, increase the mobility of the biopolymer chains, and reduce the intermolecular forces, thus providing practical benefits in terms of improved mechanical properties. Both blending and the use of plasticizers have been considered as routes to enhance the processing and physical properties of chitosan and chitosan derivatives.

In terms of blending, some researchers have combined chitosan with other biopolymers such as cellulose, starch or Konjac glucomannan in order to improve the mechanical and water-swelling properties of antimicrobial chitosan films [68, 120–125]. Many of the published studies in this field describe biocomposite materials with optimum antimicrobial activity but poor water barrier properties. Recent research performed by Fernández-Saiz *et al.* [126] showed an inhibitory effect on growth of *S. aureus* when a composite matrix gliadin/chitosonium acetate (60/40 % wt/wt) was tested and in which film integrity was maintained in the nutrient broth. However, although film disintegration was prevented, some turbidity in the nutrient medium was detected. The gliadin protein network is believed to exert a blocking effect against the dissolution of chitosonium acetate chains, albeit though some of the latter chains do still migrate and exert the required antimicrobial role. Similar findings were observed by Tanabe *et al.* [127] who studied keratin/chitosan films and observed a significant reduction of the swelling ratio as well as a significant reduction of *E. coli* (61%) when using a keratin:chitosan ratio of 5:1. In another study, the water barrier and the antimicrobial activity of high- and low-molecular weight chitosonium acetate-based solvent-cast blends with ethylene-vinyl alcohol (EVOH) copolymers were tested against *S. aureus* and *Salmonella spp* [87]. These samples showed excellent antimicrobial activity as well as enhanced water barrier when low-molecular weight chitosan was used as the dispersed phase in the blend, specifically 80/20 % (wt/wt) EVOH/chitosan. In terms of blends, the most appropriate ratio between chitosan and the water-resistant polymer will depend on their compatibility. Overall, it is possible to conclude that chitosan-based blends can be a very suitable practical means for processing chitosan while retaining or enhancing its properties, including dimensional stability, in high-humidity environments.

5.5   CONCLUDING REMARKS

Chitosan has shown excellent potential as a substrate for fabrication of a wide range of practically useful films and coatings and much research in this direction is still ongoing. While use in coatings (e.g., on food) can be considered highly suitable due to the inherent characteristics of chitosan, other applications (e.g., antimicrobial active packaging) need further careful consideration. At present, the antimicrobial activity of chitosan is attributed to migration of protonated glucosamine fractions and the processing conditions for chitosan to obtain optimum biocide properties are known. Blending chitosan with more water-resistant polymers appears to be a promising direction to enable chitosan films with sufficient dimensionally stability when in contact with high-moisture products, while simultaneously retaining biocide properties when processed from acidic solutions. A significant processing challenge still remains since the inherent biocidal performance of chitosan is strongly handicapped by heating. Much of the current research and development of product applications for chitosan in films and coatings has been carried out in laboratory or pilot plant environments. Consequently, in addition to further investigations on the characteristics of chitosan, its derivatives and blends, there is much research on standardization and reproducibility in commercial grades, including scaled-up processes and trials, which will be required in the future in order to extend the applications of this most attractive and fascinating biopolymer.

REFERENCES

[1] Arvanitoyannis, I.S., Nakayama, A., and Aiba, S. Chitosan and gelatin based edible films: state diagrams, mechanical and permeation properties. *Carbohydr Polym.* 1998; **37**: 371–382.

[2] Bough, W.A. Chitosan – A Polymer from Seafood Waste, for use in Treatment of Food Processing Wastes, and Activated Sludge. *Proc. Biochem.* 1976; **11**: 13–17.

[3] No, H.K., Meyers, S.P., and Lee, K.S. Isolation and characterization of chitin from crawfish shell waste. *J. Agric. Food Chem.* 1989; **37**: 575–579.

[4] Knorr, D. Recovery and utilization of chitin and chitosan in food processing waste management, *Food Technol* 1991; **45**: 114–122.

[5] Shahidi, F., Synowiecki, J. Isolation and characterization of nutrients and value-added products from Snow crab (Chinoecetes opilio) and Shrimp (Pandalus borealis) processing discards, *J. Agric. Food Chem* 1991; **39**: 1527–1532.

[6] Shahidi, F. and Synowiecki, J. (1992), *Advances in Chitin and Chitosan*, Brine, C.J., Sandford, P.A., and Zikakis, J.P. (Eds.), London, Elsevier, 617–628.

[7] Shahidi, F. Role of chemistry and biotechnology in value-added utilization of shellfish processing discards Can. *Chem. News* 1995; **47**: 25–29.

[8] Jeon, Y.J., Shahidi, F., and Kim, S.K. Preparation of chitin and chitosan oligomers and their applications in physiological functional foods *Food Rev. Int* 2000; **16**: 159–176.

[9] Chung, Y.C., Su, Y.P., Chen, C.C., *et al.* Relationship between antibacterial activity of chitosan and surface characteristics of cell wall. *Acta Pharmacol. Sinica* 2004; **25**: 932–936.

[10] Shahidi, F., Abuzaytoun, A.R. Chitin, chitosan and coproduce: Chemistry, production, applications, and health effects. *Adv. Food Nutr. Res.* 2005; **49**: 94–135.

[11] Yingyuad, S., Ruamsin, S., Reekprkhon, D., *et al.* Effect of chitosan coating and vacuum packaging on the quality of refrigerated grilled pork. *Packag. Technol. Sci.* 2006; **19**: 149–157.

[12] Baxter, A., Dillon, M., Taylor, K.D.A., *et al.* Improved method for IR determination of the degree of N-acetylation of chitosan, *Int. J. Biol. Macromol.* 1992; **14**: 166–169.

[13] Maghami, G.A., Roberts, G.A.F. Studies on the adsorption of anionic dyes on chitosan, *Makromol. Chem.* 1988; **189**: 2239–2243.

[14] Domard, A. Circular dichroism study on N-acetylglucosamine oligomers, *Int. J. Biol. Macromol.* 1986; **8**: 243–246.

[15] Domard, A. pH and c.d. measurements on a fully deacetylated chitosan: application to Cu –polymer interactions, *Int. J. Biol. Macromol.* 1987; **9**: 98–104.

[16] Wei, Y.C., Hudson, S.M. Binding of sodium dodecyl sulfate to a polyelectrolyte based chitosan, *Macromolecules* 1993; **26**: 4151–4154.

[17] Sashiwa, H., Saimoto, H., Shigemasa, Y., *et al.* Distribution of the acetamido group in partially deacetylated chitins, *Carbohydr. Polym.* 1991; **16**: 291.

[18] Sashiwa, H., Saimoto, H., Shigemasa, Y., *et al.* N-Acetyl group distribution in partially deacetylated chitins prepared under homogeneous conditions, *Carbohydr. Res.* 1993; **242**: 167.

[19] Raymond, L., Morin, F.G., and Marchessault, R. Degree of deacetylation of chitosan using conductometric titration and solid-state NMR, *Carbohydr. Res.* 1993; **243**: 331–336.

100   BIOPOLYMERS

[20] Niola, F., Basora, E., Chornet, E. *et al.* A rapid method for the determination of the degree of N-acetylation of chitin–chitosan sample by acid hydrolysis and HPLC, *Carbohydr. Res.* 1993; **238**: 1–9.
[21] S.H. Pangburn, P.V. Trescony, and J. Heller, in: J.P. Zikakis (Ed.), *Chitin, Chitosan and Related Enzymes*, Harcourt Brace Janovich, New York, 1984, p. 3.
[22] Rathke, T.D., Hudson, S.M. Determination of the degree of N-deacetylation in chitin and chitosan as well as their monomer sugar ratios by near infrared spectroscopy, *J. Polym. Sci., Polym. Chem. Ed.* 1993; **31**: 749–753.
[23] Wu, A.C.M., Bough, W.A., Conrad, E.C. *et al.* Determination of molecular weight distribution of chitosan by high-performance liquid chromatography, *J. Chromatog. A.* 1976; **128**: 87–99.
[24] Muzzarelli, R.A.A., Lough, C. and Emanuelli, M. The molecular weight of chitosans studied by laser light scattering, *Carbohydr. Res.* 1987; **164**: 433–442.
[25] Kumar, M.N.V.R. A review of chitin and chitosan applications. *React. Funct. Polym.* 2000; **46**: 1–27.
[26] V.F. Lee, Solution and shear properties of chitin and chitosan, PhD Dissertation, University of Washington, University Microfilms, Ann Arbor, MI, USA, 74-29 1974, p. 446.
[27] Krajewska, B. Membrane-based processes performed with use of chitin/chitosan materials. *Separ. Purif. Technol.* 2005; **41**: 305–312.
[28] Knorr, D. Use of chitinous polymers in food – a challenge for food research and development. *Food Technol.* 1984; **38**: 85–97.
[29] C. Jeuniaux. Chitosan as a tool for the purification of water. In *Chitin in nature and Technology*; Muzzarelli, R.A.A., Jeuniaux, C. and Gooday, G.W. eds.; Plenum Press: New York, 1986, 551–567.
[30] Immizi, S.A., Iqbal, J., and Isa, M. Collection of metal ions present in waste water samples of different sites of Pakistan using biopolymer chitosan. *J. Chem. Soc. Pakistan* 1996; **18**: 312–315.
[31] R.A.A. Muzzarelli. (Ed). *Chitin*; Pergamon Press: Oxford, UK, 1977.
[32] McKay, H.S., Blair, J.R., and Gardner, M. Absorption of dyes on chitin. I. Equilibrium studies. *J. Appl. Polym. Sci.* 1982; **27**: 3043–3046.
[33] Muzzarelli, R.A.A. Chitosan-based dietary foods. *Carbohyd. Polym.* 1996; **29**: 309–316.
[34] Deuchi, K., Kanauchi, O., Imasato, Y. *et al.*, Decreasing effect of chitosan on the apparent fat digestibility by rats fed on high-fat diet. *Biosc. Biochem.* 1994; **589**: 1613–1616.
[35] Ohya, Y., Takei, T., Kobayashi, H. *et al.*, Release behaviour of 5-fluorouracil from chitosan-gel microspheres immobilizing 5-fluorouracil derivative coated with polysaccharides and their cell specific recognition. *J. Microencapsul.,* 1993; **10**: 1–9.
[36] Jameela, S.R., Jayakrishnan, A. Glutaraldehyde cross-linked chitosan microspheres as a long acting biodegradable drug delivery vehicle: studies on the in vitro release of mitoxantrone and in vivo degradation of microspheres in rat muscle. *Biomaterials* 1995; **16**: 769–775.
[37] Denkbas, E.B., Seyyal, M., and Piskin, E. Implantable 5-fluorouracil loaded chitosan scaffolds prepared by wet spinning, *J. Membr. Sci.* 2000; **172**: 33–38.
[38] Gutowska, A., Jeong, B., and Jasionowski, M. Injectable gels for tissue engineering. The Anatomical Record Part A: Discoveries in Molecular, *Cellular, and Evolutionary Biology* 2001; **263**: 342–349.

[39] Chenite, A., Chaput, C., Wang, D., *et al.* Novel injectable neutral solutions of chitosan form biodegradable gels in situ. *Principio del Formulario Biomaterials* 2000; **21**: 2155–61.

[40] El Ghaouth, A., Arul, J., Ponnampalam, R., *et al.* Chitosan coating effect on storability and quality of fresh strawberries. *J. Food Sci.* 1991; **56**: 1618–20.

[41] Lee, J.W., Lee, H.H., and Rhim, J.W. Shelf life extension of white rice cake and wet noodle by the treatment with chitosan. *Korean Journal. Food Sci. Technol.* 2000; **32**: 828–33.

[42] Roller, S., Covill, N. The antifungal properties of chitosan in laboratory media and apple juice. *Int. J. Food Microbiol.* 1999; **47**: 67–77.

[43] Roller, S., Covill, N. The antimicrobial properties of chitosan in mayonnaise and mayonnaise-based shrimp salads. *J. Food Prot.* 2000; **63**: 202–209.

[44] Ha, T.J., Lee, S.H. Utilization of chitosan to improve the quality of processed milk. *Journal Korean Society. Food Sci. Nutr.* 2001; **30**: 630–34.

[45] Lee, H.Y., Kim, S.M., Kim, J.Y., *et al.* Effect of addition of chitosan on improvement for shelf-life of bread. *Journal Korean Society. Food Sci. and Nutr.* 2002; **31**: 445–50.

[46] Hein, S., Wang, K., and Stevens, W.F. Chitosan composites for biomedical applications: States, challenges and perspectives. *J. Mater. Sci. Technol.* 2008; **24**: 1053–1061.

[47] Leane, M.M., Nankervis, R., Smith, A., *et al.* Use of ninhydrin assay to measure the release of chitosan from oral solid dosage forms. *Int. J. Pharmacol.* 2004; **271**: 241–249.

[48] Torres-Giner, S., Ocio, M.J., and Lagaron, J.M. Development of active antimicrobial fiber based chitosan polysacharide structures by electrospinning. *Eng. Life Sci.* 2008; **8**: 303–314.

[49] Torres-Giner, S., Ocio, M.J., and Lagaron, J.M. Novel antimicrobial ultrathin structures of zein/chitosan blends obtained by electrospinning. *Carbohydr. Polym.* 2009; **77**: 261–266.

[50] Ralston, G.B., Racey, M.V., and Wrench, P.M. Inhibition of fermentation in baker's yeast by chitosan. *Biochim. Biophysic. Acta* 1964; **93**: 652–655.

[51] Helander, I.M., Nurmiaho-Lassila, E.L., Ahvenainen, R., *et al.* Chitosan disrupts the barrier properties of the outer membrane of gram-negative bacteria. *Int. J. Food Microbiol.* 2001; **71**: 235–244.

[52] Liu, H., Du, Y., Xiaohui Wang, X., *et al.* Chitosan kills bacteria through cell membrane damage. *Int. J. Food Microbiol.* 2004; **95**: 147–155.

[53] Mayachiew, P., Devahastin, S., Mackey, B.M., *et al.* Effects of drying methods and conditions on antimicrobial activity of edible chitosan films enriched with galangal extract. *Food Res. Int.* 2010; **43**: 125–132.

[54] Ganan, M., Carrascosa, V., and Martinez-Rodriguez, A.J. Antimicrobial activity of chitosan against Campylobacter spp. and other microorganism and its mechanism of action. *J. Food Prot.* 2009; **8**: 1735–1738.

[55] Hadwiger, L.A., Kendra, D.F., Fristensky, B.W., *et al. Chitosan both activates genes in plants and inhibits RNA synthesis in fungi.* In: Muzzarelli, R.A.A., Jeuniaux, C., Gooday, G.W. Eds.; *Chitin in Nature and Technology.* New York, USA: Plenum Press; 1985. pp 209–222.

[56] Cuero, R.G., Osuji, G., and Washington, A. N-carboxymethyl chitosan inhibition of aflatoxin production: role of zinc. *Biotechnol. Lett.* 1991; **13**: 441–444.

[57] Fernández-Saiz, P., Ocio, M.J., and Lagaron, J.M. Film forming and biocide assessment of high molecular weight chitosan as determined by combined ATR-FTIR spectroscopy and antimicrobial assays. *Biopolymers* 2006; **83**: 577–583.

[58] Lagaron, J.M., Fernández-Saiz, P., and Ocio, M.J. Using ATR-FTIR spectroscopy to design active antimicrobial food packaging structures based on high molecular weight chitosan polysaccharide. *J. Agric. Food Chem.* 2007; 55: 2554–2562.

[59] Tsai, G.J., Su, W.H., Chen, H.C., *et al.* Antimicrobial activity of shrimp chitin and chitosan from different treatments and applications of fish preservation. *Fisheries Sci.* 2002; 68: 170–177.

[60] Sagoo, S., Board, R., and Roller, S. Chitosan inhibits growth of spoilage microorganisms in chilled pork products. *Food Microbiol.* 2002; 19: 175–182.

[61] Lopez-Caballero, M.E., Gómez-Guillén, M.C., Pérez-Mateos, M., *et al.* A chitosan-gelatin blend as a coating for fish patties. *Food Hydrocoll.* 2005; 19: 303–311.

[62] Tsai, G.J., Tsai, M.T., Lee, J.M., *et al.* Effects of Chitosan and low-molecular-weight chitosan on Bacillus cereus and application in the preservation of cooked rice. *J. Food Prot.* 2006; 69: 2168–2175.

[63] Juneja, V.K., Thippareddi, H., Bari, L., *et al.* Chitosan protects cooked ground beef and turkey against Clostridium perfringens spores during chilling. *J. Food Sci.* 2006; 71: 236–240.

[64] Ye, M., Neetoo, H., and Chen, H. Control of Listeria monocytogenes on ham steaks by antimicrobials incorporated into chitosan-coated plastic films. *Food Microbiol.* 2008; 25: 260–268.

[65] Ouattara, B., Simard, R.E., Piette, G., *et al.* Inhibition of surface spoilage bacteria in processed meats by application of antimicrobial films prepared with chitosan. *Int. J. Food Microbiol.* 2000; 62: 139–148.

[66] Fernández-Saiz, P., Soler, C., Lagaron, J.M., *et al.* Effects of chitosan films on the growth of Listeria monocytogenes, Staphylococcus aureus and Salmonella spp. in laboratory media and in fish soup. *Int. J. Food Microbiol.* 2010; 137: 287–94.

[67] Coma, V., Martial-Gros, A., Garreau, S., *et al.* Edible antimicrobial films based on chitosan matrix. *J. Food Sci.* 2002; 67: 1162–1169.

[68] Moller, H., Grelier, S., Pardon, P., *et al.* Antimicrobial and physicochemical properties of chitosan-HPMC-based films. *J. Agric. Food Chem.* 2004; 52: 6585–6591.

[69] Butler, B.L., Vergano, P.J., Testin, R.F., *et al.* Mechanical and barrier properties of edible chitosan films as affected by composition and storage. *J. Food Sci.* 2006; 61: 953–956.

[70] Caner, C., Vergano, P.J., and Wiles, J.L. Chitosan films mechanical and permeation properties as affected by acid, plasticizer and storage. *J. Food Sci.* 1998; 63: 1049–1053.

[71] Park, S.Y., Marsh, K.S., and Rhim, J.W. Characteristics of Different Molecular Weight Chitosan Films Affected by the Type of Organic Solvents *J. Food Sci.* 2002; 67: 194–197.

[72] Wiles, J.L., Vergano, P.J., Barron, F.H., *et al.* Water vapor transmission rates and sorption behavior of chitosan films. *J. Food Sci.* 2000; 65: 1175–1179.

[73] Chen, R.H., Hwa, H.-D. Effect of molecular weight of chitosan with the same degree of deacetylation on the thermal, mechanical, and permeability properties of the prepared membrane. *Carbohydr. Polym.* 1996; 29: 353–358.

[74] No, H.K., Park, N.Y., Lee, S.H., *et al.* Antibacterial activity of chitosan and chitosan oligomers with different molecular weights. *Int. J. Food Microb.* 2002; 74: 65–72.

[75] Gerasimenko, D.V., Avdienko, I.D., Bannikova, G.E., *et al.* Antibacterial effects of water-soluble low-molecular-weight chitosan on different microorganisms. *Appl. Biochem. Microbiol.* 2004; 40: 253–257.

[76] Zivanovic, S., Basurto, C.C., Chi, S., *et al.* Molecular weight of chitosan influences antimicrobial activity in oil-in-water emulsions. *J. Food Prot.* 2004; 67: 952–959.

[77] Kim, K.M., Son, J.H., Kim, S.-K., *et al.* Properties of chitosan films as function of pH and solvent type. *J. Food Sci.* 2006; 71: 119–124.

[78] Kim, S.H., No, H.K., and Prinyawiwatkul, W. Effect of molecular weight, type of chitosan, and chitosan solution pH on the shelf-life and quality of coated eggs. *J. Food Sci.* 2007; 72: 44–48.

[79] Takahashi, T., Imai, M., Suzuki, I., *et al.* Growth inhibitory effect on bacteria of chitosan membranes regulated with deacetylation degree. *Biochem. Eng. J.* 2008; 40: 485–491.

[80] Hongpattarakere, T., Riyaphan, O. Effect of deacetylation conditions on antimicrobial activity of chitosans prepared from carapace of black tiger shrimp (penaeus monodon). *Songklanakarin J. Sci. Technol.* 2008; 30: 1–9.

[81] Qin, C., Li, H., Xiao, Q., *et al.* Water-solubility of chitosan and its antimicrobial activity. *Carbohydr. Polym.* 2006; 63: 367–374.

[82] Suto, S., Ui, N. Chemical crosslinking of hydroxypropyl cellulose and chitosan blends. *J. Appl. Polym. Sci.* 1996; 61: 2273–2278.

[83] Tual, C., Espuche, E., Escoubes, M., *et al.* Transport properties of chitosan membranes: Influence of crosslinking. *J. Polym. Sci.* 2000; 38: 1521–1529.

[84] Zheng, H., Du Y-M, Yu, J.-H., *et al.* The properties and preparation of crosslinked chitosan films. *Chem J. Chin. Univ.* 2000; 21: 809–812.

[85] Tang, R., Du, Y., and Fan, L. Dialdehyde starch-crosslinked chitosan films and their antimicrobial effects. *J. Polym. Sci.* 2003; 41: 993–997.

[86] Mi, F.L., Huang, C.T., and Liang, H.F. Physicochemical, antimicrobial, and cytotoxic characteristics of a chitosan film cross-linked by a naturally occurring cross-linking agent, aglycone geniposidic acid. *J. Agric. Food Chem.* 2006; 54: 3290–3296.

[87] Fernández-Saiz, P., Ocio, M.J., and Lagaron, J.M. Antibacterial chitosan-based blends with ethylene–vinyl alcohol copolymer. *Carbohydr. Polym.* 2010, 80: 874–884.

[88] Zotkin, M.A., Vikhoreva, G.A., Smotrina, T.V., *et al.* Thermal Modification and Study of the Structure of Chitosan Films. *Polym. Sci.* 2004, 46: 39–42.

[89] Ritthidej, G.C., Phaechamud, T., and Koizumi, T. Moist heat treatment on physicochemical change of chitosan salt films. *Int. J. Pharmaceut.* 2002; 232: 11–22.

[90] Kam, H.M., Khor, E., and Lim, L.Y. Storage of partially deacetylated chitosan films. *J. Biomed. Mater. Res.* 1999; 48: 881–8.

[91] Srinivasa, P.C., Ramesh, M.N., Kumar, K.R., *et al.* Properties of chitosan films prepared under different drying conditions. *J. Food Eng.* 2004; 63: 79–85.

[92] Demarger-Andre, S., Domard, A. Chitosan carboxylic acid salts in solution and in the solid state. *Carbohydr. Polym.* 1994; 23: 211–219.

[93] Osman, Z., Arof, A.K. FTIR studies of chitosan acetate based polymer electrolytes. *Electrochim. Acta* 2003; 48: 993–999.

[94] Fernández-Saiz, P., Lagarón, J.M., and Ocio, M.J. Optimization of the film-forming and storage conditions of chitosan as an antimicrobial agent, *J. Agric. Food Chem.* 2009; 57: 3298–3307.

[95] Hadwiger, L.A., Kendra, D.F., Fristensky, B.W., *et al.* 1986. Chitosan both activates genes in plants and inhibits RNA synthesis in fungi. In: Muzzarelli, R.A.A., Jeuniaux, C., and Gooday, G.W. (eds), *Chitin in Nature and Technology.* Plenum Press, New York, pp 209–214.

[96] Braun, J., Le Chanu, P., and Le Goffic, F. The immobilization of penicillin G acylase on chitosan. *Biotechnol. Bioeng.* 1989; 33: 242–246.

[97]   Siso, M.I.G., Lang, E., Carreno-Gomez, B., *et al.* Enzyme encapsulation on chitosan microbeads. *Process Biochem.* 1997; 32: 211–216.

[98]   Han, C., Zhao, Y., Leonard, S.W., *et al.* Edible coatings to improve storability and enhance nutritional value of fresh and frozen strawberries (Fragaria × ananassa) and raspberries (Rubus ideaus). *Postharvest Biol. Technol.* 2004; 33: 67–78.

[99]   Hernandez-Munoz, P., Almenar, E., Ocio, M.J., *et al.* Effect of calcium dips and chitosan coatings on postharvest life of strawberries (Fragaria×ananassa). *Postharvest Biol. Technol.* 2006; 39: 247–53.

[100] Park, S.I., Stan, S.D., Daeschel, M.A., *et al.* Antifungal coatings on fresh strawberries (Fragaria × ananassa) to control mold growth during cold storage. *J. Food Sci.* 2005; 70: M202–7.

[101] Vargas, M., Albors, A., Chiralt, A., *et al.* Quality of cold-stored strawberries as affected by chitosan-oleic acid edible coatings. *Postharvest Biol. Technol.* 2006; 41: 164–71.

[102] Soto-Peralta, N.V., Moller, H., and Knorr, D. Effects of chitosan treatments on the clarity and color of apple juice. *J. Food Sci.* 1989; 54 (2): 495–6.

[103] Imeri, A.G., Knorr, D. Effect of chitosan on yield and compositional data of carrot and apple juice. *J. Food Sci.* 1988; 53: 1707–9.

[104] Knorr, D. 1985. Utilization of chitinous polymers in food processing and biomass recovery. In: Colwell, R.R., Pariser, E.R., and Sinskey, A.J. (eds). *Biotechnology of Marine Polysaccharides.* Washington, D.C.: Hemisphere Publishing Corp. pp 313–32.

[105] No, H.K., Meyers, S.P. Application of chitosan for treatment of wastewaters. *Rev. Environ. Contam. Toxicol.* 2000; 163: 1–27.

[106] Chatterjee, S., Chatterjee, S., Chatterjee, B.P., *et al.* Clarification of fruit juice with chitosan. *Process Biochem.* 2004; 39: 2229–32.

[107] Rungsardthong, V., Wongvuttanakul, N., Kongpien, N., *et al.* Application of fungal chitosan for clarification of apple juice. *Process Biochem.* 2006; 41: 589–93.

[108] Del Blanco, L.F., Rodriguez, M.S., Schulz, P.C., *et al.* Influence of the deacetylation degree on chitosan emulsification properties. *Colloid Polym. Sci.* 1999; 77: 1087–92.

[109] Filar, L.J., Wirick, M.G. Bulk and solution properties of chitosan. In: Muzzarelli, R.A.A., Pariser, E.R. (eds). *Proceedings of the 1st International Conference on Chitin/Chitosan.* Mass.: MIT Sea Grant Program 1978; pp 169–81.

[110] Kamil, J.Y.V.A., Jeon, Y.J., and Shahidi, F. Antioxidative activity of chitosans of different viscosity in cooked comminuted flesh of herring (Clupea harengus). *Food Chem.* 2002; 79: 69–77.

[111] Kanatt, S.R., Chander, R., and Sharma, A. Effect of irradiated chitosan on the rancidity of radiation-processed lamb meat. *Int. J. Food Sci. Technol.* 2004; 39: 997–1003.

[112] Decker, E.A., Hultin, H.O. Lipid oxidation in muscle foods via redox iron. In: Angelo, A.J. (ed.). *Lipid Oxidation in Food.* Washington, D. C.: American Chemical Society. 1992; pp 33–54.

[113] Jeon, Y.J., Kamil, J.Y.V.A., and Shahidi, F. Chitosan as an edible invisible film for quality preservation of herring and Atlantic cod. *J. Agric. Food Chem.* 2002; 50: 5167–78.

[114] Tharanathan, R.N., Srinivasa, P.C., and Ramesh, M.N. (2002), A process for production of biodegradable films from polysaccharides, Indian Patent 85/DEL/2002.

[115] Collins, E.A., Bares, J., and Billmeyer, F.W. (1973), *Experimental Polymer Science*, New York, John Willey Interscience, pp 99–120.

[116] Elsabee, M.Z., Abdou, E.S., Nagy, K.S.A., *et al.* Surface modification of polypropylene films by chitosan and chitosan/pectin multilayer *Carbohydr. Polym.* 2008; 71: 187–195.

[117] Abdou, E.S., Elkholy, S.S., Elsabee, M.Z., *et al.* Improved antimicrobial activity of polypropylene and cotton nonwoven fabrics by surface treatment and modification with chitosan *J. Appl. Polym. Sci.* 2008; 108: 2290–2296.

[118] Yang, J.M., Lin, H.T., Wu, T.H., *et al.* Wettability and antibacterial assessment of chitosan containing radiation-induced graft nonwoven fabric of polypropylene-g-acrylic acid. *J. Appl. Polym. Sci.* 2003; 90: 1331.

[119] Hu, S.G., Jou, C.H., and Yang, M.C. Surface grafting of polyester fiber with chitosan and the antibacterial activity of pathogenic bacteria. *J. Appl. Polym. Sci.* 2002; 86: 2977.

[120] Pelissari, F.M., Grossmann, M.V.E., Yamashita, F., *et al.* Antimicrobial, mechanical, and barrier properties of cassava starch-chitosan films incorporated with oregano essential oil. *J. Agric. Food Chem.* 2009; 57: 7499–7504.

[121] Li, B., Peng, J., Yie, X., *et al.* Enhancing physical properties and antimicrobial activity of konjac glucomannan edible films by incorporating chitosan and nisin. *J. Food Sci.* 2006; 71: 174–178.

[122] Sebti, I., Chollet, E., Degraeve, C.N., *et al.* Water sensitivity, antimicrobial, and Physicochemical analyses of edible films based on HPMC and/or Chitosan. *J. Agric. Food Chem.* 2007; 55: 693–699.

[123] Park, S.I., Daeschel, M.A., and Zhao, Y. Functional properties of antimicrobial lysozyme-chitosan composite films. *J. Food Sci.* 2004; 69: M215–M221.

[124] Wu, Y.B., Yu, S.H., Mi, F.L., *et al.* Preparation and characterization on mechanical and antibacterial properties of chitosan/cellulose blends. *Carbohydr. Polym.* 2004; 57: 435–440.

[125] Shen, X.L., Wu, J.M., Chen, Y., *et al.* Antimicrobial and physical properties of sweet potato starch films incorporated with potassium sorbate or chitosan. *Food Hydrocoll.* 2009; DOI: 10.1016/j.foodhyd.2009.10.003.

[126] Fernández-Saiz, P., Lagaron, J.M., Hernandez-Muñoz, P., *et al.* Characterization of the antimicrobial properties against S. aureus of novel renewable blends of chitosan and gliadins of interest in active food packaging and coating applications. *Int. J. Food Microbiol.* 2008; 124: 13–20.

[127] Tanabe, T., Okitsu, N., Tachinaba, A., *et al.* Preparation and characterization of keratin-chitosan composite film. *Biomaterials* 2002; 23: 817–825.

6

Production, Chemistry and Properties of Proteins

Mikael Gällstedt

Innventia AB, Packaging Solutions, Stockholm, Sweden

Mikael S. Hedenqvist and Hasan Ture

Royal Institute of Technology, Department of Fiber and Polymer Technology, Stockholm, Sweden

6.1 INTRODUCTION

The introduction to this chapter gives a brief overview of the sources that are available to make protein-based films. Also, processing (casting, compression molding, extrusion and injection molding) of proteins into, for example, films and the associated final properties are discussed. For comprehensive references in this field, readers should consult references [1–4]. General limitations and advantages will be outlined here as well as availability and pricing where such information is available.

Biopolymers – New Materials for Sustainable Films and Coatings, First Edition.
Edited by David Plackett.
© 2011 John Wiley & Sons, Ltd. Published 2011 by John Wiley & Sons, Ltd.

6.2   PLANT-BASED PROTEINS

6.2.1   Rapeseed

We start off by discussing a source that has not had much attention until recently. The recent interest in rape or rapeseed proteins is due to the production of rapeseed methyl ester (RME) biofuel. The seeds from the rapeseed plant are 1.5–3.2 mm in diameter and weigh 2.5–6.5 g. The protein content is 15–18% and the amount of layered hull is 12–16%. The fiber content is 31–34% and is located mainly in thick cell walls where 35% is lignin. The residue after the oil has been squeezed out of the seed is a rapeseed cake (Figure 6.1). Besides residual oil the main components are fiber and protein. The essential amino acids in the rapeseed protein are arginine, isoleucine, leucine, lysine, phenylalanine, methionine, histidine, cystine, valine, tryptophan and threonine [5].

Figure 6.1 Dumbbell-shaped specimens based on plasticized rapeseed cake and produced by compression moulding. Courtesy of Sung-Woo Cho, Royal Institute of Technology, Sweden

6.2.2   Wheat Gluten

Wheat gluten, or rather vital wheat gluten, is produced by washing wheat flour extensively with water. In doing so the main part of the starch is washed out and collected. Other water-soluble substances such as albumin proteins are also, at least to some extent, removed. The final vital wheat gluten is rich in protein, but also contains water, starch, fiber, fat and ash. The protein consists of mainly two types and the distinction is easily made based on ethanol solubility. The lower molar mass gliadins are ethanol soluble whereas the higher molar mass glutenins are not. In general, the gliadin/glutenin ratio is $\sim$ 3/2. However, it should be noted that during extensive shearing or heating the protein polymerizes which implies that gliadins with intramolecular sulfur-bonds crosslinks intermolecularly and hence produces larger molecules. The inherent colour of the vital wheat gluten is brown/beige, but it is easy to colour with a dye or a pigment.

Many different methods have been used to produce plasticized vital wheat gluten and it turns out to be relatively easy to mould and shape. Casting from a solution, compression moulding, extrusion and injection moulding have all been used to produce wheat gluten films, plates and bands [6–8] (Figure 6.2).

6.2.3   Corn Zein

As with gliadin, zein prolamine is insoluble in water, except at very low or high pH, but is soluble in ethanol [9]. Corn zein consists of monomers and disulfide-linked oligomers. Because of its lack of essential amino acids and its water insolubility the main interest lately has been to use it as an industrial protein. The price was $(US) 10–40 per kg in the early 21st century [10]. Zein has an amphiphilic character where the main chain has polar amino acids but the side chains contain more than 50% nonpolar amino acids, including leucine, isoleucine, valine, alanine, phenyl alanine and glycine [11]. The most common zein aminoacids are glutamic acid (glutamine) 21–26%, leucine (20%), proline (10%) and alanine (10%) [10]. This protein can be divided in various ways,

(a)   (b)

Figure 6.2   Compression moulded (uncoloured left and coloured right) vital wheat gluten sheets plasticized with glycerol. Courtesy of Sung-Woo Cho, Royal Institute of Technology, Sweden (right figure)

and one example shows two major fractions; the α-zein which includes ~ 80% of the available prolamine and is defined as the fraction being soluble in 95% ethanol, and β-zein, which is relatively unstable, but is soluble in water.

The commercial use of corn zein includes numerous applications (e.g., in controlled release, coating, fiber and biodegradable films and plastics) [10]. Specific examples include hair fixatives, chewing gum, labels, varnishes, microspheres, coatings on confectionery and nuts [9]. In the 1970s almost 75% of the 500 tons of produced zein was used to coat tablets [10]. For several applications the yellow colour of zein, due to xanthophylls and carotenoids, is unattractive. Several ways of reducing the colour have been tested and one way to get relatively white zein is to start with waxy corn, which contains less pigment and xanthophyll [10]. A cast zein-based film is shown in Figure 6.3.

6.2.4   Soy Protein

Soy protein films are said to be clearer than films from other plant protein sources [12]. As with whey proteins, soy protein can be divided into types based on the protein content; soy protein isolate (SPI) containing more than 90% protein and soy protein concentrates containing 65–72% protein (SPC). In addition defatted soy protein flour (DSF), obtained by grinding defatted soy flakes, contains 50–59% protein. In SPI the most abundant amino acids are leucine, lysine, phenylalanine, valine and isoleucine. The concentrate is obtained from aqueous liquid extraction or acidic leaching and the isolate is obtained from aqueous or mild alkali extraction followed by isoelectric

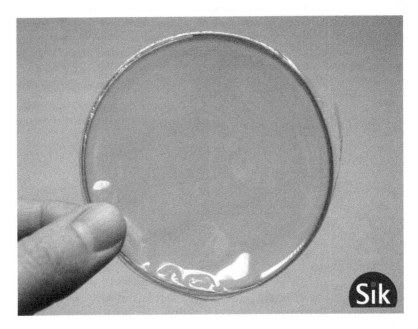

Figure 6.3   Cast plasticized zein film. Courtesy of Thomas Gillgren, SIK, Sweden

precipitation. Soy protein films can be obtained by casting from solution or by extrusion. In the latter case the protein is mixed with water and plasticizer. Extruded films of SPI, polyethylene oxide and low-density polyethylene have also been made. The properties of the final films depend on the degree of denaturation, the unfolding of protein chains exposing new groups for reactions. The properties of the films depend on the extent of thiol–disulfide interchange reactions (polymerization, cysteine content) and on hydrophobic and hydrogen bond interactions. Films can be cast from from both acidic and alkaline conditions and using heat. Lysine and cystine are involved in cross-linking reactions. Interestingly, the isoelectric point is rather low (~4.5) for the soy protein [13].

6.2.5 Kafirin (Grain Sorghum)

Kafirin is the prolamine of the staple crop sorghum grown in the semi-arid parts of Southern Africa [14]. It has properties similar to zein, but is more hydrophobic and less digestable [15]. Films can be cast from aqueous ethanol (70 wt %) solutions after heating to 70 °C. An example of a plasticizer system for kafirin is a mixture of glycerol, polyethylene glycol and lactic acid (1:1:1 on a weight basis). The glass transition temperature of the unplasticized material is, depending on the method used, 146–154 °C. For zein the T_g is higher and in the 164–184 °C range.

6.2.6 Oat Avenin

Oats are considered the sixth most important crop in the world [14]. Avenin is the oat prolamine and is most soluble in 45 wt % ethanol, hence being more hydrophilic than other prolamines. It is worth pointing out that kafirin, zein and avenin are considered safe with respect to celiac decease. Glycerol-plasticized avenin films have been cast from 45 wt % ethanol solution by first heating the solution to 70 °C for 15 min. An avenin-based pouch is shown in Figure 6.4.

6.2.7 Rice Bran Protein (RBP)

Unfractionated rice bran protein consists of a mixture of albumin, globulin, prolamin and glutelin. Rice bran is produced in large quantities as a co-product/by-product from rice milling. It has good nutritional quality and superior protein efficiency ratio. Adebiyi *et al.* [13] have cast films based on RBP at various pH, plasticizer content and with or without thermal denaturation. It was shown that the strongest films were made with heat treatment under alkaline conditions. As with other proteins, casting should be carried out some distance from the isoelectric point, which is low for RBP (~ 4.5).

6.2.8 Lupin

The seeds from lupin, a leguminous plant, can be used to make films [16]. The seeds are rich in protein, oil and nonstarch polysaccharides and oligosaccharides of the

Figure 6.4   Pouch made of plasticized avenin and containing paprika powder. Courtesy of Thomas Gillgren, SIK, Sweden

raffinose family [16]. The protein has a good balance of essential amino acids. A lupin seed protein isolate (LSPI) can be obtained by essentially grinding the seeds, removing the fat, extraction/precipitation and freeze-drying.

6.2.9   Cottonseed Proteins

Glycerol-plasticized films based on cottonseed protein have been cast [17]. The source was delipidated glandless cottonseed flour that was obtained from glandless cottonseed kernels after oil extraction. Films have also been prepared from glanded and glandless cottonseeds [18]. Films were brittle when the glycerol content was below 10 wt% and sticky when the content was higher than 30% [18]. Film puncture strength decreased in the order delipidated glandless > glanded > glandless cottonseed. The cottonseed protein consists of globulins (60%, including gossypin and congossypin) and albumins (30%) with smaller amounts of prolamins (8.6%) and glutenins (0.5%) [19]. Both gossypin ($\alpha\beta$ structure) and congossypin are oligomers with a mass average molar mass of 180–300 kDa and 127–180 kDa respectively [18]. The albumins consist of low molar mass (mass average 10–25 kDa) chains and are rich in lysine and sulfur-containing amino acids. Overall, the most common amino acid in cottonseed is glutamic acid but the protein is also rich in arginine and aspartic acid [20]. It is noteworthy that the amount of ionizable amino acids is high and the content of sulfur-containing amino acids is low, while the isoelectric point is ~ 5 [18]. In order to improve film mechanical properties (e.g., film stability and puncture strength) the protein can be cross-linked under aqueous alkaline conditions with formaldehyde, glutaraldehyde or glyoxal [17]. All three treatments lead to improved puncture strength, with use of formaldehyde being the most efficient treatment.

6.2.10   Peanut Protein

It is possible to cast bio-based films based on peanut proteins. Liu *et al.* [21] cast films with peanut protein isolate and glycerol plasticizer from a basic 90 °C solution. Whereas the peanut seed contains 45% lipid and 22~33% protein, the peanut protein isolate contains significantly more protein and less fat (protein content > 95 g/100 g). Films or solutions were exposed to different physical treatments (heat, ultrasound, UV irradiation) or chemical treatments with acetic anhydride, succinic anhydride, form-aldehyde or glutaraldehyde in order to improve the properties of the final products. It was shown that strength values in excess of 1 MPa were obtained with the heat treatment (especially at 70 °C), 24 h UV exposure, ultrasonication using a water bath or the addition of the aldehydes. It was shown that the water vapor barrier was improved by heat treatment (60–90 °C) and the use of the aldehydes. The oxygen barrier was improved using the heat treatment or UV (ultrasound was not evaluated here). Interestingly, the anhydrides did not improve the mechanical or the barrier properties.

6.3   ANIMAL-BASED PROTEINS

6.3.1   Whey Protein

Whey protein is a by-product from cheese-making and is the remaining soluble protein when casein is precipitated at pH 4.6 [2]. The four main proteins in whey are β-lactoglobulin ($\sim$ 50 wt %), α-lactalbumin ($\sim$ 20 wt %), bovine serum albumin ($\sim$ 10 wt %) and immunoglobulins ($\sim$ 10 wt %) [22]. The protein can be obtained as whey protein concentrate (WPC), having a protein content in the range 25–80 % or whey protein isolate (WPI) having a protein content above 90 %. WPC and WPI are obtained from liquid whey by ultrafiltration and spray drying. Films are typically cast from a heated solution, which enables the protein to unfold and allows the formation of intermolecular sulfur-bridges. The denaturation temperature depends on the composition and is typically 78 °C for WPI, but substantially lower for whey concentrates. The final film is water insoluble and heat-denatured WPI has a glass transition of $\sim$ 70 °C [23]. Whereas WPI is transparent and odourless, films from WPC smell of cheese and have a yellow/red colour. The WPC film shown in Figure 6.5 consists of 82% protein (mainly β-lactoglobulin) along with lactose, fat and ash. WPI composition is typically 93.5% proteins, mainly β-lactoglobulin. Films produced without heating are still water soluble since the protein still exists in the native form with mainly hydrogen bonds between molecules and the intramolecular sulfur bonds and hydrophobic groups are still hidden within the globular protein [2]. Films without plasticizer are brittle and therefore a plasticizer may have to be added.

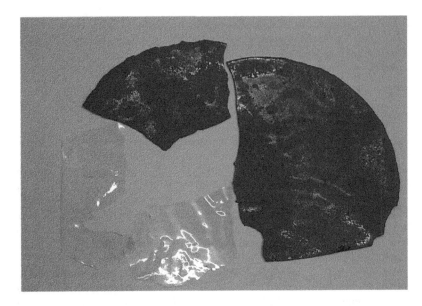

Figure 6.5   Whey protein isolate (transparent) and whey protein concentrate with 82% protein (yellow/red). Note that the whey protein isolate film is from 1999 and the other film from around the same period. Both have been kept in plastic bags and the films contain a glycerol plasticizer

Whey proteins find use as additives in pharmaceuticals, as nutrition agents and in human and animal food [22]. The international dry whey price varies with time, but is of the order of $1/kg ($1.025/kg 2010-01-09 and $0.795/kg 2010-06-12) [24].

6.3.2   Casein

Casein is the major milk protein, amounting to $\sim$ 80% of the total protein content [25]. The main amino acid in casein is glutamic acid (20.2%) [26]. Interestingly, and in contrast to many other proteins, casein has very few disulfide bonds [27]. Casein is a phosphoprotein which can be electrophorectically divided into σ_s-, κ-, β_s- and γ-casein, all of which have different molar masses and primary, secondary or tertiary protein structures. Casein from bovine milk may consist of σ_{s1} (38%), σ_{s2} (10%), β (36%), κ (13%) and γ (3%) casein fractions, all varying in molar mass in the range 19–23.9 kDa, hydrophilicity, amino acid content and isoelectric point. Casein possesses 'native casein micelles', presumably made up of aggregates with calcium phosphate bridges between the protein chains. At high pH these bridges can be cleaved, leading to destruction and restructuring of the micelles, a reduction of the high-molar mass fractions and a more transparent casein solution. Heat treatment generates stronger and stiffer, but also more brittle, films [25]. This finding is presumed to arise from the formation of isopeptide bonds between lysine and aspartic/glutamic acid residues. After extensive thermal treatment there is also the possibility that the carboxylic acid groups of the aspartic/glutamic acids can react with the hydroxyl groups of the plasticizer (i.e., glycerol). Non-food uses of rennet casein include buttons, buckles, imitation ivory (knife handles and piano keys), fountain pen barrels and shoehorns [28].

6.3.3   Egg White

The main hen egg white protein (albumen) is ovalbumin which constitutes $\sim 50\%$ of the egg white proteins [29, 30]. This protein is unique in the sense that it contains four free sulfydryl groups buried inside the molecular structure and which become available for reaction upon heat denaturation. Jerez *et al.* report the compression moulding of egg albumen, previously mixed with glycerol as plasticizer, at 60, 90 and 100 °C [29].

6.3.4   Keratin

Keratin is a protein with a relatively high content of disulfide bonds (due to cysteine). The disulfide crosslinks along with hydrogen bonding and crystallinity provide high stiffness and strength to keratin. The cysteine content varies between different keratins; for example, wool keratin contains 11–17% cysteine, whereas the corresponding value for feather keratin is 7%. Apart from cysteine, the most abundant amino acids in keratin are glycine, proline and serine, whereas there is very little histidine, lysine and methionine [3]. Chicken feathers contain a large fraction of keratin protein (91%), the rest being 1% lipid and 8% water [3]. Apart from wool and feathers, other sources of keratin include nails, hooves, epidermis, hair and horn.

Feathers are a significant by-product from the poultry industry and therefore a potentially interesting source for keratin-based materials. Likewise, $\sim 300\,000$ tons per year of human hair is cut in hairdressing saloons worldwide [31]. Keratin films can be obtained from relatively complex extraction procedures, but also from superfine powder; for example, using superfine wool powder and a plasticizer Wang *et al.* showed that it was possible to obtain compression-moulded films [32].

6.3.5   Collagen

Collagen consists of three parallel alpha chains that together form a triple-stranded superhelix [3]. The amino acid sequence is commonly glycine, proline and hydroxy-proline or alanine [3, 30]. In the triple helix the glycines are close to the central axis, whereas the prolines lie on the outside. The amino acids cystine and tryptophan are absent in collagen. Collagen is a fibrous structural protein found in bones, tendons, connective tissue, skin, ligaments and vascular systems. The most important application of collagen as a packaging material is in extruded sausage casings, where the starting material is the inner 'corium' collagen-rich layer of dehaired cattle hides [33]. Collagen and the denatured form, gelatin (see below), are used as sealants for woven polyester vascular prostheses [34].

6.3.6   Gelatin

Gelatin is produced by partial hydrolysis of collagen at low pH (gelatin A) or high pH (gelatin B) [35]. The unfolded peptide chains trap a large amount of water which then yields a hydrated material [30]. Since it originates from collagen, the most abundant amino acids in gelatin are glycine, proline and hydroxyproline [36]. The high pH treatment generates the strongest deamidation of asparagine and glutamine and the formation of free carboxyl groups. As a result, the isoelectric point is lower for gelatin B than for gelatin A. Cooling sufficiently concentrated aqueous gelatin to temperatures below 40 °C allows partial reformation/renaturation into the original collagen structure. The final gel, the properties of which depend not only on pH, but also on molecular weight, temperature, maturing time and concentration, serves as an important starting material for capsule production (Figure 6.6). Gelatin A yields a capsule that is softer and clearer than that from gelatin B. Examples of the use of gelatin include as a food stabilizer in ice cream, in pharmacy as a binder in tablets and as an adhesive in hot melt glues for case closing [37].

Compounds that are encapsulated in softgel capsules include pharmaceuticals such as anti-inflammatory agents (e.g., ibuprofen) and antihistamines (e.g., chlorpheniramine maleate) [38]. Further examples of softgel capsules are dietary supplements including phospholipids, (e.g., lecithins) and carotenoids (e.g., lycopene and lutein), personal care products (e.g., bath oils) and recreational products (e.g., paintballs).

6.3.7 Myofibrillar Proteins

In addition to connective tissue proteins, myofibrillar proteins can also be used to make films [39]. Myofibrillar proteins consist of contractive, contraction-regulatory and structural proteins and represent more than 50% of muscles, consisting mainly of myosin and actin. Both nonionized and ionized polar amino acids as well as apolar amino acids are present in myofibrillar proteins. The film making of myofibrillar proteins is usually preceded with purification/concentration steps involving

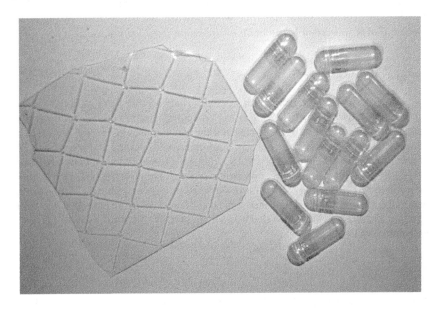

Figure 6.6   Gelatin hard-shell capsules and a gelatin sheet

mechanical separation and extensive washing processes and is referred to as the fish and beef mince process (surimi process). Films can be made directly after thawing frozen surimi or after drying and milling into powder. Films can be made either using a solvent or a dry process. As in the case of other proteins, the protein solubility is minimal near the isoelectric point (pH $\approx$ 5) and increases towards either lower or higher pH. However, the protein viscosity also increases away from the isoelectric point. Hence, a pH that yields a proper combination of high solubility and low viscosity has to be used. In addition, the protein concentration in the solution affects the viscosity. In the dry film forming/thermoforming process it is required that the protein-based material is above the T_g (which for myofibrillar protein is around 215–250 °C). The T_g can, in turn, be reduced by the presence of a plasticizer (water, glycerol, etc). As in the case of several other proteins the role of disulfide bonds seems important for the film properties [40]. It has been suggested that stabilization of films after casting and drying is supported by sulfydryl groups forming disulfide bonds.

Other proteins, or protein sources, that can be used to produce films, but which are not discussed here, include fibrinogen (blood plasma), pea, winged bean pistachio protein, cucumber pickle brine, rye secalinin (prolassmin), millet panicin (prolamin) and sunflower protein.

6.4   SOLUTION CASTING OF PROTEINS – AN OVERVIEW

Edible packaging, such as films and coatings for food ingredients, which have grease, taste or oxygen barriers, is an important area of application for protein films. In such uses, proteins can act as both a barrier and as a source of nutrition. The rather low oxygen permeability characteristics of solution-cast protein-based films have been explored in a number of studies and such new types of bio-sourced material may have interesting potential applications as films or coatings in fields as diverse as packaging, electronics and medicine [1].

The oxygen barrier properties of protein films can be attributed to the relatively high content of hydrogen bonding [41–49]. Although this intermolecular hydrogen bonding provides the excellent oxygen barrier characteristics of protein films, water resistance is usually poor.

The film-forming properties of some of the most commonly available proteins have been used particularly for encapsulating additives, enhancing the quality of cereal products and maintaining antioxidant and antimicrobial agents on the surface of food products [50]. The permselectivity (PCO_2/PO_2) properties of some protein films can also be utilized to improve the shelf-life of fresh or slightly-processed vegetables [51].

The structural and mechanical changes occurring in proteins during solution casting and heating determine the relevant process windows [52–56]. Solution-cast film processes are quite limited by factors such as temperature, time, moisture content, solvent, plasticizer concentration and pH [48, 49, 52, 57, 58].

6.4.1   Solvent Casting Procedures

Water-borne protein solutions can be readily cast into films on a laboratory scale, but can also be adapted for larger-scale coating methods as might be applied in the paper and paperboard industry or for edible packaging in the film industry. In film-making procedures, the protein is usually heated to the denaturation temperature (T_d) in order to get a cross-linked system of intermolecular disulfide bonds. Heating is applied either before or after casting, depending on the method being used.

Typically, there are several different qualities of each protein on the market and each of the proteins has different preferred processing conditions for obtaining the best possible solvent-cast films. The proteins on the market, the most common being 'concentrates' or 'isolates', differ in degree of purification. The former refer to purified proteins with some residues (e.g., starches, ash). The concentrates are commonly refined from liquid dispersions or solutions, and then spray dried and subjected to ultrafiltration to obtain protein concentrations in the range of 35–90% [59]. The latter are highly purified proteins, while the isolates are further purified through fractionation to different concentrations [59, 60]. The practical differences are that isolates normally give more transparent films with more predictable properties; however, isolates are also more expensive due to the cost of the more demanding protein purification [61].

6.4.2   Importance of pH

The importance of the solvent or liquid phase pH is pronounced for solution-cast proteins. Proteins have zwitterionic character and are therefore sensitive to the selection of pH. Many researchers have noted the importance of adjusting pH away from the isolectric point (Ip) when dealing with protein solutions or dispersions for films and coatings. With knowledge of protein properties, a pH can be chosen so as to obtain a viscosity that is most suitable for the casting method and the homogeneity of the final film. Since, without purification or fractionation, many of the plant proteins, are actually a mixture of several individual proteins, the Ip may have a broad range and the affect of pH adjustment can vary significantly depending on the difference between the selected pH and the Ip. Furthermore, film properties may also vary depending upon whether solutions are adjusted to more alkaline or acid pH, even if the pH distance from the Ip remains equal.

6.4.3   Drying Conditions

Drying conditions are a crucial factor when casting protein films. Drying rates which are too high, as a result of temperatures or air flows which are set too high or relative humidity which is too low, can result in unequal drying across or through a film with subsequent shrinkage or film fracture. Flaws such as bubbles may also arise if solvent evaporation occurs too rapidly. If temperatures are high enough, disruption of protein hydrogels may occur. Drying rates that are too low are also undesirable because processes can then become uneconomic. Therefore, laboratory evaluation of different drying conditions, easily established using conventional drying equipment (e.g., ovens, dessicators), is used to provide a good indication of the conditions that might be suitable for up-scaled casting or coating processes. In such evaluations, it is important to note that the addition of fillers or plasticizers may change the solution/dispersion properties and influence the selection of preferred drying conditions.

6.4.4   Viscosity

For the purposes of producing films or coatings, the viscosity of a protein solution is important for two reasons. First, if viscosity is too high films may be less homogeneous and may dry less efficiently. Second, if viscosity is too low the result may be undesirable absorption into a paper or paperboard substrate (e.g., if used for packaging) or into foods (e.g., if used as a food coating) Ideally, dry protein content should be as high as possible without reaching an inappropriately high viscosity since this will make the drying procedure less demanding, save on energy and maximize production. The selection of the particular protein and protein concentration will influence the preferred casting procedure and the general recipe for the formulation and again, as with drying, these are features which can be tested in the laboratory as part of formulation and process development.

6.4.5   Importance of Temperature

In terms of final protein film performance, temperature can be as important as pH. A number of protein films made without heating can have very good oxygen and grease barrier properties [1]; however, if heating is used, the mechanical properties and the structural integrity of the film or coating will depend on the selected temperatures and heating times. The denaturation temperature is sometimes referred to as the third transition temperature of peptide polymers. Most proteins do denature, at which temperature unfolding and, in many cases, cross-linking occurs. Denaturation is generally controlled by temperature and pH and therefore both should be known and controlled when preparing protein films or coatings.

6.4.6   Selection of Solvent

High protein solubility or dispersability is required to get good quality solution-cast films and many different solutions/dispersions have been studied for film casting. Protein solubility and unfolding in water–ethanol mixtures increases with a reduction of the intra- and intermolecular disulfide and hydrogen bonds and by reduction of hydrophobic interactions. The degree of unfolding is pH dependent and the reduction of disulfide cross-links depends on the amount and type of reducing agents, such as sulfites. Hydrogen bonding is decreased by using an appropriate polar solvent and the solution is normally adjusted by appropriate addition of acids or alkali.

6.4.7   Plasticizers for Protein Films and Coatings

As noted earlier, due to hydrogen bonding, protein films generally exhibit low oxygen permeability under dry conditions. However, hydrogen bonds also make the films brittle when dry, and a plasticizer is needed for the film to have desirable mechanical properties [51, 59, 62, 63]. Several factors must be taken into account when choosing between different possible plasticizers. Plasticizer and protein materials must have similar polarity to be compatible. Insufficient dispersion of plasticizer in the protein matrix results in a material with properties which depend on the plasticizer concentration gradient. The molecular weight of the plasticizer affects its diffusion properties and therefore also affects any migration from the protein matrix and thus the long-term properties of the finished material. Low-molecular weight and polar plasticizer additives increase the polymer chain mobility, with increasing permeability as a consequence.

The most commonly used plasticizer in protein films is glycerol, which is miscible in most cases; however, several other systems have also been studied [64–69]. The literature reveals studies on a variety of plasticizers including the polyfunctional alcohols such as glycerol, sorbitol, and propylene glycol, as well as di- and triethanolamine [51, 63–65, 67–77]. Both glycerol and sorbitol are frequently used as sweeteners in foodstuffs [71] and are considered harmless as plasticizers for films in contact with foodstuffs. Glycerol is renewable, water soluble, polar and nonvolatile and normally yields higher plasticization efficiency than other plasticizers at a comparable concentration. Glycerol is also rather easy to disperse in most proteins [65, 71]. The plasticizing effect of glycerol addition is mainly due to the hydroxyl groups that disrupt the hydrogen bonds in the protein matrix. However, migration of plasticizers and oxidation of free thiol groups result in aging and a decreasing fracture strain in protein films as a function of storage time and climate [68, 69, 78]. The plasticizer content should ideally be as low as possible since oxygen permeability increases in films with increasing free volume. The size and compatibility of the plasticizer is therefore of major importance. Glycerol is hygroscopic and attracts water, which in turn is also a very efficient plasticizer. This is of course a drawback for packaging film applications, and a compromise between permeability and mechanical properties has to be reached [53, 68, 79, 80]. The gas permeability of protein films generally increases with the humidity [71, 72]. Therefore, protein films need to be protected by a hydrophobic layer to be useful as barriers in packaging applications [62]. Several less hygroscopic plasticizers, such as mono-, di-, or oligosaccharides and urea, have also been studied. Amphipolar plasticizers such as octanoic and palmitic acids, dibutyl tartrate and phthalate, or mono, di- and triglyceride esters have also been studied, at least for less hydrophilic proteins such as wheat gluten and zein. However, glycerol and triethanolamine so far seem to be the most appropriate plasticizers for films of whey protein or wheat gluten [64, 68, 69].

6.4.8   Proteins as Coatings and in Composites

Various coating methods can be applicable for protein solutions or dispersions. The literature presents a number of studies in which, for example, curtain coatings or solution coatings have been applied using simple laboratory equipment. Krochta and Han studied the increase of gloss and decrease of oil absorptivity in paper with increasing thickness of WPI coating [81]. Chan and Krochta reported a considerable decrease in oxygen permeability for denatured and undenatured WPI as coatings on paperboard [82]. Trezza et al. [83] found reduced oxygen permeability when using corn zein as a paper coating. Gällstedt et al. [84] found a considerable decrease in air permeance when coating paper with whey, gluten or chitosan. Among the studies on proteins for paper coatings, grease resistance and water permeability properties have frequently been reported [74, 81, 85, 86]. A great number of studies have also been performed on whey and other proteins as oxygen barrier coatings in different food applications. Studies on protein coatings for fruits and other food products have been focused on mechanical properties, oxygen permeability, gloss, wettability and organolepticity [87–89].

6.4.9   Water Sensitivity of Protein Films

Protein films are commonly very sensitive to water, due to their high content of hydrogen bonds and polar groups. In order to be usable in packaging applications for moist or wet foods, protein films need to be protected by a water repellent material.

6.5   DRY FORMING OF PROTEIN FILMS

Fast production processes, which do not require energy for drying of solvents, are of commercial interest for manufacturing films and coatings. Dry thermoforming processes, including compression moulding, extrusion and injection moulding, are considered novel as methods to form protein-based materials into plastic items. Protein from plants (wheat gluten, soy protein, corn zein, sunflower, etc.) and animals (whey protein, myofibrillar protein, etc.) have been converted into films using dry processes.

6.5.1   Compression Moulding

Thermoplastic processing of proteins involves mixing with suitable additives under low water conditions. The resulting dough-like material is then transferred to viscoelastic melts using a combination of high temperatures, high pressure and processing times. Upon cooling, a protein-based material is formed by means of ionic, hydrophobic, and hydrophilic interactions as well as cross-linking reactions. [90, 91]. In order to improve processability and eliminate the brittleness of films due to extensive aggregation during heat treatment, an appropriate plasticizer should be incorporated into the protein matrix. Table 6.1 displays the commonly used plasticizers in compression-moulded protein-based materials. Pommet *et al.* [64] studied several types of compounds including fatty acids, water and glycerol as plasticizers for thermo-moulded wheat gluten films. Dry gluten and the compound were mixed in a two-blade counter-rotating batch mixer with 100 rpm mixing speed. The blend was thereafter compression moulded either at 100 °C for 5 min or at 130 °C for 15 min to obtain films. This work demonstrated that thermo-processing of gluten is possible with additives having low melting point, low volatility and good compability with the protein matrix. When choosing an appropriate plasticizer for thermoforming protein-based materials, its plasticizer efficiency (i.e., lowering of T_g), as well as molecular weight, polarity, amphiphilic nature and migration during aging, need to be taken into account. [64, 92–94].

Table 6.1 Commonly used plasticizers in compression-moulded protein-based materials, adapted from Hernandez-Izquiero and Krochta [91]

Plasticizer	Protein	Reference
dibuthyl phthalate	corn gluten meal	[92]
dibuthyl tartrate	corn gluten meal	[92]
diethylene glycol	sunflower protein isolate	[109]
ethylene glycol	soy protein isolate	[92]
ethylene glycol	sunflower protein isolate	[109]
glycerin	wheat gluten	[110]
glycerol	cottonseed protein isolate	[111]
glycerol	wheat gluten	[97]
gycerol	soy protein isolate	[112]
glycerol	corn gluten meal	[92]
glycerol	sunflower protein isolate	[109]
glycerol	myofibrillar protein	[39]
glycerol	whey protein isolate	[99]
lactic acid	wheat gluten	[113]
octanoic acid	corn gluten meal	[92]
octanoic acid	wheat gluten	[113]
palmitic acid	corn gluten meal	[92]
propylene glycol	sunflower protein isolate	[109]
triethylene glycol	sunflower protein isolate	[109]
water	corn gluten meal	[113]
water	wheat gluten	[113]

6.5.2 Properties of Compression-Moulded Protein-Based Films

The mechanical, barrier, thermal and solubility properties of films provide useful information for the prediction of package performance during handling and storage [91]. Table 6.2 shows the mechanical properties of various protein-based films obtained by thermoforming processes. Sothornvit *et al.* investigated the tensile properties of compression-moulded whey protein isolate sheets, plasticized with glycerol, and reported that increasing glycerol concentration from 30 to 50% reduced the tensile strength and elastic modulus and increased the extensibility of films [95]. Cuq *et al.* [96] examined the effect of thermal treatment on the mechanical properties of wheat gluten-based film. Homogeneous blends of wheat gluten and glycerol were compression moulded at 20 MPa for 10 min at different temperatures. These authors reported that increasing treatment temperature, from 80 to 135 °C, decreased the fracture strain from 468 to 236% and increased the tensile strength from 0.26 to 2.04 MPa [96]. These changes were attributed to development of heat-induced crosslinks within the protein network. Gällstedt *et al.* [97] also studied the effect of mould time (5–15 min) and relative humidity (0 and 50%) on mechanical properties of compression-moulded wheat gluten films plasticized with glycerol. It was reported that the mould time had less effect than mould temperature and glycerol

Table 6.2 Mechanical properties of various protein-based films obtained by dry processes

Film formulation	Processing method	Mould or die temperature (°C)	Test conditions	Tensile strength (MPa)	Elastic modulus (MPa)	Elongation (%)	Reference
30% Gly-WPI	Compression moulding	140	23°C/50 % RH	10	251	43	[95]
40% Gly-WPI	Compression moulding	140	23°C/50 % RH	8	144	85	[95]
50% Gly-WPI	Compression moulding	140	23°C/50 % RH	4	60	94	[95]
28.5% Gly-WG	Compression moulding	80	20°C/60-65 % RH	0.2	—	468	[96]
28.5% Gly-WG	Compression moulding	135	20°C/60-65 % RH	2	—	236	[96]
30% Gly-WG	Compression moulding	130	25°C/53 % RH	4	10	262	[98]
30% Gly-WG + 20% hemp fibers	Compression moulding	130	25°C/53 % RH	6.7	81	12	[98]
30% Gly-WG + 20% wood fibers	Compression moulding	130	25°C/53 % RH	6.5	63	17	[98]
SFPI:GLY: Water = 100:70:20	Extrusion	160	25°C/60 % RH	3.2	17	73	[103]
SPI:GLY: Water = 100:10:80	Extrusion	100–120	23°C/50 % RH	40.6	1226	3	[105]
SPI:GLY: Water = 100:30:80	Extrusion	100–120	23°C/50 % RH	15.6	374	133	[105]
SPI:GLY: Water = 100:30:80	Extrusion	100–120	23°C/0 % RH	41.1	1220	13	[105]

(continued)

Table 6.2 (*continued*)

Film formulation	Processing method	Mould or die temperature (°C)	Test conditions	Tensile strength (MPa)	Elastic modulus (MPa)	Elongation (%)	Reference
SPI:GLY: Water = 100:30:80	Extrusion	100–120	23 °C/75 % RH	4.7	70	159	[105]
45.8%GLY-WPI	Extrusion	130	23 °C/50 % RH	4.1	46	127	[91]
48.8%GLY-WPI	Extrusion	130	23 °C/50 % RH	3.5	36	121	[91]
41.9%GLY-WPI	Extrusion	130	23 °C/50 % RH	3.1	30	132	[91]
SPI:GLY:Corn starch = 50:30:50	Injection moulding	80	23 °C/11 % RH	2.9	29	89	[107]
SPI:GLY:Corn starch = 50:30:50	Injection moulding	130	23 °C/11 % RH	3.9	46	85	[107]

concentration in terms of tensile properties. Increasing relative humidity led to a decrease in fracture stress and elastic modulus and an increase in the fracture strain of samples [97]. Several attempts have been made to improve the mechanical properties of protein-based materials. In the study of Kunanopparat *et al.* [98], hemp and wood fiber were incorporated into wheat gluten/glycerol-based blends at different fiber concentrations. Addition of natural fiber in compression-moulded films increased both the tensile strength and Young's modulus, but reduced the elongation at break. Improvement in mechanical properties was reported due to the reinforcing effect of fibers and deplasticization of the matrix [98].

It is well known that protein-based films have higher water vapor permeability (WVP) values than most petroleum-derived films, as required for example in packaging. However, protein-based films show good gas barrier properties in dry conditions [43]. Table 6.3 shows the barrier properties of protein-based films manufactured by dry processes. Sothornvit *et al.* [99] examined the effect of moisture content, glycerol concentration, moulding temperature and pressure on WVP of compression-moulded whey protein isolate films and observed that changes of these parameters did not significantly affect the WVP of the films. However, films containing glycerol as plasticizer showed higher WVP values than those films plasticized with water, owing to loss of water during the compression moulding process [99]. Gällstedt *et al.* studied the effect of glycerol concentration (25–40%), moulding time (5–15 min) and temperature (90–130 °C) on the oxygen permeability (OP) and WVP of compression-moulded wheat gluten films. At the same moulding temperature and moulding time, films with 40% glycerol had higher OP and WVP values than films with 25% glycerol (Table 6.3). This result was attributed to plasticizing and the hygroscopic property of glycerol. In addition, the OP was reduced with increasing mould temperature. It was also found that the effect of moulding time and temperature on WVP varied as a function of glycerol plasticizer concentration. [97]. In a study by Foulk and Bunn [100], two commercial soy protein isolates, SUPROR 620 and SUPROR 660, were modified and acetylated to produce SY7 and SY23 thermoplastic films, formed under different moulding temperatures without using plasticizer. It was found that both the SY7 and SY23 films formed at higher compression-moulding temperatures had lower oxygen and water vapor permeability than the films formed at lower compression-moulding temperatures [100].

6.5.3 Extrusion and Injection Moulding

Compression moulding of proteins has been developed in a number of studies and this has shown the possibilities for producing high-barrier plastic sheets for future applications such as in packaging. However, even though compression moulding does not require solvents, it is a very slow production method compared with extrusion or injection moulding.

An extruder basically consists of a hopper, a heated barrel containing either one or two screws and a die. Raw materials, as granules or in powder form, are fed into a hopper, conveyed by a screw and finally pushed through a die of desired shape [91]. Feed rate, barrel temperature profile, screw speed and screw configuration, screw length-to-diameter ratio, die size and shape are considered as process variables. These greatly affect the material properties and the specific mechanical energy input, torque,

Table 6.3 Barrier properties of various protein-based films obtained by dry processes

Film formulation	Processing method	Mould or die temperature (°C)	Test conditions	WVP (g mm/d kPa m²)	OP (cm³ mm/m² day atm)	Reference
30% Gly-WPI	Compression moulding	140	25°C/0-100% RH	379	—	[99]
40% Gly-WPI	Compression moulding	140	25°C/0-100% RH	360	—	[99]
50% Gly-WPI	Compression moulding	140	25°C/0-100% RH	340	—	[99]
25% Gly-WG	Compression moulding	90	37.8°C/49% RH	1902	—	[97]
40% Gly-WG	Compression moulding	90	37.8°C/49% RH	6019	—	[97]
25% Gly-WG	Compression moulding	130	37.8°C/49% RH	2013	—	[97]
40% Gly-WG	Compression moulding	130	37.8°C/49% RH	5950	—	[97]
25% Gly-WG	Compression moulding	90	23°C/90-95% RH	—	3	[97]
40% Gly-WG	Compression moulding	90	23°C/90-95% RH	—	12	[97]
25% Gly-WG	Compression moulding	130	23°C/90-95% RH	—	1	[97]
40% Gly-WG	Compression moulding	130	23°C/90-95% RH	—	12	[97]
WG:GLY: Water = 1.65:1:1.19	Extrusion	130	22°C/0-33% RH	7	—	[106]
WG:GLY: Water = 1.65:1:1.19	Extrusion	130	22°C/0-54% RH	9	—	[106]
WG:GLY: Water = 1.65:1:1.19	Extrusion	130	22°C/0-77% RH	22	—	[106]

residence time in the barrel and the pressure at the die [91]. In injection moulding, a mould, rather than a die, is used to shape the product.

Extrusion and injection moulding have been used to make protein-based sheets or items. Redl *et al.* [101] used a co-rotating, self-wiping twin-screw extruder for extrusion of wheat gluten plasticized with glycerol. The effect of feed rate (1.9, 4.9 and 8.1 kg/h), screw speed (50, 100 or 200 rpm) and barrel temperature (40, 60 or 80 °C) on processing parameters (die pressure, product temperature, residence time and specific mechanical energy) was investigated. Results indicated that, depending on the operating conditions, either smooth-surfaced extrudates with high swell or completely disrupted samples were obtained. Extrudate break-up was attributed to the occurrence of excessive cross-linking reactions, which were induced by specific mechanical input and the high temperature reached during extrusion [101]. It can be concluded that the extrusion of wheat gluten is difficult owing to complex protein–protein interactions under shear and heat treatment, thus resulting in narrow processing windows. Ullsten *et al.* [102] added salicylic acid to wheat gluten/glycerol blends to enlarge the temperature range where wheat gluten can be extruded. These researchers reported that it was possible to extrude wheat gluten plasticized with glycerol into a flat and continuous film with a die temperature as high as 135 °C instead of 95 °C by incorporating only 1 wt % salicylic acid. The beneficial effect of salicylic acid was attributed to a lowering of protein cross-linking density due to radical scavenging effects [102]. Rouilly *et al.* [103] studied the effect of die temperature (85–160 °C), as well as water and glycerol content, on the mechanical and swelling behavior in water of sunflower protein isolate (SFPI) film produced in a single-screw extruder. A smooth and uniform film was obtained by using 20 parts of water and 70 parts of glycerol for 100 parts of SFPI, a die temperature of 160 °C and a screw speed of 20 rpm. Although the resulting film showed good mechanical properties, its sensitivity to water and tendency to swell (~ 180% after a day soaking in water) limits its potential packaging application [103]. In a study by Wang and Padua [104] corn zein film containing oleic acid as a plasticizer was manufactured using single-screw and twin-screw extruders. The tensile strength of film produced on the twin-screw extruder was found to be higher (4.2 MPa) than that of single-screw extruded film (3.1 MPa). The extensibility of twin-screw extruded films and single-screw extruded films was found to be 96.3 and 115.5% respectively. Analysis by scanning electron microscopy showed that twin-screw extruded films had fewer pinholes and more compact structure than the single-screw extruded films. The differences reflect the fact that the twin-screw extruder provided better mixing, improved conveying, and higher pressure and shearing forces during processing [104].

Soy protein-based sheet has been successfully produced on a single-screw extruder [105]. The effect of parameters such as moisture content, glycerol content and presence of cross-linking agents (zinc sulfate, epichlorohydrin and glutaric dialdehyde) on the mechanical and thermal properties of sheets was investigated. Increase in glycerol and moisture concentration led to decreases in both tensile strength and elastic modulus, while a significant increase was observed in elongation at break (Table 6.2). Use of cross-linking agents increased the elastic modulus of samples but the processibility became very poor at high concentrations. At the same glycerol content, increase in moisture concentration from 2.8 to 26% led to a decrease in glass transition temperature varying from 50 to – 7 °C, indicating the plasticizing effect of water. It was also found that the use of covalent cross-linking agents did not reduce the water absorption properties of the sheets. [105].

In the study of Hochstetter *et al.* [106], wheat gluten sheet was manufactured by using a co-rotating twin-screw extruder at a set temperature in the barrel and die of around 130 °C and a 60 rpm screw speed. These authors reported that the mechanical properties of samples measured in the machine direction were different from those measured in the cross direction. The difference was attributed to orientation and organization of denatured protein molecules in the machine direction. An increase in relative humidity caused increased water vapor permeability in gluten sheets. The measured oxygen permeability of extrudate was found to be low and comparable with that of wheat gluten film obtained by a casting process [106].

In the study of Huang *et al.* [107], soy protein isolate/acetylated high-amylose corn starch was injection moulded into Type I ASTM tensile specimens. The effects of mould temperature, storage and external lubricant on the physical and tensile properties of materials were evaluated. As the mould temperature increased, the water absorption decreased and tensile strength and stiffness increased (Table 6.2). The tensile strength of the samples stored at 11 and 50% RH for six months at room temperature increased from 11 to 38 MPa and from 5 to 16 MPa respectively. However, after being stored at the highest RH (93%) for three months at room temperature, fungal growth was observed on sample surfaces. No surface cracking was found when the samples were stored in an oven at 50 °C for four weeks. Use of beef tallow as an external lubricant significantly improved mould release during processing [107].

A mixture of soy flour, poly(ester amide), glycerol and hemp fiber (15 and 30 wt %) was pelletized and injection moulded into standard specimens. Results showed that as the hemp fiber content increased both modulus and tensile strength increased and the elongation of composites decreased. It was also found that the impact strength, flexibility and heat deflection temperature of composites was improved by using hemp fiber. However, using environmental scanning electron microscopy revealed poor fiber–matrix adhesion and inhomogeneous dispersion of fiber in the soy-based matrix [108].

6.6   CONCLUDING REMARKS

Solution casting is the most commonly used method to form protein-based films in laboratories. However, that method is an expensive process and has some limitations when it comes to fast and continuous industrial production of plastic films. Thermoplastic processes, used in the plastics industry are considered to be preferred options in the sense of avoiding the need for drying steps and therefore providing shorter processing times. Utilization of extrusion and injection-moulding technologies has the advantage of offering low cost and versatile production systems. In several studies, protein-based films have been successfully produced by dry processes. Properties of raw materials, selection of appropriate plasticizers or additives, processing and testing conditions are key variables in terms of obtaining good protein-based materials. Development of methods to produce safer foods is now and will continue to be a high-priority research area. For example, microbial contamination of meat and dairy products can occur on the surface of the food during post-process handling of food products. Such problems could be solved by antimicobial-carrying, protein-based edible films produced by compression moulding, extrusion or injection moulding processes.

REFERENCES

[1] Gennadios, A., (Ed.), Protein-based films and coatings, CRC Press, Boca Raton, Fl, USA (2002).

[2] Janjarasskul, T., Krochta J. M., *Annu. Rev. Food Sci. Technol.* 2010, 1, 415–48.

[3] Dangaran K., Tomasula, P. M., and Qi, P., Structure and Function of Protein-Based Edible Films and Coatings, in M. E. Embuscado, K. C. Huber (eds), *Edible films and coatings for food applications*, Springer, Dordrecht, Germany, pp 25–56 (2009).

[4] Rossman J. M., Commercial Manufacture of Edible Films, in M. E. Embuscado, K. C. Huber (eds), *Edible films and coatings for food applications*, Springer, Dordrecht, Germany, pp 367–390 (2009).

[5] Orlovius K., Fertilizing for High Yield and Quality Oilseed Rape, *IPI Bulletin* No. 16.

[6] Cho, S.-W., Gällstedt M., and Hedenqvist M. S. *J. Appl. Polym. Sci.* 2010, 117, 3506–3514.

[7] Wretfors, C., Cho, S.-W., Kuktaite, R., *et al.*, *J. Mater. Sci.* 2010, 45, 4196–4205.

[8] Cho, S.-W., Johansson E., Gällstedt, M., *et al.*, *Intern. J. Biol. Macromol.* 2011, 48, 146–152.

[9] Baldwin, E. A., Baker, R. A., Use of proteins in edible coatings for whole and minimally processed fruits and vegetables in Gennadios, A., (ed.), *Protein-based films and coatings*, CRC Press (2002).

[10] Shukla R., Cheryan M., *Ind. Crops Prod.* 2001, 13, 171–192.

[11] Liu, J., Lee, W. W. United States Patent (2003).

[12] Denavi G., Tapia-Blácido D. R., Añón M. C., *et al.*, *J. Food Eng.* 2009, 90, 341–349.

[13] Adebiyi, A. P., Adebiyi, A. O., Jin, D.-H., *et al.*, *Intern. J. of Food Sci. Technol.* 2008, 43, 476–483.

[14] Gillgren, T., Stading, M., *Food Biophys.* 2008, 3, 287–294.

[15] Taylor J., Taylor J. R. N., Dutton M. F., *et al.*, *Food Chem.* 2005, 90, 401–408.

[16] Pozani, S., Doxastakis, G., and Kiosseoglou, V., *Food Hydrocoll.* 2002, 16, 241–247.

[17] Marquie C., *J. Agric. Food Chem.* 2001, 49, 4676–4681.

[18] Marquié, C., Guilbert, S., Formation and properties of cottonseed protein films and coatings in Gennadios, A.,(Ed.), *Protein-based films and coatings*, CRC Press (2002).

[19] Saroso, B., *Ind. Crop Res. J.* 1989, 1, 60–65.

[20] National Cottonseed Products Association.

[21] Liu, C.-C., Tellez-Garayb, A. M., and Castell-Perez, M. E., *Lebensm. -Wiss. u.- Technol.* 2004, 37, 731–738.

[22] Gällstedt, M., Hedenqvist, M. S. *J. Appl. Polym. Sci.* 2003, 91, 60–67.

[23] Hedenqvist, M. S., Backman, A., Gällstedt, M., *et al.*, *Comp. Sci. Tech.* 2006, 66, 2350–2359.

[24] Brian Gould, Agricultural and Applied Economics, UW Madison.

[25] Ghosh A., Ali, M. A., and Dias G. J., *Biomacromol.* 2009, 10, 1681–1688.

[26] www.sci-toys.com/ingredients/casein.html. Last visited 8-Nov-2010.

[27] www.scientificpsychic.com/fitness/aminoacids1.html. Last visited 8-Nov-2010.

[28] nzic.org.nz/ChemProcesses/dairy/3E.pdf

[29] Jerez, A., Partal, P., Martinez I., *et al.*, *J. Food Eng.* 2007, **82**, 608–617.

[30] http://www.britannica.com/EBchecked/topic/180194/egg/50400/Characteristics-of-the-egg?anchor=ref501862. Last visited 8-Nov-2010.

[31] Reichl, S., *Biomaterials* 2009, **30**, 6854–6866.

[32] Wang, X., Xu, W., Li, W., *et al.*, Fibers Textiles Eastern Europe, 2009, **17**, 82–86.

[33] Osburn, W. N., Collagen casings in Gennadios, A., (ed.), *Protein-based films and coatings*, CRC Press (2002).

[34] Madaghiele, M., Piccinno, A., Saponaro, M., *et al.*, *J. Mater. Sci: Mater. Med.* 2009, **20**, 1979–1989.

[35] Bowman, B. J., Ofner III, C. M., Hard gelatin capsules in Gennadios, A., (ed.), *Protein-based films and coatings*, CRC Press (2002).

[36] www.gmap-gelatin.com/about_gelatin_comp.html. Last visited 8-Nov-2010.

[37] http://www.gelatin.co.za/gelatine-uses.html. Last visited 8-Nov-2010.

[38] Gennadios, A., Soft gelatin capsules in Gennadios, A., (ed.), *Protein-based films and coatings*, CRC Press (2002).

[39] Cuq, B., Formation and Properties of Fish Myofibrillar Protein Films and Coatings in Gennadios, A., (ed.), *Protein-based films and coatings*, CRC Press (2002).

[40] Artharn, A., Benjakul, S., Prodpran, T., *et al.*, *Food Chem.* 2007, **103**, 867–874.

[41] Lookhart, G. L., Bietz, J. A. *PBI Bulletin*, 1997, 4–6.

[42] Guilbert, S. Technology and application of edible protective films. In *Food Packaging and Preservation-Theory and Practice*, Mathlouthi, M., ed., Elsevier Applied Science Publishers, N.Y., USA, 1986, pp 371–394.

[43] Cuq, B., Gontard, N., and Guilbert, S. *Cereal Chem.* 1998, **75**, 1–9.

[44] Gennadios, A., Weller, C. L. *Food Technol-Chicago*, 1990, October, 63–69.

[45] Guilbert, S., Gontard, N. *Edible and Biodegradable Food Packaging*, In: *Foods and Packaging Materials – Chemical Interactions (Royal Society of Chemistry Special Publication, No 162)*, Ackermann, P., Jägerstad, M., Ohlsson T., eds., Royal Society of Chemistry, 1995, 159–168.

[46] Guilbert, S., Cuq, B., and Gontard, N. *Food Addit. Contam.* 1997, **14**, 741–751.

[47] Gennadios, A., Ed. *Protein-based Films and Coatings*, CRC Press LLC, USA, 2002.

[48] Herald, T. J., Gnanasambandam, R., McGuire, B. H., *et al.*, *J Food Sci*, 1995, **60**, 1147–1150.

[49] Gennadios, A., Brandenburg, A. H., Weller, C. L., *et al.*, *J. Agr. Food Chem.* 1993, **41**, 1835–1839.

[50] Redl, A, Gontard, N., and Guilbert, S., *J. Food Sci.* 1996, **61**, 116–120.

[51] Gontard, N., Marchesseau, S., Cuq, J. L., *et al.*, *Int. J. Food Sci. Tech.* 1995, **30**, 49–56.

[52] Roy, S., Weller, C. L., Gennadios, A., *et al.*, *J. Food Sci.* 1999, **64**, 57–60.

[53] Ali, Y., Ghorpade, V. M., and Hanna, M. A., *Ind. Crop. Prod.* 1997, **6**, 177–184.

[54] Kokini, J. L., Cocero, A. M., and Madeka, H., *Food Technol-Chicago*, 1995, October, 75–81.

[55] Li, M., Lee, T.–C., *J. Agr. Food Chem*, 1996, **44**, 763–768.

[56] Mujica-Paz, H., Gontard, N., *J. Agr. Food Chem*, 1997, **45**, 4101–4105.

[57] Gontard, N., Guilbert, S., and Cuq, J. L., *J. Food Sci*, 1992, **57**, 190–199.

[58] Lens, J.-P., de Graaf, L. A., Stevels, W. M., *et al.*, *Ind. Crop. Prod.*, 2003, **17**, 119–130.

[59] Anker, M., *Edible and Biodegradable Whey Protein Films as Barriers in Foods and Food Packaging*, PhD Thesis, CTH, Göteborg, Sweden, 2000.

[60] Andrews, A. T., Varley, J., eds., *Biochemistry of Milk Products*, Special Publication No. 150, The Royal Society of Chemistry, 1994.

[61] Olabarrietta, I., *Properties of chitosan and whey protein films blended with poly (ε-caprolactone)*, Thesis, KTH, Stockholm, Sweden, 2002.

[62] Muzzarelli, R. A. A., Peter, M. G., *Chitin handbook*, European Chitin Society, Grottammare, Italy, 1997.

[63] Shaw, N. B., Monahan, F. J., O'Riordan, E. D., *et al.*, *J. Food Sci.* 2002, 67, 164–167.

[64] Pommet, M., Redl, A., Morel, M.-H., *et al.*, *Polymer* 2003, 44, 115–122.

[65] Pouplin, M., Redl, A., and Gontard, N., *J. Agr. Food Chem.*, 1999, 47, 538–543.

[66] Gennadios, A., Weller, C. L., and Testin, R. F., *T ASAE*, 1993, 36, 465–470.

[67] Sánchez, A.C., Popineau Y., Mangavel C., *et al.*, *J. Food Chem.* 1998, 46, 4539–4544.

[68] Hernández-Muñoz, P., Hernández, R. J., *Effect of Storage and Hydrophilic Plasticizers on Functional Properties of Wheat Gluten Glutenin Fraction Films*. In: *Worldpak2002 Proc 13th IAPRI Conf;* CRC Press LLC, 2002, 2, 481–493.

[69] Irissin-Mangata, J., Bauduin, G., Boutevin, B., *et al.*, *Eur. Polym. J.* 2001, 37, 1533–1541.

[70] Anker, M., Stading, M., and Hermansson, A.-M., *J. Agr. Food Chem.* 1998, 46, 1820–1829.

[71] McHugh, T. H., Krochta, J. M., *J. Agr. Food Chem.* 1994, 42, 841–845.

[72] McHugh, T. H., Aujard, J.-F., and Krochta, J. M., *J. Food Sci.* 1994, 59, 416–419.

[73] Alcantara, C. R., Rumsey, T. R., and Krochta, J. M., *J. Food Process Eng.* 1998, 21, 387–405.

[74] McHugh, T. H., Krochta, J. M., *JAOCS*, 1994, 71, 307–312.

[75] Fairley, P., Krochta, J. M., and German, J. B., *Food Hydrocol.* 1997, 11, 245–252.

[76] Shellhammer, T. H., Krochta, J. M., *J. Food Sci.* 1997, 62, 390–394.

[77] Mahmoud, R., Savello, P. A., *J. Dairy Sci.* 1992, 75, 942–946.

[78] Morel, M. H., Bonicel, J., Micard, V., *et al.*, *J. Agr. Food Chem.* 2000, 48, 186–192.

[79] Gontard, N., Ring S., *J. Agr. Food Chem.* 1996, 44, 3474–3478.

[80] Gennadios, A., Brandenburg, A. H., Park, J. W., *et al.*, *Ind. Crop Prod.* 1994, 2, 189–195.

[81] Park, H. J., Kim, S. H., Lim, S. T., *et al.*, *JAOCS*, 2000, 77, 269–273.

[82] Chan, M. A., Krochta, J. M., *Tappi*, 2001, 84, 1–10.

[83] Trezza, T. A., Wiles, J. L., and Vergano, P. J., *Tappi J.* 1998, 81, 171–176.

[84] Gällstedt, M., Brottman, A., and Hedenqvist, M. S., *Packag. Technol. Sci.* 2005, 18, 161–170.

[85] Trezza, T. A., Vergano, P. J., *J. Food Sci.* 1994, 59, 912–915.

[86] Han, J. H., Krochta, J. M., *T. ASAE*, 1999, 42, 1375–1382.

[87] Park, H. J., *Trend. Food Sci. Tech.* 1999, 10, 254–260.

[88] Choi, W. Y., Park, H. J., Ahn, D. J., *et al.*, *J. Food Sci.* 2002, 67, 2668–2672.

[89] Guilbert, S., Gontard, N., and Cuq, B., *Packag. Technol. Sci.* 1995, 8, 339–346.

[90] Verbeek, C. J. R., van den Berg, L.E., *Macromol. Mater. Eng.* 2010, 295, 10–21.

[91] Hernandez-Izquierdo, V.M., Krochta, J.M., *J. Food Sci.* 2008, 73, 30–39.

[92] di Gioia, L., Guilbert, S., *J. Agric. Food Chem.* 1999, 47, 1254–1261.

[93] Chen, P., Zhang, L., and Cao, F., *Macromol. Biosci.* 2005, 5, 872–880.

[94] Tian, H., Liu, D., and Zhang, L., *J. Appl. Polym. Sci.* 2009, 111, 1549–1556.

[95] Sothornvit, R., Olsen, C.W., McHugh, T.H., *et al.*, *J. Food Eng.* 2007, 78, 855–860.

[96] Cuq, B., Boutrot, F., Redl, A., *et al.*, *J. Agr. Food Chem.* 2000, 48, 2954–2959.

[97] Gällstedt, M., Mattozzi, A., Johansson, E., *et al.*, *Biomacromol.* 2004, 5, 2020–2028.

[98] Kunanopparat, T., Menut, P., Morel, M.H., *et al.*, *Composites: Part A*, 2008, **39**, 777–785.

[99] Sothornvit, R., Olsen, C.W., McHugh, T.H., *et al.*, *J. Food Sci.* 2003, **68**, 1985–1989.

[100] Foulk, J.A., Bunn, J.M., *Ind. Crops Prod.*, 2001, **14**, 11–22.

[101] Redl, A., Morel, M.H., Bonicel, J., *et al.*, *Cereal Chem.* 1999, **76**, 361–370.

[102] Ullsten, N.H., Gällstedt, M., Johansson, E., *et al.*, *Biomacromolecules*, 2006, **7**, 771–776.

[103] Rouilly, A., Mériaux, A., Geneau, C., *et al.*, *Polym. Eng. Sci.* 2006, **10**, 1635–1640.

[104] Wang, Y., Padua, G.W., In: IFT *Annual Meeting and Food Expo*, 16–19 June 2002; Anaheim, Chicago, Ill.: Institute of Food Technologist. Poster nr 100B–37.

[105] Zhang, J., Mungara, P., and Jane, J., *Polymer*, 2001, **42**, 2569–2578.

[106] Hochstetter, A., Talja, R., Helén, H. J., *et al.*, *LWT*, 2006, **39**, 893–901.

[107] Huang, H.C., Chang, T.C., and Jane, J., *JAOCS*, 1999, **76**, 1101–1108.

[108] Mohanty, A.K., Tummala, P., Liu, W., *et al.*, *J Polym. Environ.* 2005, **13**, 279–285.

[109] Orlic, O., Rouilly, A., Silvestre, F., *et al.*, *Ind. Crops Prod.* 2003, **18**, 91–100.

[110] Pallos, F.M., Robertson, G.H., Pavlath, A.E., *et al.*, *J. Agr. Food Chem.* 2006, **54**, 349–352.

[111] Batterman-Azcona, S., Lawton, J.W., and Hamaker, B., *Scanning*, 1999, **21**, 212–216.

[112] Ogale, A.A., Cunningham, P., Dawson, P.L., *et al.*, *J. Food Sci.* 2000, **65**, 672–679.

[113] Pommet, M., Redl, A., and Morel, M.H., *Cereal Chem.* 2005, **42**, 81–91.

7

Synthesis, Chemistry and Properties of Hemicelluloses

Ann-Christine Albertsson and Ulrica Edlund
Fiber and Polymer Technology, School of Chemical Science and Engineering, Royal Institute of Technology (KTH), Stockholm, Sweden

Indra K. Varma
Centre for Polymer Science and Engineering, Indian Institute of Technology, Delhi, India

7.1 INTRODUCTION

Among the most abundant macromolecules found in nature is a family of heteropolysaccharides referred to as the hemicelluloses. Along with lignin and cellulose, hemicellulose is a key structural component of plant cell walls. Unlike cellulose, their co-component in wood and straw, the hemicellulose polysaccharides are amorphous and more or less branched, making their solubility potential markedly different. As such, and according to a classic definition, the hemicelluloses can be separated from cellulose and extracted from the primary and secondary plant cell walls via alkaline treatment [1]. The family of hemicelluloses is highly diverse in terms of their origin and their relative proportion to cellulose, but they vary even more structurally, displaying a variety of molecular weights and degree of branching. In addition, the degree of polymerization (DP) of hemicelluloses is lower than

Biopolymers – New Materials for Sustainable Films and Coatings, First Edition.
Edited by David Plackett.
© 2011 John Wiley & Sons, Ltd. Published 2011 by John Wiley & Sons, Ltd.

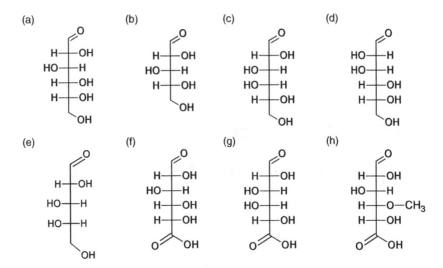

Figure 7.1 Principal sugar units of hemicelluloses: (a) D-glucose; (b) D-xylose; (c) D-galactose; (d) D- mannose; (e) L-arabinose; (f) D-glucoronic acid; (g) D-galacturonic acid; (h) 4-O-methyl-D-glucoronic acid

cellulose. The monomer units of hemicelluloses are based on hexoses and pentoses. The hexoses mainly comprise of D-glucose, D-galactose, D-mannose, D-glucoronic acid, D-galactoronic acid, and 4-O-methyl-D-glucoronic acid. The pentoses include arabinose and xylose (Figure 7.1). To a lesser extent, rhamnose and fucose sugar units are present in some hemicelluloses. A broad range of compositions from the various sugar units in different ratios and with different pendant groups increases the complexity of these heteropolysaccharides [2]. Their relative content and structural composition also varies within a species depending on the location in the plant or tissue of origin.

For more than a century, the isolation, recovery and structural analysis of hemi-celluloses has been a continuous challenge. An in-depth knowledge about their structures and function in the native and extracted states is now available. In recent years, it is with renewed and escalating interest that the utilization of these poly-saccharides as a source of low-molecular weight decomposition products and/or as a biopolymer material is being investigated. Replacing oil-derived materials and chemicals with products generated from an abundant and green resource presents an appealing alternative for future material production and fits the drive towards a more sustainable society. In this chapter, the chemistry and properties of hemicellu-loses is described.

7.2   STRUCTURE

The hemicelluloses are fairly complex heteropolysaccharides with backbones composed of one or a combination of pyranose and furanose sugar units, the predominant ones [3] being depicted in Figure 7.1. Hemicelluloses mainly comprise

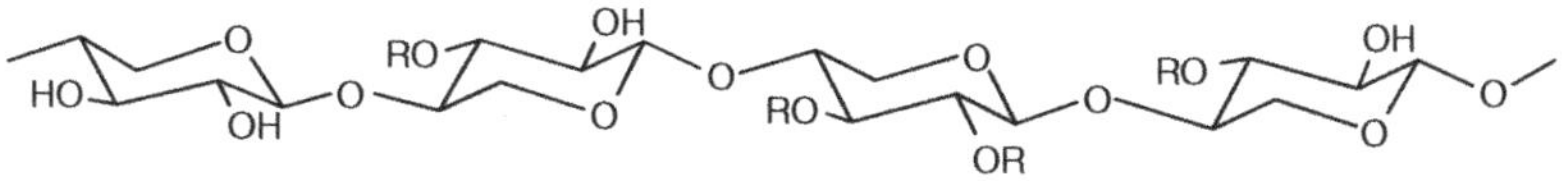

Figure 7.2 Representative structure of xylan from annual plants. R denotes the main sites of side groups or branches

of (1 → 4)-β-linked xylopyranose units and are widely distributed in various parts of the plants such as roots, stems, leaves and seeds [1]. Hemicellulose polysaccharides also occur in algae [4]. Hardwoods which are obtained from angiosperms (e.g., rye and oats) consist mainly of xylan (Figure 7.2) whereas the softwoods obtained from gymnosperms (e.g., wheat and barley) contain substantial amounts of mannose-rich polysaccharides in addition to xylans. The branching as side groups in positions C-2 and C-3 further alters the structure of hemicelluloses. Typical examples of side groups include arabinofuranose, xylopyranose, rhamnose, glucuronic acid and acetyl groups, and phenolic acids. The latter, such as ferulic acid and coumaric acid, are esterified to arabinofuranosyl units and are known as potential antioxidant, antimicrobial and anti-inflammatory agents.

The principal sugar building blocks make up a plethora of different structures that are usually divided into four main categories of hemicelluloses (Figure 7.3):

1. Xylans or xyloglucans
2. Mannans or mannoglycans
3. Xyloglucans
4. β-glucans

Xylans are the most abundant type of hemicelluloses and are characterized by having a xylopyranosyl backbone with primarily (1 → 4)-β-linkages as shown in Figure 7.2. Side chains are typically formed by branching in the C-2 and/or C-3 position. Xylans found in straw carry side chains where branching units are typically formed of L-arabinofuranose [5–7]. In fact, substituted arabinoxylans are very typical in cereal endosperms. In addition, there are acidic xylans found in some plants in which the acidic groups mainly stem from 4-O-methyl-D-glucoronic acid [8], as for instance in hardwoods such as aspen, beech, and birch, in which the backbone is substituted with such uronic acid units approximately every tenth xylopyranose. Acetylated xylans bear occasional acetyl side groups at the O-2 and/O-3 positions [9]. There are also other xylans with more complex branching structure, carrying for instance arabino-furanose, xylopyranose, galactopyranose and/or uronic acid side groups. Based on their branching characteristics, the xylans are often divided into several sub-groups. The homoxylan group comprise the structurally most simple chains, being linear with the xylopyranosyl backbone units linked through (1 → 3), (1 → 4), or mixed (1 → 3, 1 → 4) bonds. The glucuronoxylans bear D-glucoronic acid residues, which may or may not be 4-O-methylated, as substituents on the main chain. In addition, some glucuronoxylans are O-acetylated. Glucuronoxylans that in addition carry L-arabi-nofuranosyl substituents are usually considered to constitute a separate sub-group of xylans: the (arabino)glucuronoxylans. The arabinoxylan sub-group differs from the former in bearing L-arabinofuranosyl substituents, but none of the D-glucoronic acid

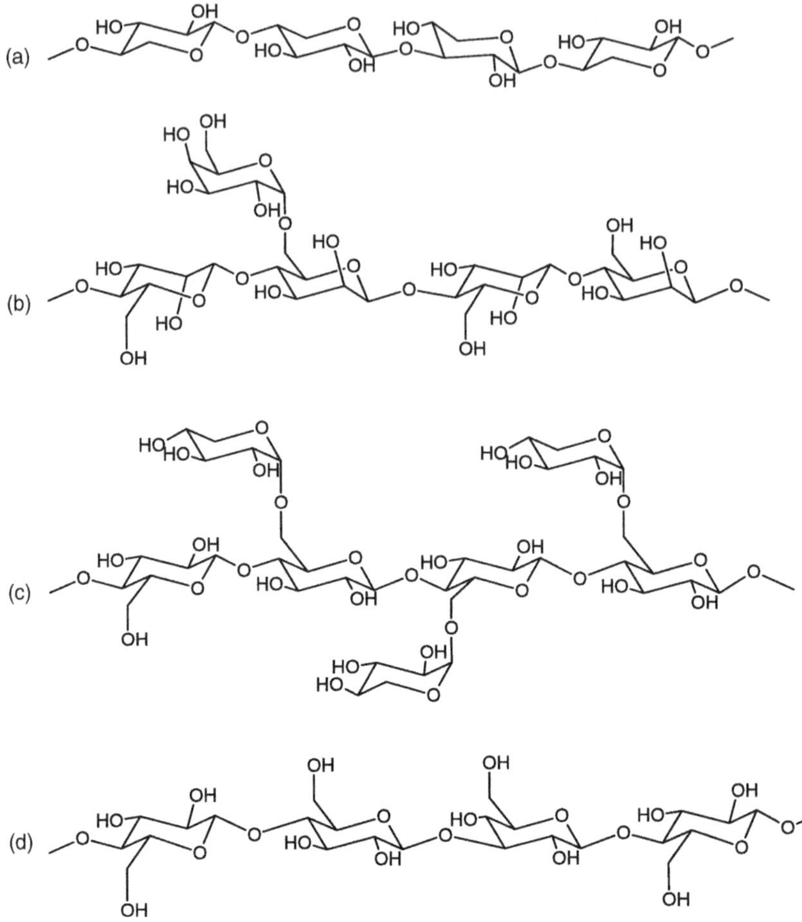

Figure 7.3 Representative structures of: (a) homoxylan with mixed β-$(1 \rightarrow 3, 1 \rightarrow 4)$ linkages; (b) D-galacto-D-mannan; (c) D-xylo-D-glucan; (d) mixed $(1 \rightarrow 3, 1 \rightarrow 4)$ linkage β-D glucan

type. The most complex xylans, in which a range of different sugar substituents may be found, are denoted as heteroxylans.

Mannans are generally divided in two types: galactomannans and glucomannans. The former type consists of a D-mannopyranosyl backbone with $(1 \rightarrow 4)$-β-linkages and with short branches of D-galactopyranosyl units linked to some of the mannan units at position C-6. The glucomannans have a D-mannopyranosyl backbone in which β-D-glucopyranosyl units are randomly distributed in the backbone chain connected via $(1 \rightarrow 4)$-β-linkages. Some glucomannans in addition carry D-galactopyranosyl substituents, just like the galactomannans, and are then referred to as (galacto)glucomannans. Mannans are the major constituents of softwood hemicelluloses [2]. Such mannans are glucomannans and the mannan:glucose ratio differs depending on the origin. Most often, the softwood mannans are acetylated at the mannosyl C-2 or C-3 position [10]. Some mannans are frequently utilized in the food industry as emulsifiers and thickeners (e.g., guar gum and locust bean gum, water-soluble galactomannans with a fairly high degree of branching) [11]. Another such mannan is the slightly O-acetylated konjac glucomannan [12], which has a mannose:glucose ratio of 1.6 : 1 and a higher molecular weight than the mannans derived from either guar or locust bean gum.

Xyloglucans vary according to the side group structure. All share the backbone structure of (1 → 4)-β-linked D-glycopyranose rings widely occurring in the primary cell walls of higher plants such as angiosperms and grasses, as well as in fruity plants such as tamarind, tomato or yellow passion fruit [13,14]. There are α-linked xylopyranosyl groups at the C-6 position of some glucopyranosyl residues and, based on the extent of xylosylation, the xyloglucans are often divided into two groups. The XXXG group contains repetitive units of four glucopyranosyl units of which three carry xylopyranosyl substituents while the XXGG group backbones have repetitive groups of two non-substituted glucopyranosyl units followed by two xylosylated glucopyranosyl units. The xylopyranosyl substituents may further be functionalized with units stemming from galactose, arabinose and/or fucose. In addition, some xyloglucans are O-acetylated.

β-glucans, also known as mixed β-glucans, are linear hemicelluloses with backbones consisting of β-D-glycopyranosyl units, resembling the cellulose structure. However, while the cellulose chain is strictly regular with solely (1 → 4)-β-linkages, the β-glucan backbone has monomers attached in a mixture of (1 → 3) and (1 → 4) linkages, hence the name mixed β-glucans [15]. Two major linkage patterns may be identified. A sequence of two consecutive (1 → 4) linkages followed by a (1 → 3) bond is referred to as a cellotriosyl sequence while a cellotetraosyl segment comprises three repetitive (1 → 4) linkages separated by a (1 → 3) bond. Even longer (1 → 4) linked chain segments have been identified, but generally the cellotriosyl and cellotetraosyl blocks make up at least 90% of the backbone [15]. The relative occurrence and distribution of cellotetraosyl and cellotriosyl building blocks can vary from species to species and even from one plant to another, depending on the growing environment. For instance, cereal plants are generally rich in β-glucans, but while some, such as barley, are rich in β-glucans, others, like wheat, have a hemicellulose fraction dominated by arabinoxylan. Furthermore, the ratio of cellotriosyl:cellotetraosyl building blocks differs between these two, with a higher value for wheat ($\sim$ 4.5) than barley ($\sim$ 3) [16].

7.3   SOURCES

As previously discussed, carbohydrate components comprising of cellulose, hemicellulose and pectins constitute 90% of primary cell walls. Depending on their origin (i.e., *Gramineae* (grass and cereals), *Gymnosperms* (softwoods) and *Angiosperms* (hardwoods)), the distribution of the constituents varies. For example, the cell walls of barley straw and maize stems consist mainly of rigid cellulose microfibrils (37–38%), an amorphous matrix of noncellulosic heteropolysaccharides (28–35%) and associated lignin (15–16%). The hemicelluloses of cereal stalks from the *Graminae* family

Table 7.1   Constituents of some lignocellulosic fibers [17]

Fiber	Cellulose (wt %)	Hemicellulose (wt %)	Lignin (wt %)	Pectin (wt %)
Flax	71	18.6–20.6	2.2	2.3
Hemp	70–74	17.9–22.4	3.7–5.7	0.9
Jute	61–71.5	13.6–20.4	12–13	0.2
Ramie	68.6–76.2	13.1–16.7	0.6–0.7	1.9
Sisal	66–78	10–14	10–14	10
Banana	63–64	10	5	
Oil palm EFB	65		19	
Coir	32–43	0.15–0.25	40–45	3–4
Cereal straw	38–45	15–31	12–20	8

have a backbone of $(1 \rightarrow 4)$-linked β-linked β-D xylopyranosyl residues to which are attached a number of L-arabinofuranose and/or D-glucuronic acid units.

Timell has reviewed the literature pertaining to wood hemicelluloses [2]. Amongst the gymnosperm group the hemicelluloses of the families of Pinaceae (genera more commonly encountered commercially such as *Pinus* and *Picea*) were investigated in details almost 50 years ago. The major hemicellulose in softwood is acetylated galactoglucomannan, which can constitute up to 20 wt/wt % of the dry wood. Smaller amounts of arabino-4-O-methylglucuronoxylan (5–10 wt/wt%) are also present. The molar ratios for galactose:glucose:mannose are approximately $1:1:3$ and $0.1:1:3$ in said mannans. Acetylation occurs at the C-2 and C-3 positions (one acetyl per 3–4 backbone hexose units). The molecular weight is around 16 000–24 000 Da [18]. The distribution of various polysaccharides in some plant lignocellulosic fibers is summarized in Table 7.1. In most of these plant fibers, hemicellulose is the second most abundant polysaccharide and only in coir fibers is the lignin content significantly higher.

7.3.1   Species

Hemicellulose composition in terms of particular sugar constituents and amounts varies widely from one plant species to another. Often, there are two or several types of hemicelluloses co-existing in the same species with composition variations in the different macroscopic parts of the plant (e.g., branches, core, fibers, knots and roots). Furthermore, hemicellulose composition may vary within the wood structure (e.g., between earlywood, latewood, sapwood, heartwood, and compression wood) and there may also be structural differences within the different layers in the cell wall. Table 7.2 presents an overview of the typical sugar composition of the hemicelluloses found in a range of plants.

The hemicellulose composition of hardwoods such as aspen, beech, and birch is quite similar, comprising a major portion of xylan and smaller amounts of mannan. The xylans are fairly linear with occasional uronic acid substituents. Hemicellulose isolated from beech is a typical example, consisting of a backbone of $1 \rightarrow 4$-linked β-D-xylopyranose units and, for every 10 units in the backbone, there is one

Table 7.2   Relative occurrence (%) of monosaccharides in different hemicelluloses

Plant source	Xylose	Arabinose	Glucose	Galactose	Mannose	Uronic acids	Ref.
Spruce mannan			15	1	62		18
Spruce arabino-galactan		11	8	35	30	8	19
Larch arabino-galactan		16	1	79	1	1	19
Pine arabino-galactan		14	3	69	7	4	19
Birch xylan	92.6	1.2	2			4.1	20
Birch mannan	1	2	28	1	68		21
Beech xylan	91.6	1, 2	2			5.2	20
Aspen xylan	95		0.7	0.6		3.8	22
Aspen mannan	2	1	41	1	55		21
Sugarcane bagasse	55	13	28	2.6	1.5	3	23
Wheat straw xylan	56.9	17.1	12.1	5.6		8.3	5
Rye straw xylan	78.3	12.4	5.9	2.5	0.4	5	24
Passion fruit rind	29	1	42	15	9	<0.5	13
Date palm leaf	75	6.3	0.3	1		17	25
Flax xylan	83.2	1	2.8	3.4	0.6	9	26
Flax mannan	7.4	3.9	42.2	8.8	37.8		26

4-O-methyl-D-glucuronic acid substituent [20, 27]. In birch, the structure consists of one 4-O-methyl-D-glucuronic acid for approximately every 15 xylopyranose units [8, 20]. Aspen xylan is typically acetylated at the C-2 and/or C-3 position and seems not to be arabinosylated [22]. Mannans found in these hardwoods are of the glucomannan type with a backbone of 1 → 4-linked β-D-mannopyranose and β-D-glucopyranose units. The glucomannans isolated from aspen and birch woods are acetylated at the C-2 or C-3 position [21].

The xylans found in annual plants such as maize, rice, or cereals are often more diverse and complex than those found in hardwoods. In this case, hemicelluloses may be acetylated, more or less branched and typically contain uronic acid and arabinose residues. Straw hemicelluloses have been thoroughly isolated and studied with respect to structural composition. Shives of flax, as an example, contain mainly acetylated 4-O-methylglucuronoxylan and smaller amounts of O-acetylglucomannan [26]. As for many other species, hemicellulose composition varies in different parts of the flax plant, with a higher xylan:mannan ratio in the woody core as compared with the bast fibers [28]. The O-acetyl-4-O-methylglucuronoxylan chain has a xylose:uronic acid ratio of around 13 : 1 while the glucomannan component has a glucose:mannose ratio of 1 : 1. Another example is corn where the glucoronoarabinoxylans isolated from corn cob and corn stover respectively differ in their degree of acetylation and substitution pattern [22]. The xylan hemicelluloses in cereals such as oat, wheat, barley and rye, consist of a 1 → 4 linked β-D-xylopyranosyl backbone, as described earlier. Typically, these xylans carry L-arabinofuranosyl groups linked by (1 → 3) or (1 → 2) bonds and uronic acid substituents, mainly 4-O-methyl-D-glucuronic acid. Such a structure is characteristic for rye straw xylans, having a xylose:arabinose: uronic acid ratio of around 85 : 12 : 3, indicating that there is an arabinose residue on every seventh xylose residue [24]. The xylan isolated from barley is reported to be similar but in addition contains non-terminal L-arabinofuranose units, namely 2-O-D-xylopyranosyl-L-arabinofuranose substituents attached to the xylopyranosyl main chain through the C-3 position [29]. Wheat straw xylans are also reported to consist of a β-D-xylopyranosyl backbone with L-arabinofuranosyl, D-xylopyranosyl, and uronic acid side groups where approximately 1 of 15 xylose residues carry an arabinose group through position 3 and 1 of 19 xylose residues carry a xylose pendant group through position 2 [30]. Somewhat in contradiction, wheat xylan has been reported to lack L-arabinofuranosyl side groups, consisting only of the β-D-xylopyranosyl backbone with D-xylopyranosyl and a small amount of uronic acid side groups [31]. In addition to arabinoxylans, cereal hemicelluloses contain β-glucans [15]. As mentioned earlier, the xylan:β-glucan ratio is different for different cereal species, as is the cellotriosyl:cellotetraosyl ratio in the β-glucan portion. Wheat has a significantly higher share of the trisaccharide units in its β-glucan than barley or oat.

In softwoods the major type of hemicellulose is glucomannan, constituting up to 20 wt/wt % of the dry wood, while xylans and arabinogalactans are present in smaller amounts. The glucomannan of common softwoods such as pine and spruce is an acetylated galactoglucomannan, composed of an O-acetylated β-(1 → 4) linked glucomannan backbone with side groups of D-galactose α-(1 → 6) linked to some of the mannosyl units [2, 18]. A relative sugar ratio of mannose:glucose: galactose:acetyl of $1:0.3:0.2:0.4$ has been reported. Around 5–10% of the softwood may consist of arabinoglucoronoxylan having a 1 → 4-linked β-D-xylopyranosyl backbone with L-arabinofuranose and 4-O-methyl-D-glucuronic acid substituents. Arabinogalactans are structurally fairly similar in pine, larch and spruce. Their backbones are composed of β-D-(1 → 3)-galactopyranose units carrying side chains of galactose, arabinose or glucuronic acid units attached to the C-6 position [19].

Xyloglucans are found in a wide range of locations in higher land plants, such as grasses, seeds, softwoods, and fruits [13, 14]. Tamarind, apricot, cherry, peaches, tomato and pumpkin are a few examples of fruit with a documented portion of xyloglucans [14, 32, 33]. The seeds of tamarind tree are known to be a rich source of xyloglucan and are used for industrial recovery through alkaline extraction.

7.3.2   Distribution

The structure, amount and distribution of hemicellulose in different plant species and within the same species and different parts of the plant varies significantly. Thus, pineapple fiber contains an arabino-4-O-methylglucuronoxylan as the main hemicellulose component in contrast to bast and leaf fibers such as jute, roselle, sisal, flax and sansevieria that contain 4-O-methylglucuronoxylans or sunn hemp which contains a glucomannan [34]. Similarly, hemicelluloses in wood species of commercial importance for pulping, such as birch, beech, hemlock, and spruce, have been investigated for more than five decades [2, 3, 35, 36]. Some detailed analyses of these studies have been reported earlier in this text. The glucuroarabinoxylan fine structure in the cell walls of *Aechmea* leaf chlorenchyma indicates different degrees of arabinose substitution which affects the wall strength [37]. Olive stone and olive seeds are lignocellulosic materials having significant differences in cellulose and hemicellulose contents. In the stone, cellulose is the main component (29.79–34.35 g/100 g dry matter) while in seeds, hemicellulose is predominant [38].

7.3.3 Co-Constituents

The carbohydrate mass in plant cell walls consists of cellulose, pectins and hemicelluloses. Cellulose is a structurally simple polysaccharide consisting of β-(1 → 4)-linked D-glucopyranosyl units, being linear, non-water-soluble and of high molecular weight. Pectins are oligo- and polysaccharides and, as for the hemicelluloses, this group is heterogeneous with saccharide chains of various lengths, constituents and degree of branching [39–41]. Pectins are highly water soluble and may form gels. Rhamnose, galactose and arabinose are common constituents of the pectic substances. Homogalacturonan, for instance, is a linear polymer with β(1 → 4)–linked D-galactopyranose units while rhamnogalacturonan I is branched with a backbone consisting of β-pyranose units of D-galactose and L-rhamnose linked with (1 → 4) and (1 → 2) linkages respectively. Homogalacturonan is the most abundant pectic substance and may, depending on its source, be methyl esterified at C-6 and/or acetylated at C-2 or C-3. The role of pectins in the cell wall is diverse. Pectins are believed to actively influence the ion transport, plant growth and development, defence, pH, water retention and cell wall porosity among other factors, and to have health-promoting bioactivity in humans, such as an immune-regulatory effect [42, 43]. Industrially, pectin is mainly utilized as a food additive (e.g., as a thickener or gelling agent).

In contrast to its carbohydrate co-constituents, lignin is a complex, and highly aromatic biopolymer consisting of monolignol units, the major three being sinapyl alcohol, coniferyl alcohol, and p-coumaryl alcohol [44]. These propylphenol adducts are linked through ether and alkyl bridges in an irregular fashion. Lignin in the native state is closely associated with the plant polysaccharides, and typically much effort is spent in the removal of lignin in the recovery of purified polysaccharide components, such as the delignification strategies dealt with in the next section of this chapter. Recent studies have established the presence of covalent bonds between lignin and polysaccharides in the native state, in so-called lignin–carbohydrate complexes (LCCs) [45], which may account for the difficulties in isolation and recovery of hemicelluloses in full yield. Although naturally occurring in massive amounts, there are few applications for lignin at present. Following the recent worldwide efforts to find and develop non-oil-derived materials for a sustainable future, there is now a renewed interest in lignin as a green resource.

7.4 EXTRACTION METHODOLOGY

Hemicelluloses have been isolated and characterized from a broad variety of plants and agricultural materials. In recent years the focus has been to examine residues remaining after the production of different agricultural products (e.g., wheat straw, brewery's spent grain, jute and luffa fruit fibers, rice straw, red gram husks, flax fibers, corn stover, etc.) [22, 26]. The closely integrated structure of carbohydrate components and lignin in the cell walls of plants and wood makes the isolation and recovery of hemicelluloses a challenge. A number of methods have been presented, typically being multi-step procedures involving mechanical and chemical treatments. There is always a risk of backbone hydrolysis in the process, causing the extracted product to be lower in molecular weight and/or less branched than in the native state [46]. The yield may be modest due to decomposition of the hemicelluloses into free sugar units. For the same reason, the hemicelluloses released in process liquids during pulping may not be representative of the native form.

The initial approach to extraction is to expose the cell walls to liquid penetration, achieved by mechanically processing the wood or plant material into chips, shives or a ground mass, which facilitates the exposure and availability of cell walls to liquid penetration for subsequent leaching of soluble substances. Pectic substances are highly soluble and are easily removed (e.g., by water solutions and ammonium oxalate). A next step is typically to remove lignin fractions. Delignification can be done through acid chlorite treatment [47]. A classical procedure involves $NaClO_2$ and glacial acetic acid at temperatures around 60 °C, but there is unfortunately a risk of degradation of the hemicelluloses in the process. The intermediate produced after delignification is referred to as holocellulose.

A range of treatments of the holocellulose to recover the hemicellulose fraction have been suggested. Extraction with acidic hot water is an effective way of releasing the hemicelluloses into solution, but chain degradation and the formation of furfural discourages this approach. A much more common method is extraction in alkaline media. The holocellulose is thereby successively extracted with hot water and sodium and/or potassium hydroxide solutions. A similar approach involves fractional extraction in hot ethanol/benzene. Fractionation is usually completed in ethanol by precipitation. Typically, a higher molecular weight fraction is precipitated while oligomer fractions remain in solution. Degradation from glycoside bond cleavage may occur under basic conditions. It has been shown that acetyl side groups are readily cleaved off at elevated pH and elevated temperatures, affecting for instance the acetylated glucomannan recovery from softwood. Alkaline extraction is commonly combined with other fractionation methods (e.g., gel filtration). For example, the soluble fractions of hemicelluloses may be further upgraded by ultrafiltration and/or diafiltration in which polysaccharide fractions are concentrated and purified by means of a pressurized membrane system where monomers and salts are removed to the permeate and hence separated from saccharides of higher molecular weight remaining in the retentate [48]. An alternative route to alkaline extraction is the steam explosion treatment of wood chips or shives. The chips are impregnated with water and exposed to hot steam for a few minutes, generating a slurry from which lignocellulosic material as well as soluble polysaccharide fractions can be separated and recovered.

Hemicelluloses from straws and grasses can effectively be recovered using an alkaline peroxide treatment, where the hydroxyperoxide anions formed mediate both delignification and solubilization of the polysaccharide chains. As an example, a sequential treatment of sugarcane bagasse with sodium hydroxide and hydrogen peroxide gave soluble fractions where a high yield of xylans could be recovered by subsequent precipitation in ethanol [49].

Dimethylsulfoxide-solubilized polysaccharides from delignified corn stover and aspen have recently been isolated and characterized [22]. The biomass was delignified by two different techniques, a standard acid chlorite and a pulp and paper technique (comprising of chelation, peroxide and acid chlorite treatment). Xylan acetylation was intact after the chlorite delignification process but deacetylation was observed when the pulp and paper technique was used. Thus, the delignification process may alter the xylan structure. Constitutional components of the ball-milled cereal straw (barley straw and maize stem), avoiding the classical delignification step, have been fractionated by sequential extraction with 90% neutral dioxane, 80% acidic dioxane, DMSO and 8% KOH. Some degradation of macromolecules was observed when acidic and alkaline solvents were used due to cleavage of glycosidic linkages [50].

Heat fractionation using microwave irradiation is another alternative for hemicellulose extraction from wood and straw. To allow for softwood hemicellulose extraction without major concurrent deacetylation, a fractionation technique involving exposure to microwaves has been used, albeit the products were partly degraded in the process. Spruce chips were soaked in water or a slightly basic medium and heat-fractionated for up to 20 minutes to give O-acetylated galactoglucomannan in decent yields [18, 51]. In a related study, microwave irradiation of water-impregnated spruce chips at 200 °C gave yields up to 70% [52], while steam treatment of the same type of spruce chips at 200 °C gave lower yields. The recovery of hemicelluloses from flax shives was also possible by microwave extraction. The use of water-soaked shives treated at 200 °C enabled the recovery of approximately half of the xylan contents and most of the less common glucomannan component [26]. Water-soaked shives treated at 180 °C gave lower yields than when a solution with 30% ethanol was used, but the molecular weights of the high-molecular weight fractions retrieved did not differ significantly [53].

7.5 MODIFICATIONS

The hemicellulose chains have hydroxyl groups in each repeating unit and lend themselves to chemical or enzymatic modification. Thus, for more than 40 years, chemical modification of hemicelluloses has been investigated as a means to introduce new groups and new properties (e.g., to alter the solubility or to achieve thermoplastic derivatives). Reports on hemicellulose modification are, however, fairly few compared to the related work on cellulose or starch, although analogous modifications are applicable to each of these polysaccharides. The most common reactions applied to modify hemicelluloses have been acetylation and alkylation. A schematic illustration of common covalent modifications presented in the literature is shown in Figure 7.4.

7.5.1 Esterification

Esterification of the hydroxyl groups of the sugar units is readily conducted by conventional carboxylic acid chemistry. A classic example is the treatment with an acid anhydride in the presence of a mildly basic catalyst (sodium acetate or pyridine for instance). This route was followed to prepare acetate, propionate, butyrate, caprate, laurate, myristate, and palmitate esters from xylan [54]. Direct esterification with succinic acid anhydride in basic media also generated an esterified xylan [55]. Aromatic esters of xylan can be obtained by benzoylation, for instance via reaction with benzoyl chloride and aqueous sodium hydroxide at 80 °C. N,N-dimethylformamide (DMF) is a common solvent in these processes. In fact, the use of DMF has

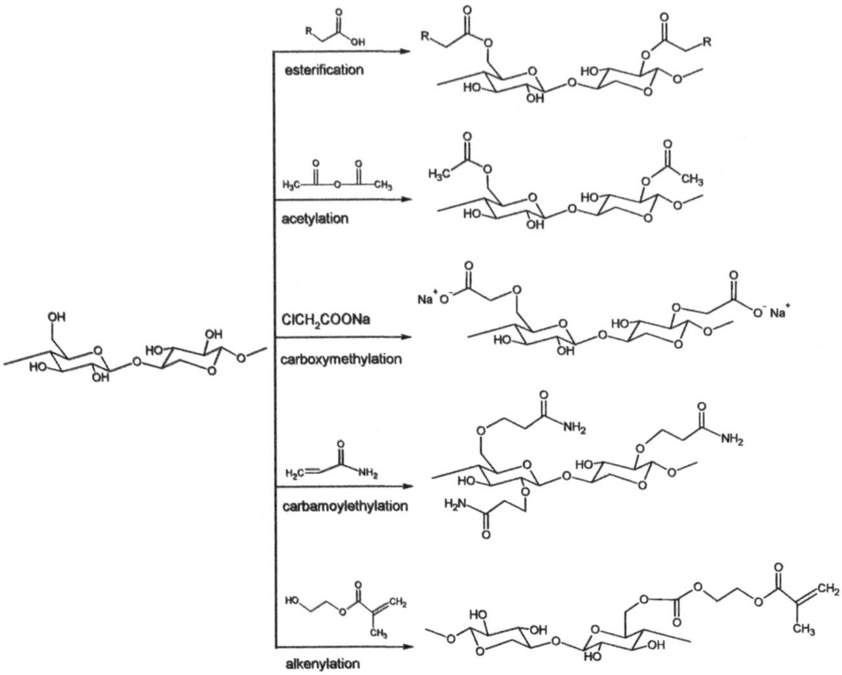

Figure 7.4   A schematic view of possible modifications of hemicelluloses

been shown to be appropriate in several cases, explained by solvation of the hydrophilic hemicellulose chains, preventing aggregation and facilitating the reaction between the hemicellulose hydroxyl groups and the acylating agent. For instance, xylans from rye straw were esterified with a range of acyl chlorides in a DMF/LiCl system in the presence of triethylamine and 4-(dimethylamino)pyridine, yielding high degrees of substitution after relatively short reaction times with only minor degradation effects on the hemicellulose chains [24]. Just like cellulose, hemicellulose chains may be converted to tosyl esters. Moreover, as an example, tosylated galactomannan can be treated with an alkali thiocyanate, yielding a thiocyano-substituted chain with degrees of substitution ranging from 0.44 to 0.52 [56]. The classic nitration of cellulose is applicable to hemicelluloses as well, a treatment with HNO_3 and H_2SO_4, yielding a partially degraded polysaccharide with hydroxyl groups converted to nitrate esters.

7.5.2   Etherification

A Williamson ether synthesis is a facile route to the replacement of sugar unit hydroxyl groups with alkoxy moieties. There are a range of examples in the literature, one of the most common being carboxymethylation. Carboxymethylation is frequently applied to cellulose in order to increase the solubility and form a material with plastic-like properties. A common method, carboxymethylation by sodium mono-chloroacetate in NaOH solution, was applied to hemicelluloses isolated from sugar-cane bagasse and shown to give carboxymethylated hemicelluloses in ethanol suspension with degrees of substitution ranging from 0.1 to 0.56, depending on the reaction conditions [23]. However, the hemicelluloses were markedly degraded in the process. Hemicelluloses from sugarcane bagasse have also been derivatized by etherification with 2,3-epoxypropyltrimethylammonium chloride in a alkaline me-dium, resulting in substitution preferably in the C-3 position with yields ranging from around 35 to 42%, depending on the reaction conditions. The product in this case is a cationic hemicellulose with potential use as an additive in paper making [57]. Methylation of hemicelluloses is another straightforward etherification process. Methylation procedures described in the literature include dimethyl sulfate treatment in a sodium hydroxide medium and methyl iodide/silver oxide treat-ment [35, 36].

The reaction of hemicelluloses isolated from wheat straw with acrylamide in alkaline solution was carried out to obtain products with different degrees of substitution [58]. Structural characterization confirmed the grafting of carbamoy-lethyl groups onto the main chain of the hemicelluloses as shown in Figure 7.4. It was also shown that a fraction of the formed carbamoylethyl groups were converted to carboxyethyl groups due to the alkaline conditions so that, in effect, a bifunctional hemicellulose derivative was formed in the process. Another thermoplastic xylan adduct was prepared by reacting low-molecular weight xylan with propylene oxide under basic conditions [59].

7.5.3   Miscellaneous Treatments

In addition to the conventional modification of hemicellulose, with chemistries analogous to those carried out on cellulose, a range of other substitutions have been reported. Such modifications may serve a number of purposes, including hydropho-bization or the creation of a thermoplastic or a hydrogel.

Covalent cross-linking of hemicelluloses has been mediated by partial substitution of the hydroxyl groups on spruce-derived acetylated galactoglucomannan with 2-hydroxylethylmethacrylate (HEMA), yielding a hemicellulose with pendant vinyl functionalities. In a subsequent step involving radical polymerization in the presence of free HEMA a highly hydrophilic cross-linked network with elastic and transparent properties was formed [60]. By further development of the coupling strategy, it was possible to form a number of alkenyl derivatives of this hemicellulose and a library of hydrogels therefrom [61]. Sulfation of hemicelluloses has been described for oat-derived xylan in two alternative reaction routes: either homogeneously by amidosulfuric acid in dimethyl sulfoxide solution or by reaction with chlorosul-furic acid in DMF solution, both procedures giving appreciable degrees of substitution [55].

An alternative to the use of hemicellulose as a polymeric material is partial or full decomposition into oligo- or monosaccharide units for subsequent use as chemicals. Recently, following increasing awareness of the climate situation, there has been a markedly increased interest in the utilization of non-edible green resources for chemical or material production. From this perspective, the degradation of hemicelluloses into oligo- and/or monosaccharides shows potential as an alternative source of chemicals. Conversion of hemicellulose chains into monomeric sugars is readily accomplished by acid hydrolysis, typically using sulfuric acid [62]. Enzymatic cleavage of glycosidic bonds is another option that offers potential high selectivity and a controlled degradation pattern. There are a range of enzymes available with known activities toward various pyranose and/or furanose polysaccharide structures [50, 63, 64]. The products generated from hemicellulose hydrolysis may either be used directly in a range of processes, or fermented into high-value adducts. A process already of commercial importance is the production of xylitol from hemicelluloses through yeast-mediated fermentation [65]. Hemicellulose-derived biofuel is an anticipated future product. A process for the production of ethanol has been presented in which hemicelluloses from sugarcane bagasse were hydrolyzed in sulfuric acid, electrodialyzed, and the resulting monosaccharides fermented to ethanol using a xylose-fermenting yeast, *Pachysolen tannophilus DW06*, yielding 0.36 g ethanol per g sugar [66]. Butanediol has also been produced from aspen wood xylans in a combined hydrolysis and fermentation process in which enzymatic hydrolysis yielded mainly xylose and xylobiose with a subsequent turnover equal to 0.30 g butanediol per g sugar [67].

7.6   APPLICATIONS

Hemicellulose is an attractive resource for green chemicals, potentially yielding a range of valuable products (e.g., sugars and alcohols) upon degradation and possibly subsequent fermentation. In the chemical pulping industry, hemicelluloses are typically released into the cooking liquor, removed by washing of the pulp fibers, and later burned for energy recovery. Various conversions to low-molecular weight products were discussed in the previous section of this chapter. In addition, hemicelluloses lend themselves to material design by direct utilization of the biopolymers. Considering the abundance of hemicelluloses in nature, there are still very few material applications of hemicelluloses presented and commercial products are even more scarce. Yet, in research terms, chemical modification of hemicelluloses is increasingly being exploited for the development of novel environmentally friendly and biocompatible advanced materials A short overview of the state-of-art for applications of pure and modified hemicelluloses is given below.

Hemicelluloses have been used in the past as thickeners and adhesives, in wound management [68] and for other medicinal purposes [8, 69]. Many polysaccharides, including a range of hemicelluloses, have been known for a great many years to be useful film formers and, the interesting barrier properties of hemicelluloses have also been acknowledged. The film-forming properties of arabinoxylans have been utilized to protect protein foam against thermal disruption [70] and arabinoxylans have also been explored as film constituents for potential packaging applications [71]. Likewise, hemicelluloses from the mannan family, such as konjac glucomannan or spruce-derived galactoglucomannan have been used in the preparation of packaging films [72, 73]. The research and development related to use of hemicelluloses in films and coatings was recently reviewed [74].

Within the biomedical field, hemicellulose-based materials have been of interest for their gel- and film-forming properties. Films have been explored for wound-dressing purposes [75] and xyloglucan gels for controlled drug delivery [76]. Covalent hydrogels designed from alkenyl-functionalized galactoglucomannan have been shown to offer sustained drug delivery of incorporated substances, mediated by diffusion [77] and enzymatic hydrolysis [64]. Microsphere formulations have been designed and investigated for several decades because of their potential use for incorporation of bioactive agents, predominantly within the field of controlled drug delivery. Recently, a method for the preparation of hemicellulose hydrogel microspheres has been presented, in which O-acetyl galactoglucomannan from spruce was functionalized and microspheres were formed in an oil-water emulsion during *in situ* covalent cross-linking into a hydrogel. The microspheres thus obtained displayed sustained delivery of incorporated model drugs [78]. Sulfated hemicelluloses mimic the structure of sulfonated polysaccharides like heparin and, as such, may play a role as anticoagulants [55].

Within the food sector, some hemicelluloses have been utilized for a long time due to their nontoxicity, water solubility and ability to form gels. Typical examples are the mannans such as guar gum, locust bean gum, and konjac glucomannan, which are commercial emulsifiers and thickening agents [12]. Upon addition of an alkaline solution, preferably at elevated temperature, konjac glucomannan forms a stable gel for dietary use. Also xyloglucans and β-glucans may be gel formers and all these components have been investigated, not only from a consistency-regulating perspective, but also for their potential bioactive effects on human metabolism, such as an effect on the glycemic response [79]. β-glucans isolated from barley, rice, and oats have been shown to have a promising cholesterol-regulating effect [80]. A commercial product is Glucagel™, a β-glucan gel with fat-like consistency [81]. Phenolic acids (e. g., ferulic acid and p-coumaric acid) which are esterified to arabinofuranosyl residues in hemicelluloses have been used as natural food preservatives because they inhibit peroxidation of fatty acids [82].

Hemicellulose derivatives have been reported to improve the quality of certain papers. The influence of quaternized hemicelluloses and carboxymethyl hemicellulose as wet-end additives on the physical properties of hand sheets was recently reported [83]. The physical properties of hand sheets were significantly improved by adding cationic hemicellulose (DS = 0.37) and carboxymethyl hemicellulose (DS = 0.35), each at 1.0% (based on dry pulp weight). The properties of sheets prepared from old corrugated container (OCC) pulp were improved by addition of quaternized xylan-rich sugarcane bagasse hemicelluloses [84].

7.7   CONCLUDING REMARKS

Hemicelluloses account for 20–30% of the biomass of annual and perennial plants, thereby representing an important class of polysaccharides and an almost inexhaustible natural resource for the development of eco-friendly and biocompatible materials. The utilization of hemicelluloses from annual agricultural residual biomass – constituting a secondary resource for the production of useful biopolymers and other chemicals – has very significant potential in terms of new bio-based materials which can contribute to sustainable growth. The secondary resources that have been explored to some extent in the past are wheat straw, sugarcane bagasse, sunflower husk, oat spelts, flax shives, olive stones, red gram husks, corn cobs, brewer's spent grain, and rice bran. More efforts now have to be expended for exploration of other types of residual biomass, such as herbaceous (tall grass) and dedicated short-rotation woody crops, which are very promising resources in terms of chemicals, biopolymers, and eco-friendly materials. In summary, hemicelluloses need to be continuously explored as a highly promising resource for future materials and chemicals.

REFERENCES

[1] G. O. Aspinall, *Adv. Carbohydr. Chem.*, 1959, 14, 429–468.

[2] T. E. Timell, *Wood Sci. Technol.*, 1967, 1, 45–70.

[3] G. O. Aspinall, *Ann. Rev. Biochem.*, 1962, 31, 79–102.

[4] C. T. Bishop, G. A. Adams, and E. O. Huges, *Can. J. Chem.*, 1954, 32, 999–1004.

[5] G. A. Adams, A. E. Castagne, *Can. J. Chem.*, 1951, 29, 109–122.

[6] G. O. Aspinall, E.G. Meek, *J. Chem. Soc.*, 1956, 3830–3834.

[7] Y. Ghali, A. Youssef, E. A. E. Mobdy, *Phytochem.*, 1974, 13, 605–610.

[8] A. Ebringerova, T. Heinze, *Macromol. Rapid Commun.*, 2000, 21, 542–556.

[9] A. Teleman, J. Lundqvist, F. Tjerneld, *et al.*, *Carbohydr. Res.*, 2000, 329, 807–815.

[10] S. Willför, R. Sjöholm, C. Laine, *et al.*, *Carbohydr. Polym.*, 2003, 52, 175–187.

[11] J. J. Gonzalez, *Macromolecules*, 1978, 11, 1074–1085.

[12] H. Zhang, M. Yoshimura, K. Nishinari, *et al.*, *Biopolymers*, 2001, 59, 38–50.

[13] S. C. Fry, *J. Exp. Bot.*, 1989, 40, 1–11.

[14] B. M. Yapo, K. L. Koffi, *J. Sci. Food Agric.*, 2008, 88, 2125–2133.

[15] S. W. Cui, Q. Wang, *Struct. Chem.*, 2009, 20, 291–297.

[16] M. S. Izydorczyk, L. J. Macri, and A. W. MacGregor, *Carbohydr. Polym.*, 1998, 35, 249–258.

[17] Adapted from Bismarck, A., Mishra, S. and Lampke, T., Plant fibers as reinforcement for green composites, in A. K. Mohanty, M. Misra, and L. T. Drzal (eds.), *Natural Fibers, Biopolymers and Biocomposites*, CRC Press, Taylor and Francis, NW, FL, USA, pp 37–108 (2005).

[18] J. Lundqvist, A. Teleman, L. Junel, *et al.*, *Carbohydr. Polym.*, 2002, 48, 29–39.

[19] S. Willför, R. Sjöholm, C. Laine, *et al.*, *Wood Sci. Technol.*, 2002, 36, 101–110.

[20] A. Teleman, M. Tenkanen, A. Jacobs, *et al.*, *Carbohydr. Res.*, 2002, 337, 373–377.

[21] A. Teleman, M. Nordström, M. Tenkanen, *et al.*, *Carbohydr. Res.*, 2003, 338, 525–534.

[22] R. Naran, S. Black, S. R. Decker, *et al.*, *Cellulose*, 2009, 16, 661–675.

[23] J.-L. Ren, R.-C. Sun, and F. Peng, *Polym. Degr. Stab.*, 2008, 93, 786–793.

[24] R.-C. Sun, J. M. Fang, and J. Tomkinson, *J. Agric. Food Chem.*, 2000, 48, 1247–1252.

[25] A. Bendahou, A. Dufrense, H. Kaddami, *et al.*, *Carbohydr. Polym.*, 2007, 68, 601–608.

[26] A. Jacobs, M. Palm, G. Zacchi, *et al.*, *Carbohydr. Res.*, 2003, 338, 1869–1876.

[27] G. O. Aspinall, E. L. Hirst, and R. S. Mahomed, *J. Chem. Soc.*, 1954, 1734–1738.

[28] D. E. Akin, G. R. Gamble, W. H. Morrison, *et al.*, *J. Sci. Food Agric.*, 1996, 72, 155–165.

[29] G. O. Aspinall, R. J. Ferrier, *J. Chem. Soc.*, 1957, 4188–4194.

[30] J. M. Lawther, R.-C. Sun, and W. B. Banks, *Int. J. Polym. Mater.*, 1997, 36, 53–64.

[31] G. O. Aspinall, R. S. Mahomed, *J. Chem. Soc.*, 1954, 1731–1734.

[32] N.B. Shankaracharya, *J. Food Sci. Technol.*, 1998, 35, 193–208.

[33] C. Kurz, R. Carle, and A. Schieber, *Food Chem.*, 2008, 106, 421–430.
[34] S. K. Bhaduri, S. K. Sen, *Carbohydr. Res.*, 1983, 121, 211–220.
[35] G. G. S. Dutton, T. G. Murata, *Can. J. Chem.*, 1961, 39, 1995–2000.
[36] G. G. S. Dutton, S. A. McKelvey, *Can. J. Chem.*, 1961, 39, 2582–2589.
[37] J. Ceusters, E. Londers, K. Brij, *et al.*, *Phytochemistry*, 2008, 69, 2307–2311.
[38] G. Rodriguez, A. Lama, R. Rodriguez, *et al.*, *Biores. Technol.*, 2008, 99, 5261–5269.
[39] M. C. Jarvis, *Plant Cell Environ.*, 1984, 7, 153–164.
[40] G. Brigand, A. Denis, M. Grall, *et al.*, *Carbohydr. Polym.*, 1990, 12, 61–77.
[41] A. G. J. Voragen, G. J. Coenen, R. P. Verhoef, *et al.*, *Struct. Chem.*, 2009, 20, 263–275.
[42] B. S. Paulsen, H. Barsett, *Adv. Polym. Sci.*, 2005, 186, 69–101.
[43] D. Mohnen, *Curr. Op. Plant. Biol.*, 2008, 11, 266–277.
[44] E. Adler, *Wood Sci. Technol.*, 1977, 11, 169–218.
[45] M. Lawoko, G. Henriksson, and G. Gellerstedt, *Biomacromolecules*, 2005, 6, 3467–3473.
[46] P. Hoffmann, R. Patt, *Holzforschung*, 1976, 30, 124–132.
[47] W. G. Campbell, I. R. C. McDonald, *J. Chem. Soc.*, 1952, 2644–2650.
[48] A. Andersson, T. Persson, G. Zacchi, *et al.*, *Appl. Biochem. Biotechnol.*, 2007, 136–140, 971–984.
[49] J. X. Sun, X. F. Sun, R.-C. Sun, *et al.*, *Carbohydr. Polym.*, 2004, 56, 195–204.
[50] A. X. Jin, J. L. Ren, F. Peng, *et al.*, *Carbohydr. Polym.*, 2009, 78, 609–619.
[51] H. Stålbrand, J. Lundqvist, A. Andersson, *et al.*, *ACS Symp. Ser.*, 2004, 864, 66–78.
[52] M. Palm, G. Zacchi, *Biomacromolecules*, 2003, 4, 617–623.
[53] A. U. Buranov, G. Mazza, *Carbohydr. Polym.*, 2010, 79, 17–25.
[54] J. F. Carson, W. D. Maclay, *J. Am. Chem. Soc.*, 1948, 70, 293–295.
[55] K. Hettrich, S. Fischer, and N. Schröder, *Macromol. Symp.*, 2006, 232, 37–48.
[56] J. F. Carson, W. D. Maclay, and W. Dayton, *J. Am. Chem. Soc.*, 1948, 70, 2220–2223.
[57] J.-L. Ren, R.-C. Sun, and C.F. Liu, *J. Appl. Polym. Sci.*, 2007, 105, 3301–3308.
[58] J.-L. Ren, F. Peng, and R.-C. Sun, *Carbohydr. Res.*, 2008, 343, 2776–2782.
[59] R. K. Jain, M. Sjöstedt, and W. G. Glasser, *Cellulose*, 2000, 7, 319–336.
[60] M. S. Lindblad, E. Ranucci, and A.-C. Albertsson, *Macromol. Rapid Commun.*, 2001, 22, 962–967.
[61] J. Voepel, U. Edlund, and A.-C. Albertsson, *J. Polym. Sci., Part A: Polym. Chem.*, 2009, 47, 3595–3606.
[62] B. Saha, R. J. Bothast, *Appl. Biochem. Biotechnol.*, 1999, 76, 65–77.
[63] J. P. Vincken, G. Beldman, and A. G. J. Voragen, *Carbohydr. Res.*, 1997, 298, 299–310.
[64] A. Andersson Roos, U. Edlund, J. Sjöberg, *et al.*, *Biomacromolecules*, 2008, 9, 2104–2110.
[65] L. F. Chen, C. S. Gong, *J. Food Sci.*, 1985, 50, 226–228.
[66] K.-K. Cheng, B.-Y. Cai, J.-A. Zhang, *et al.*, *Biochem. Eng. J.*, 2008, 38, 105–109.
[67] E. K. C. Yu, L. Deschatelets, L. U. L. Tan, *et al.*, *Biotechnol. Lett.*, 1985, 7, 425–430.
[68] L. L. Lloyd, J. F. Kennedy, P. Methacann, *et al.*, *Carbohydr. Polym.*, 1998, 37, 315–322.
[69] M. Hashi, T. Takeshita, *Agric. Biol. Chem.*, 1979, 43, 961–967.
[70] A. Ebringerova, Z. Hromadkova, *Genet. Eng. Rev.*, 1999, 16, 325–346.
[71] P. Zhang, R. L. Whistler, *J. Appl. Polym. Sci.*, 2004, 93, 2896–2902.
[72] J. Hartman, A.-C. Albertsson, and J. Sjöberg, *Biomacromolecules*, 2006, 7, 1983–1989.
[73] K. S. Mikkonen, M. P. Yadav, and P. Cooke, *Bioresources*, 2008, 3, 178–191.
[74] N. M. L. Hansen, D. Plackett, *Biomacromolecules*, 2008, 9, 1493–1505.

[75] D. Melandri, A. De Angelis, R. Orioli, *et al.*, *Burns*, 2006, **32**, 964–972.

[76] S. Miyazaki, F. Suisha, N. Kawasaki, *et al.*, *J. Control. Rel.*, 1998, **56**, 75–83.

[77] J. Voepel, J. Sjöberg, and A.-C. Albertsson, *J. Appl. Polym. Sci.*, 2009, **112**, 2401–2412.

[78] U. Edlund, A.-C. Albertsson, *J. Bioact. Compat. Polym.*, 2008, **23**, 171–186.

[79] N. Latha, R. Kapoor, *J. Food Sci. Technol.*, 2004, **41**, 83–85.

[80] T. S. Kahlon, F. I. Chow, *Cereal Foods World*, 1997, **42**, 86–92.

[81] K. R. Morgan, C.J. Roberts, S. J. B. Tendler, *et al.*, *Carbohydr. Res.*, 1999, **315**, 169–179.

[82] G. Mandalari, C. B. Faulds, and A. I. Sancho, *J. Cereal Sci.*, 2005, **42**, 205–212.

[83] J.-L. Ren, F. Peng, R.-C. Kang, *et al.*, *Carbohydr. Polym.*, 2009, **75**, 338–342.

[84] J. L. Ren, R.C. Sun, F. Peng, *et al.*, *China Pulp Paper*, 2007, **26**, 14–16.

8

Production, Chemistry and Properties of Cellulose-Based Materials

Mohamed Naceur Belgacem
Grenoble INP-Pagora, St. Martin d'Hères, France

Alessandro Gandini
CICECO and Chemistry Department, University of Aveiro, Aveiro, Portugal

8.1   INTRODUCTION

Virtually from its inception, humanity has had a very rewarding interaction with cellulose, first as a pristine source of energy, shelter, clothing and other daily needs, then as a source of writing substrates, hence a determining factor in the spreading of culture. The Industrial Revolution spurred a major technological revamping of papermaking and a historical event with the production of the *first* thermoplastic material, well before the fossil-based polymers, in the form of cellulose nitrate, soon followed by the acetate homologue. As the most abundant biopolymer on earth, with some 10^{12} tons in essentially vegetable biomass, readily renewed thanks to solar energy, and as a macromolecular material possessing unique properties and chemical

Biopolymers – New Materials for Sustainable Films and Coatings, First Edition.
Edited by David Plackett.
© 2011 John Wiley & Sons, Ltd. Published 2011 by John Wiley & Sons, Ltd.

reactivity, cellulose continues to represent an invaluable resource for mankind, with an incessantly growing number of novel applications, including a vast array of derivatives [1]. As opposed to the very uneven worldwide distribution of fossil resources, the ubiquitous nature of cellulose, albeit in different fibrous morphologies, makes it available to communities worldwide for local exploitation.

Cellulose is a linear homopolysaccharide consisting of β-D-glucopyranose units linked by glycosidic $\beta(1\text{-}4)$ bonds in a $^4C_1$ conformation (1). Strictly speaking therefore, its repeat element is constituted by two anhydroglucose units (AGU). The two end-groups of this polymer are not chemically equivalent, since one bears the 'normal' C4–OH group (non-reducing end), whereas the other has a C1–OH moiety in equilibrium with the corresponding aldehyde function (reducing end).

1

This structure is responsible for the peculiar properties of cellulose, namely its highly hydrophilic character and the correspondingly high surface energy, its biodegradability and relative thermal fragility and of course its marked reactivity associated with the three OH groups present in each AGU. Whereas these structure–property relationships are also common to other polysaccharides such as starch, hemicelluloses and chitin, cellulose displays a stronger aptitude to crystallize through the establishment of a regular network of intra- and inter-molecular –O–H···H–O– hydrogen bonds. The high cohesive energy ensuing from these physicochemical interactions explains why cellulose does not possess a liquid state, since its melting temperature is well above that at which chemical degradation takes place.

The DP of cellulose varies considerably from hundreds to thousands of AGUs, both as a function of the species and of the isolation and purification procedures. The latter operations often introduce other functionalities, such as carbonyl and carboxyl groups, into the macromolecular backbone, which are known to provide specific features to cellulose, including surface charge and additional chemical reactivity.

The natural manifestations of cellulose are invariably in the shape of fibers, whose role is predominantly that of providing mechanical strength to wood and annual plants, as the reinforcing elements in the diverse cell wall composite morphologies incorporating lignins as matrices and hemicelluloses as compatibilizing interface elements. The isolation of cellulose fibers from these vegetable structural composites has been the traditional task of the pulping industry and the different approaches implemented today, including novel strategies that comply with the realization of the biorefinery concept, fall outside the scope of this chapter, but are of course very thoroughly documented elsewhere [2, 3]. The same applies to cotton science and technology [4].

The hierarchical morphology of natural cellulose fibers is illustrated in Figure 8.1 and the details of this supramolecular assembly and crystal domain formations are well established thanks to extensive X-ray diffraction, electron microscopy, solid-state ^{13}C- NMR and neutron diffraction studies [1].

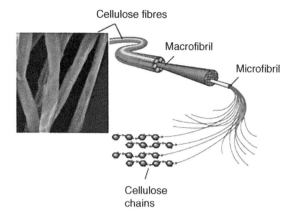

Figure 8.1   Hierarchical morphology of a plant cellulose fiber

The proportion of crystalline and amorphous domains in natural cellulose can vary considerably at the macroscopic level of a fiber assembly, but tend to favour the ordered region as the constitutive elements are progressively refined down to the microfibrils, with major consequences in terms of enhanced mechanical properties and reduced chemical reactivity, as discussed below.

Certain bacteria, fungi and algae are also responsible for the biosynthesis of cellulose, sometimes with very original morphologies and a high purity (because of the absence of the other characteristic vegetable components). Cellulose from these sources often exhibits a high degree of crystallinity.

The different wood and plant fibers that have been used in papermaking for centuries are hollow, a morphology that suits quite adequately the properties expected from such applications. The first production of continuous cellulose materials, such as fibers (rayon) and films (cellophane), dates back to the end of the 19th century, when the viscose process was developed. This technology, still operational today, converts cellulose into an alkali-soluble xanthogenate derivative, which is spun or extruded in an acidic medium, producing a regenerated cellulose structure [5]. As briefly reviewed later, over the years other technologies have been developed, mostly based on the use of cellulose solvents (i.e., without the need for chemical modifications and regeneration).

The realm encompassing the numerous aspects related to cellulose and its derivatives is monumental and thoroughly covered by a rich fundamental and applied bibliography [1, 5, 6]. This chapter purports to tackle both some of its most essential and basic notions and the recent research and technological developments that have marked it because of their originality and contributions to the progress of materials science. The present survey appears justified however in the light of the fact that the last few decades have witnessed a spectacular revival of interest in cellulose chemistry and processing, arguably the most dynamic period of applied research that this renewable resource has known.

8.2   PRISTINE CELLULOSE AS A SOURCE OF NEW MATERIALS

Notwithstanding the predominance of papermaking [2] and cotton textile manufacturing [4] as the fundamental users of cellulose fibers, new domains of exploitation are being actively sought, calling upon both classical morphologies and nanocrystalline counterparts. This section reviews advances in the specific context of materials based on essentially unmodified fibers, as far as their chemical nature is concerned.

8.2.1   All-Cellulose Composites

The idea of constructing a composite material in which both the matrix and the reinforcing elements are the same polymer was originally put forward by Ward and Hine [7] who applied it to poly(ethylene). Nishino's group was the first to report the application of this concept to cellulose [8] by impregnating a cellulose solution into uniaxially aligned cellulose fibers and later by selectively dissolving the surface of the cellulose fibers and then compressing and drying the system. Similar approaches were implemented by other groups, all based on embedding highly crystalline fibers into a matrix of regenerated cellulose, albeit using different solvents and fibers [9]. In some instances, the modulus of elasticity and the strength of these original composites reached values well above those of the best cellulose-reinforced thermoplastics, namely more than 10 GPa and 100–500 MPa respectively.

A recent extension of these concepts to the preparation of cellulose aerogels ('aerocellulose') describes the partial and controlled dissolution of microcrystalline cellulose in LiCl/DMAc and precipitation of the ensuing gels, followed by freeze drying to preserve their open morphology [10]. These highly porous materials had densities of 100–350 kg m^{-3}, flexural strengths as high as 8 MPa and a maximum stiffness of 280 MPa. Other types of composites prepared from cellulose, but involving a chemical modification of the fibers' outer sleeve, are discussed below in the appropriate section.

8.2.2   Cellulose Nano-Objects

The fundamental and applied research related to the isolation, characterization and exploitation of nanoscopic cellulose fibers has reached exhilarating new heights in the last several years. Broadly speaking, two different types of nanocellulose are relevant here, nanofibrils isolated from lignocellulosic vegetable structures [11] and nanofilaments produced by certain bacteria [1, 11d, 12]. Other cellulose nanocrystalline fibers found in some exotic species in a very pure form, but in unexploitable quantities, are not reviewed here.

Plant-based cellulose nanocrystals (CNs) and microfibrillated cellulose (MFC) have been known for decades, but the recent outburst of academic and industrial interest in macromolecular materials from renewable resources [13], coupled with major advances in the techniques used for their isolation, has brought about a very lively revival in studies dealing with them [11].

The basic operation in the isolation of CNs from much larger fibrous morphologies is acid-catalyzed hydrolysis of the amorphous and para-crystalline cellulose macromolecules, which ultimately leaves only the highly ordered and regular rod-like nanocrystals suspended in the resulting aqueous medium. Figure 8.2 shows schematically these amorphous and crystalline regions in a fiber.

The optimization of CN isolation processes has led to efficient procedures which call upon the use of strong acids under carefully monitored conditions of temperature, reaction time and stirring. Repeated centrifugation and dialysis against distilled water delivers the final CN suspension. Figure 8.3 shows a typical TEM image of these nanocrystals ($L \sim 210\,\text{nm}$, $d \sim 5\,\text{nm}$, $L/d \sim 40$) extracted from sugar-beet cellulose [11c].

Research related to the isolation of MFC has concentrated on combining pretreatments based on enzymatic hydrolysis and mechanical beating with disintegration of

microfibril

fiber

Figure 8.2 Schematic view of the regular and irregular assemblies of cellulose macromolecules in a fiber

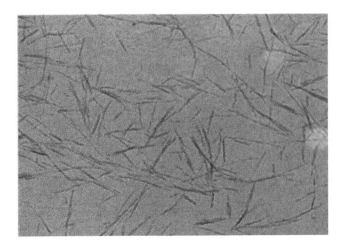

Figure 8.3   TEM image of cellulose whiskers isolated from sugarbeet [11c]. Reprinted from M. A. S. Azizi, F. Alloin, A. Dufresne, Review of recent research into cellulosic whiskers, Their properties and their application in nanocomposite field, *Biomacromolecules*, 6, 612. Copyright (2005) with permission from American Chemical Society

nanofibrils from the cell wall by high-shear homogenization of dilute aqueous suspensions of the fibers. An example of the ensuing web-like nanomorphology is shown in Figure 8.4. An alternative approach calls upon the use of TEMPO to oxidize a proportion of the cellulose primary OH groups and thus append carboxylate moieties which, because of the electrostatic repulsive forces induced by the surface negative charges, greatly facilitates fiber disintegration through simple shearing [14].

Bacterial cellulose (BC) is the other remarkable form of nano-sized fibers, which are produced by strictly aerobic and non-photosynthetic gram-negative bacteria. These bacteria, usually found in fruit, vegetables, vinegar and alcoholic beverages, are capable of converting glucose, glycerol and other organic substrates into cellulose within a period of a few days [12]. As an example, the cellulose biosynthesized by *G. xylinus* is identical to that produced by plants, but displays a higher crystallinity and is free of lignin, hemicelluloses and the other natural components normally associated with cellulose from plants. Figure 8.5 shows the fiber web secreted by the bacteria, and Figure 8.6 illustrates the pure cellulose morphology after the removal of microorganisms by washing with an aqueous alkaline solution. The extremely high water affinity of BC produces gel-like morphologies, such as the object shown in Figure 8.7, which is 99% water, but has surprisingly good mechanical stability. Figure 8.8 shows a BC film obtained by oven drying a gelled membrane such as that displayed in Figure 8.7. BC can be readily fragmented by mechanical shearing with a kitchen mixer or reduced to nanocrystals by acid treatment [15].

Of course, vegetable cellulose nanofibers, CN and MFC, also display an extreme hydrophilicity, which gives rise to highly viscous aqueous suspensions, even at low cellulose contents.

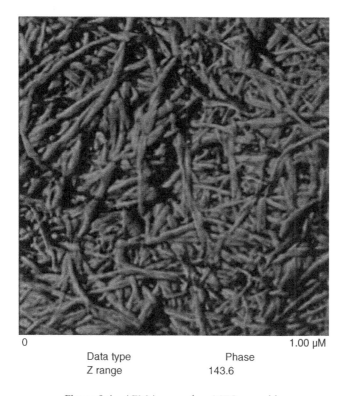

0   1.00 µM

Data type   Phase
Z range   143.6

Figure 8.4 AFM image of an MFC assembly

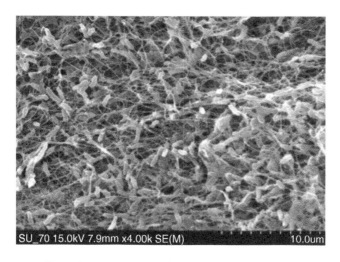

Figure 8.5 SEM image of secreted bacterial cellulose

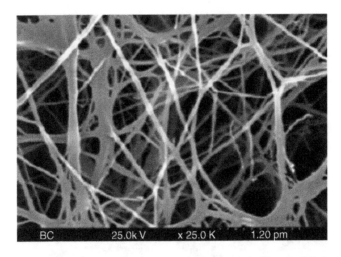

Figure 8.6 SEM image of a bacterial cellulose web after purification

These beautiful nano-objects can be turned into equally stunning materials [11] through self-assembly, such as nanopapers displaying very high toughness ($W_A = 15\,MJ/m^3$) with a Young's modulus of 13 GPa and a tensile strength of more than 200 MPa [16, 17], transparent films with high gas barrier properties [18] and a temperature-independent refractive index [19]. As for BC, its applications in the biomedical domain are widening with particularly promising results [1, 20]. The use of cellulose nanofibers in composite materials, as such, or after suitable surface modification, is reviewed below together with the corresponding materials incorporating more traditional

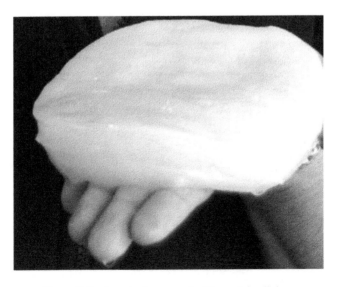

Figure 8.7 A typical water gel of bacterial cellulose

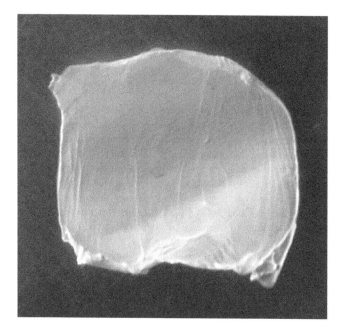

Figure 8.8   A typical dried membrane of bacterial cellulose

fibers. Whereas up to a few years ago these cellulose nanofibers were relegated to little more than research curiosities, substantial progress has since been made in terms of new technologies which should render their production economically viable.

8.2.3   Model Cellulose Films

Considerable advances have been reported lately in the preparation and characterization of model cellulose films using different techniques such as Langmuir–Blodgett (LB) deposition and spin coating from solution or suspension [21]. Apart from the very rewarding type of imagery these films provide, their properties have been and are being assessed with a variety of techniques and for a wide range of both fundamental studies and application purposes, with special emphasis on papermaking.

Cellulose coatings, with thickness varying from less than ten to tens of nm, have been used to study the adsorption of polyelectrolytes, surfactants, dyes and enzymes [21]. The results of these pioneering investigations have provided an improved understanding of theoretical issues and the associated technological opportunities. Equally important are the measurements of surface forces, namely cellulose/cellulose, cellulose/polyelectrolyte, and cellulose/inorganic–surface interactions [21], as well as adhesion forces [22]. The study of the swelling behavior of model cellulose surfaces, including wetting and drying cycles, has been most informative for both fundamental and practical (e.g., fiber hornification) aspects [22, 23].

Two recent publications have extended the context of these investigations by reporting the preparation and characterization of cellulose nanocrystal LB films [24] and cellulose nanofibril model surfaces [25] (i.e., studies involving the cellulose nanoobjects currently in (justified) fashion).

8.3   NOVEL CELLULOSE SOLVENTS

The dissolution of cellulose is conditioned by the basic requirement to replace the intermolecular hydrogen bonds by thermodynamically more favorable solvent–OH interactions. Two approaches have been traditionally applied to implement these conditions and thus form cellulose solutions: (i) derivatization through a readily reversible reaction; and (ii) physicochemical interaction which does not alter the cellulose structure. In approach (i), the chemical strategy, first implemented with the xantogenate process mentioned earlier, requires acid media such as H_3PO_4, $HCOOH/H_2SO_4$ and $CF_3COOH/(CF_3CO)_2O$, or the nitrating mixture N_2O_4/DMF. These systems have lost much of their utility with the advent of more viable and simpler physical alternatives. In approach (ii), the list of suitable solvents is now quite rich and includes both aqueous and non-aqueous combinations [26], among which solutions of LiCl in DMA or NMP have been the most successful. N-methylmorpholine-N-oxide (NMMO) was until recently the only non-derivatizing single-component solvent for cellulose, albeit associated with rather laborious handling. Over the last decade, new cellulose solvents have been discovered and put to good use for such purposes as spinning and chemical modification. For example, Zhang's group has perfected a set of simple aqueous solvents based on mixtures of strong bases and urea derivatives [27], typically LiOH/urea and NaOH/thiourea, which readily dissolve cellulose with degree of polymerization (DP) as high as 1300.

In a different vein, Heinze's group has perfected a novel non-aqueous solvent combination consisting of DMSO and ammonium fluorides [28] and has shown that high-molecular weight cellulose, such as cotton linters with DP close to 4 000, could be dissolved within one minute.

The ionic liquid revolution has also led to new cellulose treatments. Swatloski et al. (2002) reported the first such successful dissolution of cellulose [29]. Since then, the search has gained momentum and numerous publications have widened the number of ionic liquids that are good cellulose solvents, even for highly crystalline BC with a DP of 6500. In these comprehensive investigations, the contributions of Heinze's group [26, 28], and, more recently, that of Ohno's laboratory [30], have been substantial. Figure 8.9 shows the structure of some of the most common ionic liquids.

Chloride anion-based ionic liquids are the most efficient in this context because the strong Cl$^-$ hydrogen bonding basicity can induce efficient replacement of the strong intermolecular cellulose –OH...HO– bonds by the ionic liquid molecule, as sketched in Figure 8.10 [31].

The dissolution of cellulose by ionic liquids has been the subject of hundreds of publications, dealing with dissolution in all its aspects [32], the rheology of the ensuing solutions [33], electrospinning [34] and, cellulose modification in these media [26, 28]. A useful review recently covered some of these topics with stimulating reflections [35]. This state of affairs, coupled with similar investigations on lignocellulosic materials from annual plants and wood [35], suggests

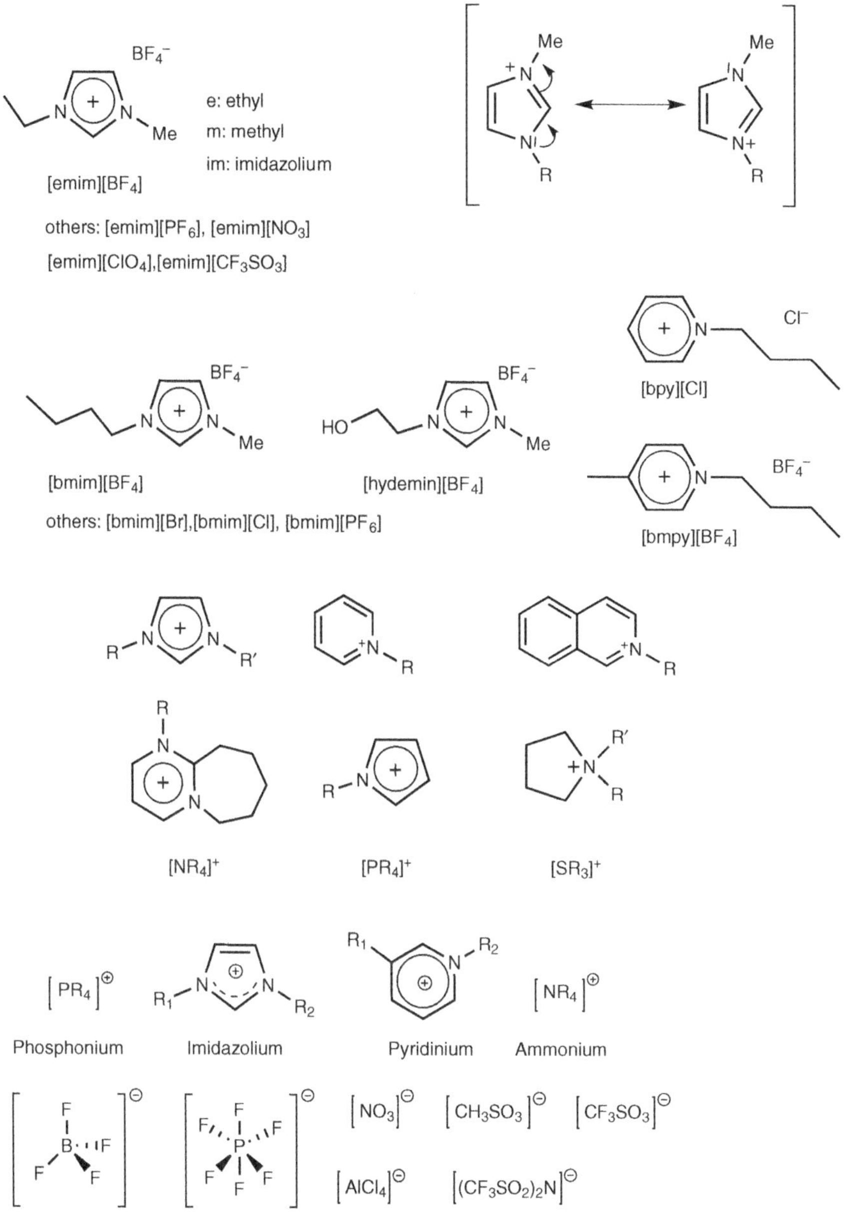

BF$_4^-$

[emim][BF$_4$]

e: ethyl
m: methyl
im: imidazolium

others: [emim][PF$_6$], [emim][NO$_3$]
[emim][ClO$_4$],[emim][CF$_3$SO$_3$]

BF$_4^-$

[bmim][BF$_4$]

others: [bmim][Br],[bmim][Cl], [bmim][PF$_6$]

BF$_4^-$

[hydemin][BF$_4$]

Cl$^-$

[bpy][Cl]

BF$_4^-$

[bmpy][BF$_4$]

[NR$_4$]$^+$ [PR$_4$]$^+$ [SR$_3$]$^+$

$\left[\text{PR}_4\right]^\oplus$ $\left[\text{NR}_4\right]^\oplus$

Phosphonium   Imidazolium   Pyridinium   Ammonium

$\left[\text{NO}_3\right]^\ominus$ $\left[\text{CH}_3\text{SO}_3\right]^\ominus$ $\left[\text{CF}_3\text{SO}_3\right]^\ominus$

$\left[\text{AlCl}_4\right]^\ominus$ $\left[(\text{CF}_3\text{SO}_2)_2\text{N}\right]^\ominus$

Figure 8.9 Some of the most common ionic liquids

that a new era in cellulose technology might be looming on the horizon, in which fiber regeneration, the separation of biomass components in biorefineries and the synthesis of cellulose derivatives will also be conducted in 'green' ionic liquid media.

Figure 8.10 A possible mechanism of cellulose dissolution in a Cl⁻ ionic liquid

8.4 CELLULOSE-BASED COMPOSITES AND SUPERFICIAL FIBER MODIFICATION

The incorporation of cellulose fibers in macromolecular materials is another field of scientific and technological development that has witnessed a strong increase in momentum since the beginning of the third millennium [11a, 11c, 17, 36]. In most instances, the fiber surface requires some type of modification in order to provide an optimized interfacial interaction with the matrix polymer (i.e., good wetting, adhesion and/or chemical bonding) [37]. Given the abundant literature in the form of comprehensive monographs on these topics, the present discussion is restricted to the most relevant recent contributions.

The impressive growth of interest in cellulose-based composites can be attributed to the general trend towards investigations on materials from renewable resources [13] and, in more specific terms, to the numerous advantages provided by cellulose fibers, in all their sizes and shapes, as reinforcing elements in composites. Their positive features include very good mechanical properties, amplified in the nanofibers discussed above, renewable character, ubiquitous availability at low cost, biodegradability or recyclability, low density and modest abrasivity. In other words, cellulose fibers constitute a very serious alternative to glass counterparts. Composites with both thermoplastic and cross-linked matrices have been prepared and tested with fibers which cover the whole spectrum of morphologies and chemical compositions, from wood flour to pure whiskers.

The by-now classical drawbacks associated with the use of cellulose fibers are, on the one hand, the poor interfacial compatibility between their polar surface and nonpolar, or poorly polar polymers like polyolefins and, on the other hand, the aptitude of cellulose to absorb moisture, related to its richness in hydroxyl groups, and thereby lose some of its mechanical strength. Hence, the need for surface modification based ideally on applying a single simple treatment capable of solving both problems. A multitude of strategies and specific approaches have been studied and many of them have been successfully implemented [36, 37]. Composites without any specific preliminary fiber treatment are however also relevant [11, 17, 36, 37], particularly with very polar matrices, and some recent examples are the first to be discussed here.

8.4.1 Composites with Pristine Fibers

Aqueous suspensions of bacterial cellulose are particularly suited for mixing with water-soluble polymers or polymer emulsions before drying to form nanocomposites.

Brown and Laborie [38] introduced different proportions of high-DP poly(ethylene oxide) (PEO) into the growth medium of *Acetobacter xylinum*, thereby inducing the (partial) association of the PEO chains with the surface of the nascent cellulose nanofibers. The isolated composites showed an enhancement of the mechanical properties of the PEO matrix, which also acquired a higher resistance to thermal decomposition, while its melting temperature decreased. A careful examination of their morphology revealed that BC associated into nanofiber bundles whose size (75–770 nm width) increased with PEO content, but whose surface became correspondingly smoother.

When BC was generated in the presence of hydroxyethyl cellulose (HEC), the individual fibrils were coated by it and exhibited smaller diameters than those produced without it [39]. The ensuing self-assembled nanocomposites had tensile strengths close to 300 MPa (i.e., some 20% higher than that of BC or wood-based nanopapers) without any appreciable decrease in strain to failure.

The incorporation of BC into thermoplastic starch [40] and chitosan [41] has also provided evidence of excellent interface compatibility, clearly associated with the common polysaccharide structures, as in the similar situation of MFC–chitosan nanocomposites [42]. As expected, the films prepared with these two chitosan combinations displayed a strong increase in mechanical properties as the cellulose nanofiber content was increased. As shown in Figure 8.11, because the average width of the fibers was lower than the range of visible wavelengths, high transparency was maintained at loadings up to 10% BC. Figure 8.12 confirms the homogeneous distribution of the BC fibers (10%) within the chitosan film.

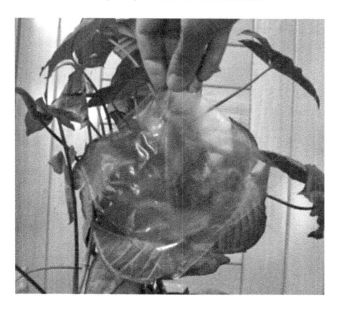

Figure 8.11   Chitosan film with 10% of BC nanofibers [41]. Reprinted from S. C. M. Fernandes, *et al.*, Novel transparent nanocomposite films based on chitosan and bacterial cellulose, *Green Chem.* 11, 2023. Copyright (2009) with permission from Royal Society of Chemistry. DOI http://dx.doi.org/10.1039/B919112G

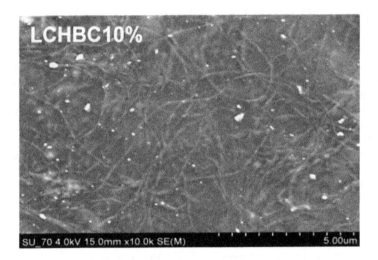

Figure 8.12 SEM visualization of the BC fiber network inside a chitosan film [41]. Reprinted from S. C. M. Fernandes, *et al.*, Novel transparent nanocomposite films based on chitosan and bacterial cellulose, *Green Chem.* 11, 2023. Copyright (2009) with permission from Royal Society of Chemistry. DOI http://dx.doi.org/10.1039/B919112G

In another direction, films were prepared from commercial acrylic latexes and shredded BC fibers [43] in order to enhance the mechanical properties of the ensuing composites. Again, the distribution of the nanofibers within the polymer sheet was quite uniform, despite its relatively poor polarity. This observation was rationalized by the compatibilizing role of the surfactant(s) present in the latexes. Figure 8.13

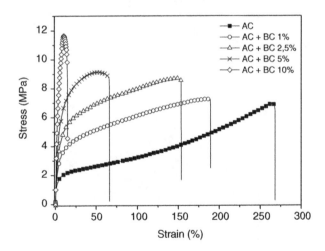

Figure 8.13 Stress–strain plots of nanocomposites of an acrylic copolymer (AC) containing an increasing amount of shredded BC [43]. Reprinted from C.S. R. Freire, *et al.*, Novel bacterial cellulose–acrylic resin nanocomposites, *Compos. Sci. Technol.* 70, 1148. Copyright (2010) with permission from Elsevier

illustrates the enhancement of strength and the corresponding reduction in strain to failure.

The rigidity of shape-memory polyurethanes was enhanced by the incorporation of well-dispersed cellulose nanocrystals, with a modulus increase of about 50% with 1% of fibers, without any appreciable loss of the recovery features over several cycles [44].

Cellulose nanocomposites with a polar thermoplastic poly(ethylene oxide–co–epichlorohydrin) matrix were prepared by solvent casting and characterized in terms of the reinforcement induced by formation of a three-dimensional percolating network of nanofibers [45]. The application of a papermaking technique to the preparation of MFC-reinforced poly(lactic acid) allowed the formation of tough sheets from an aqueous suspension of both polymers [46].

8.4.2 Superficial Fiber Modification

Both physical and chemical approaches have been successfully adopted to modify cellulose fiber surfaces [37], although the latter has received much more attention. The present survey covers all types of modification, including those which were not followed by the incorporation of the fibers into a matrix.

The use of plasma treatment has recently provided some interesting results, including a strong enhancement of hydrophobicity associated with both the incorporation of fluorine atoms and surface etching, following the application of a low-pressure SF_6 plasma [47]. Belgacem's group recently published two papers dealing with the treatment of cellulose samples with cold plasma in the presence of several coupling agents, namely vinyl trimethoxysilane (VTS) and γ-methacrylopropyl trimethoxysilane (MPS) on the one hand [48], and myrcene (MY) and limonene (LM) on the other [49]. Contact angle and XPS measurements showed that the surface cellulose chains had been chemically grafted, with a decrease in the polar component of the surface energy from about $23\,mJ/m^2$ to practically zero for all the treated samples. This was amply confirmed by the XPS spectra, which also indicated that MY had been coupled more efficiently than LM, as expected, given its higher degree of unsaturation. As for the siloxane coupling agents, MPS was found to be more reactive than VTS.

VTS   **MPS**   **LM**   **MY**

The chemical modifications recently reported can be subdivided into different categories, based on both the adopted approach and the specific purpose of the treatment.

8.4.2.1 Surface Hydrophobization

The reaction of fatty acid chlorides with conventional cellulose fibers was recently studied by two groups [50, 51] in an approach that aims at giving a complete exploitation of renewable resources. The ensuing esterified fibers were thoroughly characterized and, among other relevant properties, an increase in their hydrophobic character was readily attained. When the reaction was carried out in a non-swelling medium such as toluene, the esterification was essentially confined to the fiber surfaces and this was quite effective for achieving good interface adhesion when fibers were incorporated into a poly(ethylene) matrix [52, 53]. However, when the modification was carried out in a swelling medium such as DMF [50], the esterification proceeded further into the fiber thickness, thus generating fibers with an outer sleeve of thermoplastic material. This process enabled the preparation of a material which could be converted to a composite simply by hot pressing. The outer layer of the esterified fibers melted together to form a matrix, within which the inner core of the unmodified fibers played the role of reinforcing element. Figure 8.14 illustrates the morphology obtained with this in-depth plasticization.

The coupling of fluorinated moieties at the surface of a material is a well-known method for reducing its hydrophilic character and this approach has given positive results when applied to cellulose fibers. Three reports dealing with this issue were recently published [54]. These methods involved trifluoroacetic anhydride (TFA), 3,3,3-trifluoropropanoyl chloride (TFP) and pentafluorobenzoyl choride (PFB) as the reagents for appending hydrophobic moieties through condensation reactions with the surface OH groups of different cellulose fibers.

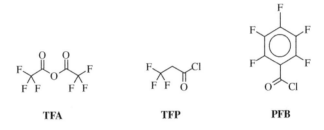

TFA   **TFP**   **PFB**

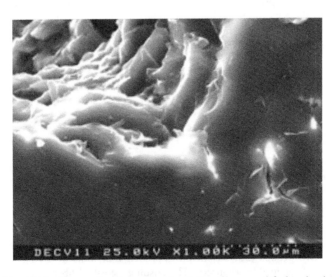

Figure 8.14   SEM micrograph of hot-pressed cellulose fibers modified in depth by a fatty acid chloride [50]

The success of the esterifications using TFA, TFP or PFB was assessed by FTIR spectroscopy and thereafter all the modified fibers were characterized exhaustively to determine the extent of surface coverage, the ensuing surface energy and the resistance of the modification to hydrolytic conditions. Highly biphobic surfaces were obtained in all instances, as witnessed by high contact angles with both water (> 120°) and diiodo-methane (70–100°). These findings were coupled with a significant decrease in the dispersive component of the surface energy and values of the polar component approaching zero. Whereas fibers esterified with TFP and PFB retained their surface biphobicity after prolonged contact with water in liquid or vapor form, their TFA counterparts showed high sensitivity to hydrolysis typical of trifluoroacetates [54a]. The latter behavior could be exploited in paper applications such as packaging, where the biphobic character should only be a temporary property related to the specific life cycle of the material.

A superhydrophobic cellulose surface was achieved by combining the two essential requirements associated with that property (loosely defined as a surface that gives a contact angle higher than 140°), namely highly non-polar chemical moieties, such as perfluoro groups, coupled with a micro- or nano-roughness that impedes water spreading. In this study [55], successive treatments involving the deposition of silica nanoparticles, followed by their perfluorination, as sketched in Figure 8.15, created precisely this combination of chemical and physical features on the final cellulose surfaces and resulted in contact angles close to 150°.

A different strategy for preparing cellulose–inorganic hybrid materials with hydrophobic properties was implemented by converting some of the surface OH groups into urethane moieties by reaction with (3-isocyanatopropyl)triethoxysilane. This was followed by different modifications of the siloxane moieties, including their grafting with long perfluorinated groups [56]. Figure 8.16 illustrates these surface reactions. The micro/nano asperities thus generated, coupled with the presence of the perfluoro moieties, gave water and diiodomethane contact angles as high as 140 and 134° respectively.

An original, straightforward and industrially viable approach to the preparation of hydrophobic cellulose surfaces was positively assessed very recently [57]. The idea is to flow a mixture of trichloromethylsilane (TCMS) and moisture of known composition in air at room temperature onto a cellulose substrate (e.g., filter paper) where a series of reactions take place rapidly to form various hybrid structures and hence different nanomorphologies on the fiber surfaces. Figure 8.17 shows schematically the chemical constructs that are likely to be generated. The preliminary results gathered using different experimental conditions indicated that a substantial hydrophobization (water contact angles around 130°) can be attained in less than a minute and with very modest amounts of reagent. This is a very promising system, thanks to its simplicity and green connotations, which could in principle be scaled up to a technological level. This process has given an equally interesting outcome when applied to other OH-bearing substrates like starch, chitosan and poly(vinyl alcohol) [58].

The preparation of hydrophobic cellulose surfaces has been critically examined in a more systematic fashion in a recent review [59].

8.4.2.2 *Surface-Initiated Polymerization and Polymer Grafting*

The modification of cellulose by polymer grafting is a classical topic which has interested chemists for decades, as already thoroughly discussed in a book published

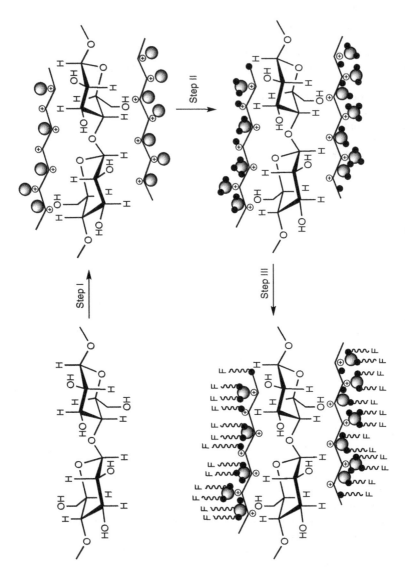

Figure 8.15 Creating a nano-roughness and appending perfluoro moieties to prepare superhydrophobic cellulose fibers [55]. Reprinted from G. Gonçalves, *et al.*, Superhydrophobic cellulose nanocomposites, *J. Coll. Interface Sci.* 324, 42. Copyright (2008) with permission from Elsevier

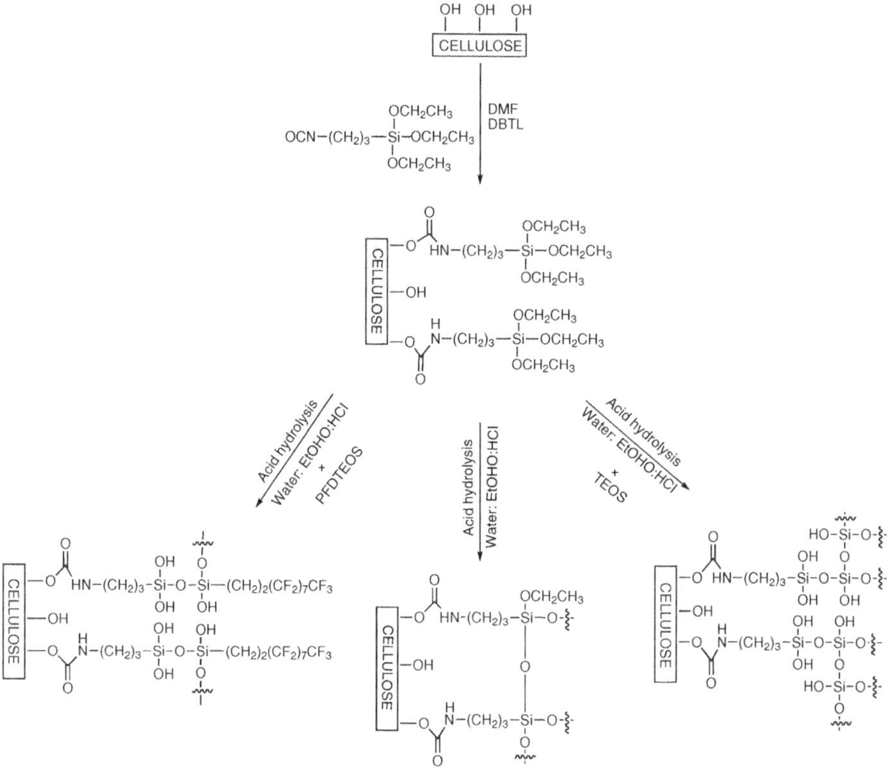

Figure 8.16 Reaction paths related to the preparation of different cellulose–inorganic hybrid materials, including hydrophobic surfaces [56]. Reprinted from A. G. Cunha, *et al.*, Preparation and characterization of novel highly omniphobic cellulose fibers organic–inorganic hybrid materials. *Carbohydr. Polym.* 80, 1048. Copyright (2010) with permission from Elsevier

some thirty years ago [60]. Renewed activity in this field within the last decade reflects the importance attached to materials derived from cellulose fibers, in which surfaces have been modified to widen their applications, notably in the realm of composites. As already emphasized, a highly adhesive interface between fibers and matrix is an essential prerequisite for ensuring the adequate response of a composite to mechanical stress. The presence of macromolecular grafts at the fiber surface can contribute in a substantial way to achieving this goal, either by establishing chain entanglements with the matrix macromolecules or, better still, through continuous covalent bonds between fibers and matrix [37]. The contributions to this rejuvenated area of cellulose chemistry have been reviewed recently [37d, 61], with particular emphasis on the application of both RAFT (reversible addition–fragmentation transfer) and ATRP (atom transfer radical polymerization) living systems, and therefore only a choice of salient features related to the latest advances is discussed here.

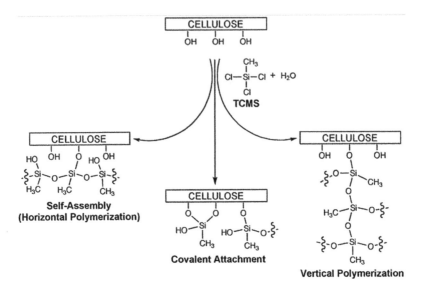

Figure 8.17 Possible mechanisms associated with the interactions of TCMS and moisture with the surface OH groups of cellulose fibers [57]. Reprinted from A. G. Cunha, *et al.*, Preparation of highly hydrophobic and lipophobic cellulose fibers by a straightforward gas–solid reaction, *J. Coll. Interface Sci.* 44, 588. Copyright (2010) with permission from Elsevier

An interesting extension of a novel mode of cellulose chemical activation put forward a decade ago [62] consisted in reacting cellulose fibers with difuctional coupling agents (i.e., pyromellitic dianhydride (PMDA), benzophenone-3,3',4,4'-tetracarboxylic dianhydride (BPDA), 1,4-phenylene diisocyanate (PPDI), methylene-bis-diphenyl diisocyanate (MDI), γ-mercaptopropyltriethoxysilane (MRPS), and γ-methacrylopropyltriethoxysilane (MPS)) [63]. The strategy here is based on the fact that only one of the functionalities reacts with the cellulose surface OH groups, either because of steric hindrance [62], or because the other functionaility simply does not react, so that the resulting surface bears a moiety available for further modification.

PMDA   BPDA

PPDI   MDI

MRPS   **MPS**

The incorporation of the anhydride- and isocyanate-modified fibers into a cellulose ester matrix gave rise to coupling reactions with the residual OH groups of the matrix, thus forming continuous covalent bonds between the two components of the composites. The same occurred when the siloxane-modified fibers were incorporated into a natural rubber matrix.

In a totally different vein that resembles the in-depth modification with fatty acid chains discussed above, a whole variety of cellulose substrates were partially oxypropylated in order to generate a thermoplastic sleeve around the fibers. This sleeve functioned as the matrix upon hot pressing, thus giving rise to composite materials entirely derived from cellulose [64]. The gas–solid oxypropylation of OH-bearing natural polymers, through the 'grafting-from' anionic polymerization of propylene oxide initiated by surface oxyanions, has received a good deal of attention, particularly in terms of the total conversion of the substrate into a viscous polyol [65].

The application of click chemistry to the grafting of cellulose fibers was first reported by Krouit et al. [66] who introduced acetylenic moieties by esterification with undecenoic acid and thereafter appended azido-terminated poly(caprolactone) (PCL) chains. Cellulose nanocrystals were linked to give an organized gel using the same click chemistry after attaching N_3 and $C \equiv C$ moieties to two different batches and mixing them [67]. An alternative way to graft PCL diol onto cellulose called upon the use of 2,4-toluene diisocyanate as a coupling mediator [68], and this successful approach was then extended to polyether branches of different DP and polarity [69]. Interestingly, all these modified fibers maintained their bio-degradability, albeit with slower kinetics compared with the pristine cellulose substrate [68, 69].

An original 'grafting-from' mechanism was recently put forward by Hong et al. [70] who appended benzophenone moieties to cotton fibers before irradiating the fabrics with UV light in the presence of acrylamide. Attachment of poly(acrylamide) chains to the cellulose backbone was amply corroborated by a thorough characterization of the modified materials.

8.4.2.3 Miscellaneous Chemical Modifications

Among the numerous reports on the coupling of different chemical species to the surface of cellulose fibers [37], a selection of recent original publications is mentioned here. Heinze and co-workers [71] synthesized a dendritic cellulose architecture bearing polyamidoamine structures, which reached the third generation with a DS of 0.69 (i.e., with nearly 70% of the C-6 primary OH involved in the modification). As discussed below, water solubility of these carefully characterized materials was one of the notable features in terms of their applications.

Alila *et al.* [72] grafted cellulose with *N,N*-carbonyldiimidazole and then appended numerous amines, among which long aliphatic moieties terminated with multiple primary amino groups provided promising materials for further exploitation (e.g., by reactive incorporation of these modified fibers in an epoxy matrix in order to prepare composites with continuous covalent linkages at the interface). A variety of chemical modifications have been applied to MFC both in aqueous and organic media [73], including reactions with glycidyl methacrylate, diisocyanates and cyclic anhydrides. The partial acetylation of cellulose whiskers was carried out at the same time as the actual acid-catalyzed hydrolysis used to prepare the whiskers from cotton linters [74]. Both modifications were aimed at either stabilizing these nanofibers against spontaneous association or at their incorporation into polymeric matrices as reinforcing elements.

8.5   CELLULOSE COUPLED WITH NANOPARTICLES

Within the last decade, much research has been devoted to attaching metal and oxide nanoparticles to cellulose fibers [37*d*]. Arguably, the most promising aspects of these investigations are related to the deposition of silver nanoparticles for antibacterial applications and of titania counterparts for photocatalyic materials. Examples of the former include the use of waterborne fluorinated siloxanes, nano-sized Ag particles and an organic–inorganic binder to prepare biphobic cotton fibers with antimicrobial properties [75], a systematic study of the quantitative antibacterial role of cellulose/Ag nanocomposites [76], the *in situ* synthesis of Ag nanoparticles on ZnO whiskers incorporated into a paper matrix [77] and the grafting of cellulose with poly (acrylamide) followed by the generation of Ag nanoparticles on the modified substrate [78].

An alternative approach describes the surface-initiated ATRP of *t*-butylacrylate onto filter paper followed by the hydrolysis of the grafted polymers and the chelation of silver ions by the ensuing carboxylic groups [79], which gave cellulose materials with a strong antibacterial action against *E. coli*. Ferrairia *et al.* [80] described the preparation of ultrathin cellulose films containing immobilized silver nanoparticles following the grafting of amino groups discussed above [72]. The nanoparticles were in a size range of 7–30 nm and their location could be visualized both at the surface and within the thin films.

An interesting study of cellulose fibers decorated with TiO_2 nanoparticles [81] called upon a preliminary modification with different siloxanes. These cellulose nanohybrids have potential application because of self-bleaching properties through solar photoactivity.

8.6   ELECTRONIC APPLICATIONS

Coating cellulose fibers with a semiconducting polymer has recently motivated researchers with studies on poly(pyrrole) [82], poly(3,4-ethylene dioxythiophene [83] and poly(3-octylthiophene) [84], among other conventional conjugated polymers. On the whole, since these publications only point to its obvious interest and potential development, this field seems to require more systematic investigations in order to move into maturity. The latest contributions are moving in that direction, with two studies on well-characterized materials which displayed, in particular, high conductivities from poly(pyrrole)-coated MFC [85a] and from poly(aniline) deposited on microporous cellulose ester membranes [85b].

Kim *et al.* [86] were the first to report that cellulose in the form of a paper sheet could be turned into smart electroactive devices producing large bending displacements with modest actuation voltages and power consumptions. Other electronic applications of modified cellulose include the development of an organic memory device based on carbazole-bearing fibers [87], photocurrent-generating films based on porphyrin-bearing cellulose [88] and an electroactive actuator based on bacterial cellulose [89].

8.7 BIOMEDICAL APPLICATIONS

Hoenich [90] and more recently Gatenholm and Klemm [20b] have updated the literature on the biomedical potential of bacterial cellulose. These are incisive reviews and the interested reader can therefore find the most relevant information in those monographs. Without repeating the issues put forward there, it is important to emphasize that BC has become a definitive material in terms of applications related to implants and biofabrication, among others, for which it is providing extremely useful outcomes.

Specific cellulose derivatives are regularly synthesized with the purpose of fulfilling a given task related to biological activities. Three very different examples are cited here to illustrate this point, without any pretense to represent a comprehensive view of this vast topic. First, the type of dendronized cellulose quoted earlier [71] has been used to develop biofunctionalized surfaces by first introducing a multitude of primary amino groups at the shell of the dendrons and then covalently attaching enzymes [91]. Second, improved blood compatibility has been achieved by grafting poly(*p*-vinyl sulfobetaine) onto the polysaccharide backbone of cellulose membranes using the ATRP mechanism [92]. Third, bioactive papers (i.e., paper-based materials), with potential analytical functions such as pathogen detection have been investigated [93]. The enormous potential of the latter stems primarily from its extreme simplicity and economy, compared with other more intricate and costly laboratory or field devices.

8.8 CELLULOSE DERIVATIVES

Esters and ethers represent the dominant cellulose derivatives on the chemical commodity market [5, 6, 94]. Some of these derivatives are indeed the oldest thermoplastic materials, prepared by man some 150 years ago. Throughout the 20th century, the technologies related to their manufacture were regularly improved without, however, any substantial qualitative innovation. The state of the art concerning novel strategies to prepare cellulose derivatives was provided by Heinze and Petzold in a recent monograph [26] which sets the stage for possible radical innovations. The obvious trends in this domain, as indeed in any domain of industrial chemistry, have to do with improving the green connotation of the processes. In support of this strategy, numerous recent publications propose novel approaches to the synthesis of cellulose derivatives, including the use of ionic liquids [26, 28, 95], microwave irradiation [96] or solvent-free systems [97].

Whereas the production of 'minor' derivatives required for specific applications such as the biomedical and optoelectronic fields should undergo this important process reorganization within a short time, the heavy industry associated with the fabrication of massive quantities of derivatives such as cellulose acetate, is likely to be slower in responding.

8.9   CONCLUDING REMARKS

Cellulose has returned in full swing into the limelight as a source of novel high-tech materials, quite apart from its traditional utilization in paper products, cotton textiles and thermoplastic polymers in the form of esters and ethers. Within the general trend toward a growing exploitation of renewable resources using green processes, cellulose has all the qualifications to respond adequately to both the improvement of existing technologies and its intervention as a key element in the construction of original razoredge products and devices. This chapter hopefully provides clear indications regarding the validity of this assessment.

REFERENCES

[1] D. Klemm, B. Heublein, H.-P. Fink, et al., Angew. Chem. Int. Ed. 2005, 44, 3358.
[2] J. Gullichsen, H. Paulapuro, Eds., Papermaking Science and Technology, Fapet Oy, Helsinki, Vols. 1–19 1999. Roberts J. C., Paper Chemistry, 2nd ed., Chapman & Hall, London, 1996. M. Ek, G. Gellesteddt, G. Eriksson, Eds., Pulp and Paper Chemistry and Technology, Volumes 1–4, de Gruyter, Berlin, 2009.
[3] B. Kamm, P. R. Gruber, and M. Kamm, Biorefineries–Industrial Processes and Products., Volumes 1 and 2, Wiley VCH, Weinheim, 2006.
[4] S. Gordon, Y.-L. Hsieh, Eds., Cotton: Science and Technology, Woodhead, Cambridge, 2007.
[5] H. A. Krässig, Cellulose-Structure, Accessibility and Reactivity, Gordon & Breach, Iverdon, 1993.
[6] D. Klemm, B. Philipp, T. Heinze, et al., Comprehensive Cellulose Chemistry, Vols 1 and 2, Wiley VCH, Weinheim, 1998. D. N.-S. Hon N. Shiraishi, Eds., Wood and Cellulose Chemistry, 2nd edn. Marcel Dekker, New York, 2001.
[7] I. M. Ward and P. J. Hine, Polym. Eng. Sci. 1997, 37, 1809 and Polymer, 2004 45, 1413.
[8] T. Nishino, I. Matsuda, and K. Hirao, Macromolecules 2004, 37, 7683. T. Nishino, N. Arimoto, Biomacromolecules, 2007, 8, 2712. N. Soykeabkaew, N. Arimoto, T. Nishino, et al., Compos. Sci. Technol. 2008, 8, 2201. N. Soykeabkaew, C. Sian, S. Gea, et al., Cellulose. 2009, 16, 435.
[9] W. Gindl, J. Keckes, Polymer 2005, 46, 10221.W. Gindl, T. Schoberl, and J. Keckes, J. Appl. Phys. A: Mater. Sci. Process 2006, 68, 19. B. Duchemin, R. Newman, and

M. Staiger, *Cellulose* 2007, **14**, 311. B. J. C. Duchemin, R. H. Newman, and M. P. Staiger, *Compos. Sci. Technol.* 2009, **69**, 1225.

[10] B. J. C. Duchemin, M. P. Staiger, N. Tucker, *et al.*, *J. Appl. Polym. Sci.* 2010, **115**, 216.

[11] *a.* A. Dufresne, Cellulose-Based Composites and Nanocomposites, in M.N. Belgacem, A. Gandini (eds), *Monomers, Polymers and Composites from Renewable Resources*, Elsevier, Amsterdam, 2008. Chapter 19, p. 401. *b.* M. M. de Souza Lima, R. Borsali, *Macromol. Rapid Comm.* 2004, **25**, 771. *c.* M. A. S. Azizi, F. Alloin, and A. Dufresne, *Biomacromolecules* 2005, **6**, 612. *d.* I. Siró, D. Plackett, *Cellulose* 2010, **17**, 459. *e.* Y. Habibi, L. A. Lucia, and O. J. Rojas, *Chem. Rev.* 2010, **110**, 3479.

[12] É. Pecoraro, D. Manzani, Y. Messaddeq, *et al.*, Bacterial Cellulose from Glucanacetobacter xylinus: Preparation, Properties and Applications, in M.N. Belgacem, A. Gandini (eds), *Monomers, Polymers and Composites from Renewable Resources*, Elsevier, Amsterdam, 2008, Chapter 17, p. 369.

[13] M.N. Belgacem, A. Gandini (eds), *Monomers, Polymers and Composites from Renewable Resources*, Elsevier, Amsterdam, 2008. A. Gandini, *Macromolecules*, 2008, **41**, 9491.

[14] T. Saito, M. Hirota, N. Tamura, *et al.*, *Biomacromolecules* 2009 **10**, 1992. M. Hirota, N. Tamura, T. Saito, *et al.*, *Cellulose* 2010, **17**, 279.

[15] A. Hirai, O. Inui, F. Horii, *et al.*, *Langmuir* 2009, **25**, 497.

[16] M. Henriksson, L. A. Berglund, P. Isaksson, *et al.*, *Biomacromolecules* 2008, **9**, 1579.

[17] L. A. Berglund, T. Peijs, *MRS Bull.* 2010, **35**, 201.

[18] H. Fukuzumi, T. Saito, T. Iwata, *et al.*, *Biomacromolecules* 2009, **10**, 162.

[19] A. N. Nakagaito, M. Nogi, and H. Yano, *MRS Bull.* 2010, **35**, 214.

[20] *a.* W. K. Czaja, D. J. Young, M. Kawecki, *et al.*, *Biomacromolecules*, 2007, **8**, 1. *b.* P. Gatenholm, D. Klemm, *MRS Bull.*, 2010, **35**, 208.

[21] E. Kontturi, T. Tammelin, and M. Österberg, *Chem Soc. Rev.* 2006, **35**, 1287. See also *ACS Symp. Ser.* 2010, **1019**, an issue entirely devoted to model cellulosic surfaces.

[22] R. Sczech, H. Riegler, *J. Coll. Interf. Sci.* 2006, **301**, 376.

[23] C. Aulin, A. Shchukarev, J. Lindqvist, *et al.*, *J. Coll. Interf. Sci.* 2008, **317**, 556.

[24] Y. Habibi, L. Foulon, V. Aguié-Béghin, *et al.*, *J. Coll. Interf. Sci.* 2007, **316**, 388.

[25] S. Ahola, J. Salmi, L.-S. Johansson, *et al.*, *Biomacromolecules*, 2008, **9**, 1273.

[26] T. Heinze, K. Petzold, Cellulose Chemistry: Novel Products and Synthesis Paths, in M.N. Belgacem, A. Gandini (eds), *Monomers, Polymers and Composites from Renewable Resources*, Elsevier, Amsterdam, 2008, Chapter 16, p. 343.

[27] J. Cai, L. Zhang, *Macromol. Biosci.* 2005, **5**, 539. D. Ruan, L. Zhang, A. Lue, *et al.*, *Macromol. Rapid Commun.* 2006, **5**, 1495. H. Qi, X. Sui, J. Yuan, *et al.*, *Macromol. Mater. Eng.* 2010, **295**, 695.

[28] S. Köhler, T. Heinze, *Macromol. Biosci.* 2007, **7**, 307. M. Schobitz, F. Meister, and T. Heinze, *Macromol. Symp.* 2009, **280**, 102. S. Köhler, T. Liebert, and T. Heinze, *Macromol. Biosci.* 2009, **9**, 836.

[29] R. P. Swatloski, S. K. Spear, J. D. Holbrey, *et al.*, *J. Am. Chem. Soc.* 2002, **124**, 4974.

[30] Y. Fukaya, K. Hayashi, M. Wada, *et al.*, *Green Chem.* 2008, **10**, 44. H. Ohno, Y. Fukaya, *Chem. Lett.* 2009, **38**, 2. K. Fujita, N. Nakamura, K. Igarashi, *et al.*, *Green Chem.* 2009, **11**, 351.

[31] L. Feng, Z. Chen, *J. Mol. Liq.* 2008, **142**, 1.

[32] J. Vitz, T. Erdmenger, C. Haensch, *et al.*, *Green Chem.* 2009, **11**, 417. J. Zhang, H. Zhang, J. Wu, *et al.*, *Phys. Chem. Chem. Phys.* 2010, **12**, 1941. H. Liu, K. L. Sale, B. M. Holmes, *et al.*, *J. Phys. Chem. B* 2010, **114**, 4293.

[33] M. Gericke, K. Schlufter, T. Liebert, et al., Biomacromolecules 2009, 10, 1188. X. Chen, Y. Zhang, L. Cheng, et al., J. Polym. Environ. 2009, 17, 273.

[34] S. Xu, J. Zhang, A. He, et al., Polymer 2008 49, 2911. S.-L. Quan, S.-G. Kang, and I.-J. Chin, Cellulose 2010, 17, 233.

[35] A. Pinkert, K. N. Marsh, S. Pang, et al., Chem. Rev. 2009, 109, 6712; Ind. Eng. Chem. Res. 2010, 49, 11809.

[36] A. K. Bledzki, J. Gassan, Progr. Polym. Sci. 1999, 24, 221. A. Mohanty, M. Misra, and L. T. Drzal, eds, Natural Fibers, Biopolymers and Biocomposites, CRC Press, Boca Raton, 2005. M. J. John, R. D. Anandjiwala, Polym. Compos. 2008, 29, 187. T. Sabu, S. Pothan, eds, Cellulose Fiber Reinforced Polymer Composites, Old City Publishing, Philadelphia, 2009.

[37] a. M. N. Belgacem, A. Gandini, Surface Modification of Cellulose Fibers, in M.N. Belgacem, A. Gandini, eds, Monomers, Polymers and Composites from Renewable Resources, Elsevier, Amsterdam, 2008, Chapter 18, p. 385 and references therein. b. M. N. Belgacem, A. Gandini, Compos. Interf. 2005, 12, 41. c. A. Gandini, C. S. R. Freire, Cellulose Chem. Technol. 2006, 40, 691. d. A. Gandini, M. N. Belgacem, Physical and Chemical methods of fiber surface modification, in N. E. Zafeiropoulos, ed., Interface Engineering in Natural Fiber Composites, Woodhead Publishing, Cambridge (UK), 2011, Chapter 1, p. 3.

[38] E. E. Brown M.-P. Laborie, Biomacromolecules 2007, 8, 3074.

[39] Q. Zhou, E. Malm, H. Nilsson, et al., Soft Matter 2009, 5, 4124.

[40] C.S.R. Freire, A.J.D. Silvestre, C. Pascoal Neto, et al., Composite Sci. Technol. 2009, 69, 2163.

[41] S. C. M. Fernandes, A. L. Oliveira, C. S. R. Freire, et al., Green Chem. 2009, 11, 2023.

[42] C.S.R. Freire, A.J.D. Silvestre, C. Pascoal Neto, et al., Carbohydr. Polym. 2010, 81, 394.

[43] C.S. R. Freire, A. J. D. Silvestre, A. Gandini, et al., Compos. Sci. Technol. 2010, 70, 1148.

[44] M. L. Auad, V. S. Contos, S. Nutt, et al., Polym. Int. 2008, 57, 651.

[45] J. R. Capadona, K. Shanmuganathan, S. Trittschuh, et al., Biomacromolecules 2009, 10, 712.

[46] A. N. Nakagaito, A. Fujimura, T. Sakai, et al., Compos. Sci. Technol. 2009, 69, 1295.

[47] R. Barni, S. Zanini, D. Beretta, et al., EPJ Appl. Phys. 2007, 38, 263.

[48] C. Gaiolas, A. P. Costa, M. Nunes, et al., Plasma Process. Polym. 2008, 5, 444.

[49] C. Gaiolas, M. N. Belgacem, L. Silva, et al., J. Coll. Interface Sci. 2009, 330, 298.

[50] C. S. R. Freire, C. Pascoal Neto, A.J.D. Silvestre, et al., J. Appl. Polym. Sci. 2006, 100, 1093. C. S. R. Freire, A.J.D. Silvestre, C. Pascoal Neto, et al., J. Coll. Interf. Sci. 2006, 301, 205.

[51] D. Pasquini, M.N. Belgacem, A. Gandini, et al., J. Coll. Interface Sci. 2006, 295, 79.

[52] C. S. R. Freire, A.J.D. Silvestre, C. Pascoal Neto, et al., Compos. Sci. Technol. 2008, 68, 3358.

[53] D. Pasquini, E. M. Teixera, A. A. S. Curvelo, et al., Compos. Sci. Technol. 2008, 68, 193.

[54] a. A. G. Cunha, C. S. R. Freire, A. J. D. Silvestre, et al., J. Coll. Interface Sci. 2007, 316, 360. b. Biomacromolecules 2007, 8, 1347. c. Langmuir 2007, 23, 10801.

[55] G. Gonçalves, P. A. A. P. Matques, T. Trinidade, et al., J. Coll. Interface Sci. 2008, 324, 42.

[56] A. G. Cunha, C. S. R. Freire, A. J. D. Silvestre, et al., Carbohydr. Polym., 2010, 80, 1048.

[57] A. G. Cunha, C. S. R. Freire, A. J. D. Silvestre, *et al.*, *J. Coll. Interface Sci.* 2010, 344, 588.
[58] A. G. Cunha, C. S. R. Freire, A. J. D. Silvestre, *et al.*, unpublished results.
[59] A. Gandini, A. G. Cunha, *Cellulose*, 2010 17, 875, 1045.
[60] A. Hebeish, J. T. Guthrie, *The Chemistry and Technology of Cellulosic Copolymers*, Springer-Verlag, Berlin, 1981.
[61] D. Roy, M. Semsarilar, J. T. Guthrie, *et al.*, *Chem. Soc. Rev.* 2009, 38, 2046.
[62] A. Gandini, V. R. Botaro, E. Zeno, *et al.*, *Polym. Int.* 2001, 50, 7.
[63] B. Ly, W. Thielemans, A. Dufresne, *et al.*, *Compos. Sci. Technol.* 2008, 68, 3193.
[64] A. J. de Menezes, D. Pasquini, A. A. da Silva Curvelo, *et al.*, *Carbohydr. Polym.* 2009, 76, 437 and *Cellulose* 2009, 16, 239.
[65] A. Gandini, M. N. Belgacem, Partial or Total Oxypropylation of Natural Polymers and the Use of the Ensuing Materials as Composites or Polyol Macromonomers, in M.N. Belgacem, A. Gandini, eds, *Monomers, Polymers and Composites from Renewable Resources*, Elsevier, Amsterdam, 2008, Chapter 12, p. 273.
[66] M. Krouit, J. Bras, and M. N. Belgacem, *Europ. Polym,. J.* 2008, 44, 4074.
[67] I. Filpponen, D. S. Argyropoulos, *Biomacromolecules* 2010 11, 1060.
[68] O. Paquet, M. Krouit, J. Bras, *et al.*, *Acta Mater.* 2010, 58, 792.
[69] E. B. Ly, J. Bras, P. Sadocco, *et al.*, *Mater. Chem. Phys.* 2010, 120, 438.
[70] K. H. Hong, N. Liu, and G. Sun, *Europ. Polym. J.* 2009, 45, 2443.
[71] M. Pohl, J. Schaller, F. Meister, *et al.*, *Macromol. Rapid Commun.* 2008, 29, 142.
[72] S. Alila, A. M. Ferraria, A. M. Botelho do Rego, *et al.*, *Carbohydr. Polym.* 2009, 77, 553.
[73] P. Stenstad, M. Andersen, B. S. Tanem, *et al.*, *Cellulose* 2008, 15, 35.
[74] B. Braun, J. R. Dorgan, *Biomacromolecules* 2009, 10, 334.
[75] B. Tomsic, B. Simoncic, B. Orel, *et al.*, *J. Sol-Gel Sci. Technol.* 2008, 47, 44.
[76] R. J. B. Pinto, P. A. A. P. Marques, C. Pascoal Neto, *et al.*, *Acta Biomater.* 2009, 5, 2279.
[77] H. Koga, T. Kitaoka, and H. Wariishi, *J. Mater. Chem.* 2009, 19, 2135.
[78] R. Tankhiwale, S. K. Bajpai, *Colloid Surf. B: Biointerf.* 2009, 69, 164.
[79] F. Tang, L. Zhang, Z. Zhang, *et al.*, *J. Macromol. Sci. Part A: Pure Appl. Chem.* 2009, 46, 989.
[80] A. M. Ferraria, S. Boufi, N. Battaglini, *et al.*, *Langmuir* 2010 26, 1996.
[81] G. Gonçalves, P. A. A. P. Marques, R. J. B. Pinto, *et al.*, *Compos. Sci. Technol.* 2009, 69, 1051.
[82] D. Beneventi, S. Alila, S. Boufi, *et al.*, *Cellulose* 2006, 13, 725.
[83] I. Wistrand, R. Lingström, and L. Wågberg, *Europ. Polym. J.* 2007, 43, 4075.
[84] P. Sarrazin, D. Chaussy, O. Stephan, *et al.*, *Colloid Surf. A: Physicochem. Eng. Asp.* 2009, 349, 83.
[85] *a*. G. Nyström, A. Mihranyan, A. Razaq, *et al.*, *J. Phys. Chem. B* 2010, 114, 4178. *b*. A. A. Qaiser, M. M. Hyland, and D. A. Patterson, *J. Phys. Chem. B* 2009, 113, 14986.
[86] J. Kim, J. S. Yun, and Z. Ounaies, *Macromolecules* 2006, 39, 4202.
[87] M. Karawaka, M. Chikamatsu, Y. Yoshida, *et al.*, *Macromol. Rapid Commun.* 2007, 28, 1479.
[88] K. Sakakibara, Y. Ogawa, and F. Nakatsubo, *Macromol. Rapid Commun.* 2007, 28, 1270.
[89] J.-H. Jeon, I.-K. Oh, C.-D. Kee, *et al.*, *Sens. Actuators B: Chem.* 2010, 146, 307.
[90] N. Hoenich, *Biores.* 2006, 1, 270.

[91] M. Pohl, N. Michaelis, F. Meister, *et al.*, *Biomacromolecules* 2009, **10**, 382.

[92] P.-S. Liu, Q. Chen, X. Liu, *et al.*, *Biomacromolecules 2009* **10**, 2809.

[93] R. Pelton, *Trends Anal. Chem.* 2009, **28**, 925.

[94] E. Doelker, *Adv. Poly. Sci.* 1993, **107**, 199.

[95] S. Barthel, T. Heinze, *Green Chem.* 2006, **8**, 301. M. Gericke, T. Liebert, and T. Heinze, *Macromol. Biosci.* 2009, **9**, 343. W. Mormann, M. Wezstein, *Macromol. Biosci.* 2009, **9**, 369.

[96] S. Possidonio, L. C. Fidale, and O. A. El Seoud, *J. Polym. Sci.: Part A: Polym. Chem.* 2009, **48**, 134. L. Crépy, L. Chaveriat, J. Banoub, *et al.*, *ChemSusChem* 2009, **2**, 165.

[97] J. Li, L.-P. Zhang, F. Peng, *et al.*, *Molecules* 2009, **14**, 3551. J. C. P. Melo, E. C. da Silva Filho, S. A. A. Santana, *et al.*, *Colloid Surf. A: Physicochem. Aspects* 2009, **346**, 1338.

9

Furan Monomers and their Polymers: Synthesis, Properties and Applications

Alessandro Gandini

CICECO and Chemistry Department, University of Aveiro, Aveiro, Portugal

9.1 INTRODUCTION

Most of the chapters of this section deal with *natural polymers* and the materials they can provide by suitable physical and/or chemical manipulations, whereas polylactides are synthesized from naturally occurring monomers. In all these contexts, the *basic* macromolecular structure is determined by the renewable resource at stake. The properties of ensuing materials will hence be largely dependent on source, albeit with important modulations arising from appropriate chemical modifications. In other words, the latitude in the shaping of materials' features cannot deviate very far from the imprint imposed by the starting natural resource (e.g., starch polymer properties are determined to a large degree by the starch raw material and the properties of polylactides cannot be modified beyond the inevitable chemical restrictions imposed by the structure of their monomer units). This is also true of other monomers and polymers from renewable resources not discussed in this book, such as terpenes and lignins [1], which are subject to the same limitations. Although these considerations

Biopolymers – New Materials for Sustainable Films and Coatings, First Edition.
Edited by David Plackett.
© 2011 John Wiley & Sons, Ltd. Published 2011 by John Wiley & Sons, Ltd.

are by no means reductive as to the remarkable potential of all the materials described in this section and highlighted in concrete terms in the second section of the book, they do introduce an intrinsic qualitative boundary, which is not encountered in the case of furan polymers.

Two fundamental aspects make the realm of furan polymers qualitatively different from the above domains. First, although furan derivatives are ubiquitous in nature, none of these interesting molecules are likely to be exploited as monomers. The real monomer *precursors* within the furan family are instead prepared from sugars and polysaccharides. Second, these precursors are a source of a wide variety of monomers with very different chemical properties. This translates into an array of polymers covering an entire spectrum of macromolecular materials, potentially capable of simulating the properties and applications of presently available polymer commodities, predominantly derived from fossil resources. This strategy constitutes a possible alternative to petrochemistry through a very promising specific application of the biorefinery paradigm. The purpose of this chapter is to illustrate this working hypothesis with a choice of systems and materials possessing a serious chance of success.

Furan (1), like its two major homologues pyrrole (2) and thiophene (3), is a five-membered unsaturated heterocycle bearing two conjugated double bonds. Its chemistry [2] and the interest of its derivatives [3] as organic synthons and as major commodities in important areas like medicine, liquid crystals and specific additives, are very well documented, but fall mostly outside the scope of this monograph.

(1) 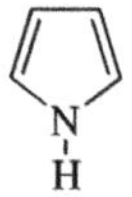   **(2)**    **(3)**

The relevant aspects in the present context concern thoroughly studied features [4] related to: (i) the reactivity of the furan ring with free radicals, since its dienic character gives rise to relatively well-stabilized entities (Scheme 9.1), the consequences of which can affect the aptitude of certain furan monomers such as vinyl furoate to undergo regular radical polymerization, rather than playing the role of retarders; (ii) the ease with which the furan ring participates in electrophilic substitutions at its alpha (C2 or C5) positions (Scheme 9.2) and hence promotes branching reactions in the cationic polymerization of furan monomers like 2-vinylfuran; (iii) the high susceptibility of hydrogens bound to a carbon linking two furans to be abstracted as protons, hydride ions or atoms, depending on the chemical environment of a given system, thus

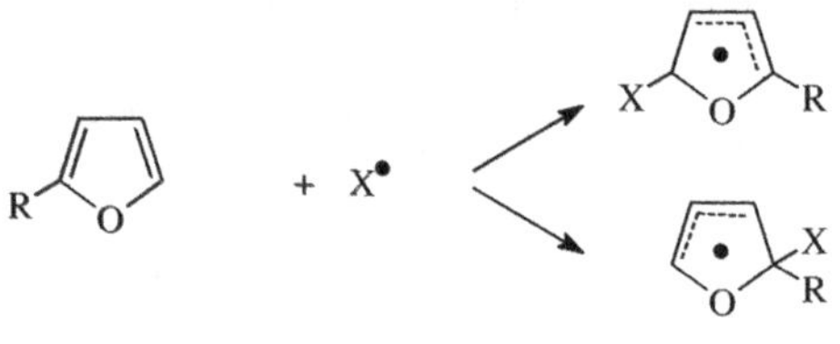

Scheme 9.1

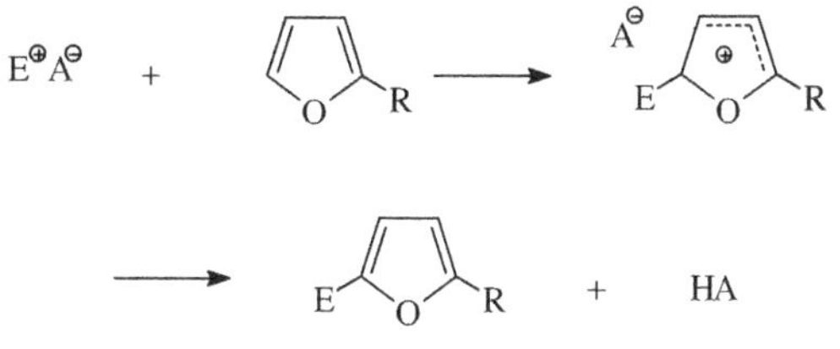

Scheme 9.2

favouring often undesired polymer destabilization, as in the case of polyfurfuryl alcohol discussed below; and (iv) the pronounced aptitude of furans to participate in the Diels–Alder (DA) reaction because of their pronounced dienic properties, which favors thermally reversible coupling with dienophiles like maleimide, as illustrated in Scheme 9.3.

9.2   PRECURSORS AND MONOMERS

It is remarkable that both pentose- and hexose-type saccharidic structures can be converted into furan derivatives by an acid-catalyzed dehydration mechanism of their sugar units. Pentose hemicelluloses such as xylan are very common components of certain woods and annual plants, so that agricultural and forestry wastes constitute very profitable biomass residues for the production of furfural (F). The mechanism, following the initial depolymerization of the polysaccharide, is illustrated for xylose and rhamnose in Scheme 9.4, the latter as a precursor to 5-methylfurfural (MF).

The variety of renewable resources available for preparing F includes, among others, sugarcane bagasse, oat and rice hulls, corn cobs, olive husks and cotton seeds. As a result, it is possible to build an industrial facility virtually anywhere in the world, considering moreover that such an initiative does not require high investment. It is therefore not surprising that F has been an industrial commodity for about a century, with a present yearly production of more than 300 000 tonnes at a price close to 1$ per kg. Likewise, certain C6 sugars and polysaccharides can be used to prepare hydroxymethylfurfural (HMF) through an analogous mechanism. However, HMF has not yet reached the status of an industrial commodity because of its relative chemical fragility, which tends to provoke important losses during its synthesis (low yields) and to make it difficult to store. Solutions are being actively sought in numerous laboratories [4f] through intensive research on novel synthetic procedures

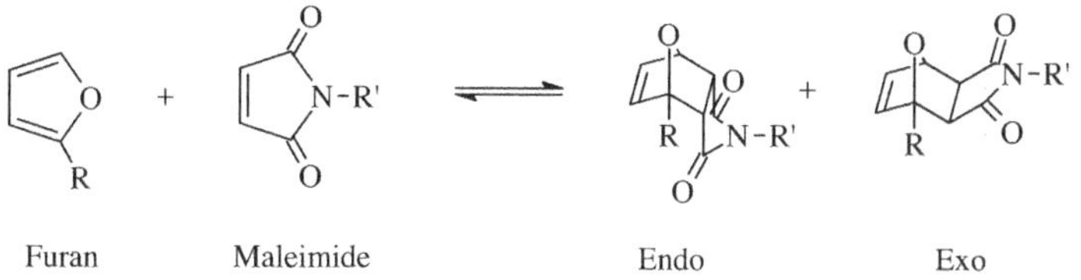

Scheme 9.3

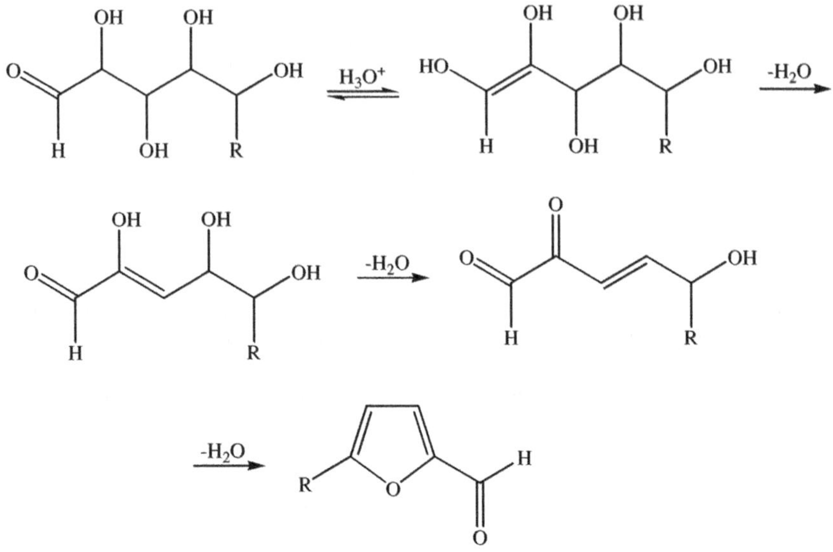

R = H, F; R = CH₃, MF

Scheme 9.4

and on ways to stabilize HMF or to prepare *in situ* highly stable derivatives, such as the corresponding dialdehyde (FCDA) and diacid (FDCA).

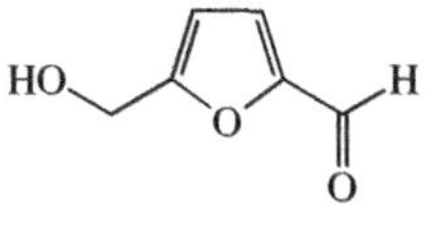

HMF 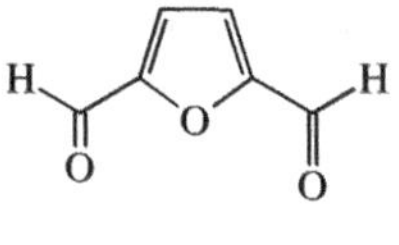   **FCDA**

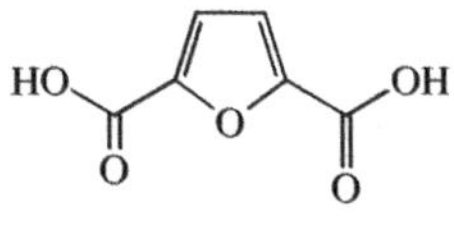

FDCA

Given the advances achieved by these investigations and the general consensus on the urgent need for HMF and derivatives, there seems to be little doubt that these compounds will become commercially available in large quantities within a few years.

Considering the availability of F and HMF as first-generation industrial furan compounds, the working hypothesis that follows naturally for a macromolecular scientist interested in polymers from renewable resources is to consider the types of furan monomers that can be prepared from each of these precursors. This has been largely planned and executed in the laboratory, thanks to decades of wide-ranging research [4], which should be considered as essentially exploratory in terms of both monomer syntheses and the study of their polymerization, including of course a preliminary assessment of the polymers' properties and potential applications. The keyword here is work in progress, since all these issues are being pursued with the intent of extending the range of materials and of optimizing the various steps leading to them [4f].

F and MF have been used as starting synthons for the preparation of a whole host of furan monomers [4] bearing a moiety suitable for chain polymerization(s) (Scheme 9.5). Depending on their specific reactivity, these monomers have been polymerized and copolymerized via free radical, cationic and anionic initiation [4].

Monofunctional furans from F have been converted into difunctional difuran mono-mers through the acid-catalyzed coupling reaction shown in Scheme 9.6 [4]. Here the context changes in that the ensuing monomers bear typical step-growth structures.

Likewise, but in a more straightforward approach, HMF was used as the precursor to a variety of step-growth monomers [4], as illustrated in Scheme 9.7.

None of the monomers described above have reached an industrial status yet, although some are certainly on the way. The exception is furfuryl alcohol (FA), readily prepared by the reduction of F, which has been a commercially available compound for decades. Indeed, FA is the most important industrial furan derivative on the market, since some 85% of the F production is converted into FA. The reasons for this are explained in the next section devoted to furan polymers.

FA

The points raised in the introduction concerning the unique qualitative aspects related to furan chemistry at the service of macromolecular science should now be clear. The condensed overview presented above shows that a choice of cheap and ubiquitous renewable resources provides straightforward access to two first-gener-ation furan compounds. Thus, using a strategy that simulates the equivalent petro-chemical approach to the present variety of industrial monomers, a panoply of monomers can be prepared.

It is important to emphasize however that, should any of these monomers become a valuable precursor to economically viable macromolecular materials with realistic technological applications, all the structures shown in the above schemes involve synthetic pathways that can be improved in terms of optimization, economics and green character.

9.3   POLYMERS

A brief but systematic summary is given here of the numerous different polymer structures that have been or can be prepared from the furan monomers discussed above. This survey is accompanied by objective/subjective assessments of the potential interest of these macromolecules as real materials, capable of substituting fossil-derived

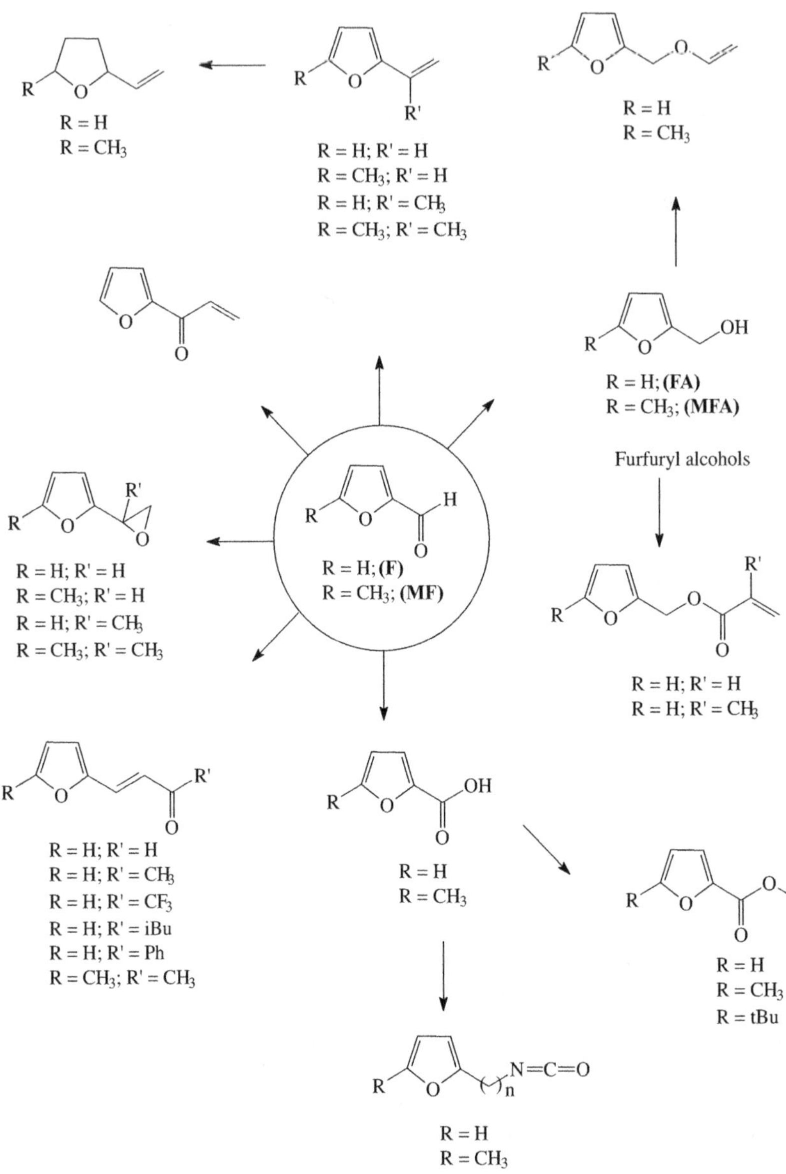

Scheme 9.5

counterparts, or displaying novel properties associated with high-tech applications. A critical appraisal is therefore presented here in which only the most relevant systems are selected and reasons given for their superiority in terms of perspective development, compared with other less-convincing investigation outcomes.

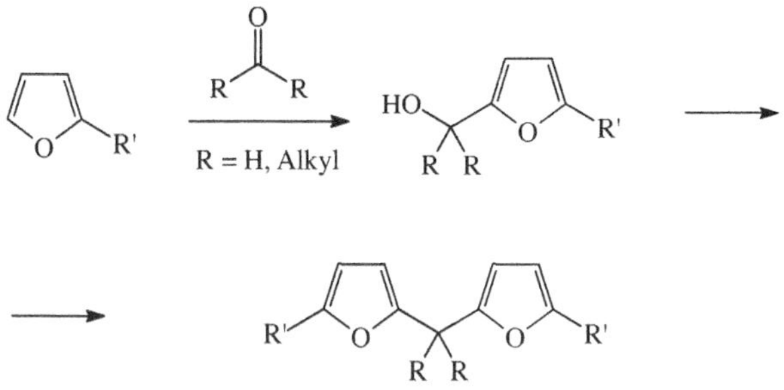

R' = CH$_2$-NH$_2$, COOH, COOAlkyl, etc.

Scheme 9.6

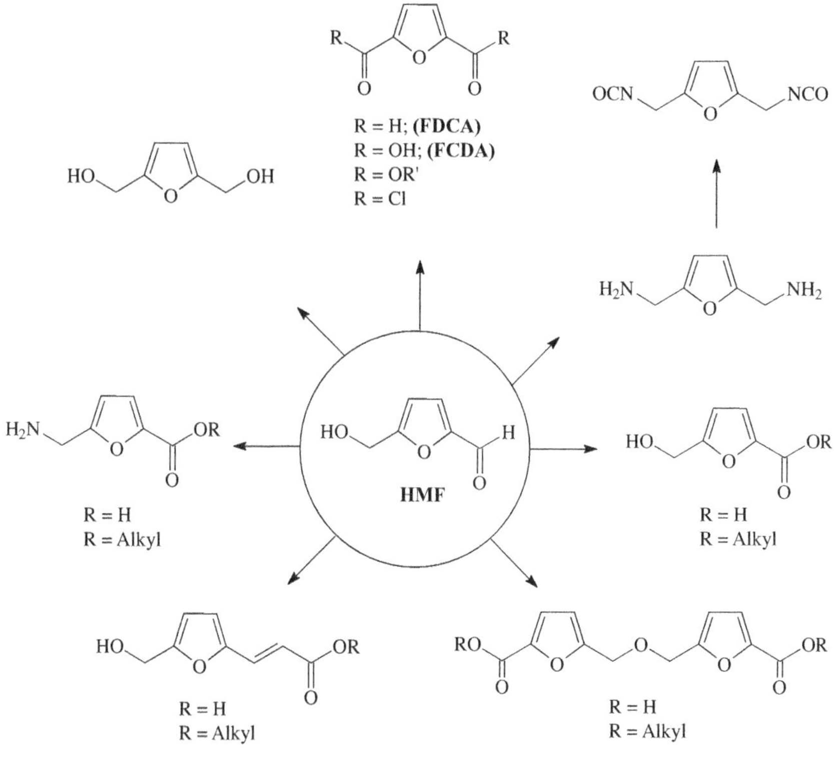

Scheme 9.7

9.3.1   Chain-Growth Systems

9.3.1.1   Free Radical Polymerization

The behavior of furan monomers in radical polymerizations [4c, 4e] can be discussed in terms of three situations: (i) the structure of the monomer strongly favours the formation of a stabilized furyl radical rather than the propagating alkenyl coun-terpart and hence self-inhibition takes place, as with 2-vinylfuroate (see Scheme 9.5), because the ester moiety does not provide any significant delocalization of the unpaired electron; (ii) the reverse situation occurs when stabilization is quite effective at the expense of the formation of the furyl radical, in which case polymerization takes place in a conventional fashion, as with furfuryl acrylate and methacrylate (see Scheme 9.5); (iii) intermediate instances characterized by a balanced competition between the two possible radical structures, resulting in retardation and sometimes limited polymer yields caused by the radical trapping role of the furan rings borne by the accumulating polymer chains, as with 2-vinylfuran (see Scheme 9.5). Copolymerizations of furan monomers with conven-tional comonomers display similar features.

Based on thorough experimental evidence, only those furan monomers that allow the formation of conventional propagating radicals, following both initiation and propagation steps, deserve attention as a source of homopolymers and copolymers. Furfuryl methacrylate has been employed successfully in the synthesis of copolymers in which the pendant furan heterocycles were used for reversible cross-linking through the DA reaction, as discussed below. An additional positive feature is that furfuryl acrylates and methacrylates are readily obtained from FA by esterifi-cation or transesterification, as sketched in Scheme 9.5. Interestingly, these studies were also useful in pointing out that the radical trap efficiency of some furan structures could be turned into a profitable application precisely by looking into a way of amplifying this role, rather than reducing it, and thereby developing very powerful radical inhibitors [4c, 4e]. Increasing the degree of conjugation between the hetero-cycle and the moiety appended to it at C2 achieved this objective, because molecules such as 4,

$$R = \overset{\text{O}}{\overset{\|}{\text{C}}}-R' \text{ or } C \equiv N$$

4

readily synthesized from F, were shown to play a very efficient role as inhibitors in polymerizations, oxidations, photolyzes, and other systems governed by free radicals. This was because the power of **4** in trapping free radical intermediates exceeds that of even such strong classical inhibitors as hydroquinone, a feature arising from the very high delocalization of the unpaired electron shown in the thoroughly stabilized radical structure **5**, in which X represents the trapped radical moiety.

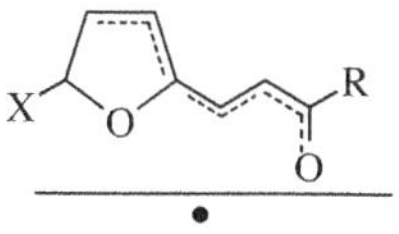

5

The valorization of simple furan molecules in the important role of inhibiting free radical chain mechanisms was also applied successfully to the stabilization of poly (propylene) against degradation during its melt-processing maleation in the presence of peroxides, by using furanacrylic moieties [5].

9.3.1.2 Anionic Polymerization

Among the numerous furan monomers tested for their aptitude to respond to anionic initiation, 2-furyl oxirane (see Scheme 9.5) is undoubtedly the most interesting [4c, 4e, 6]. Conventional initiators such as tBuOK and Al(iPro)$_3$ are efficient initiators, the latter providing a regiospecific propagation through the α-opening of the epoxide. The ensuing polyether 6 has a T_g close to room temperature.

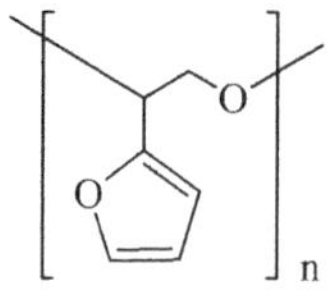

6

The most salient feature associated with 2-furyl oxirane is its extreme reactivity to nucleophilic initiators, extending to amines and alcohols which are notoriously incapable of activating aliphatic and aromatic oxiranes. This peculiarity can be turned into a very useful tool for the straightforward synthesis of block and star copolymers (e.g., from linear and branched macropolyols) as well as graft macro-molecules (e.g., from polyvinyl alcohol or cellulose) without the use of a catalyst. The furan polyether blocks and grafts were however found to be limited in DP to about 10 by transfer reactions, although this aspect was not optimized and it seems therefore likely that longer chains could be obtained.

The unique reactivity of 2-furyl oxirane in simple anionic systems, coupled with its ready availability from a one-step synthesis using F [6], make this furan monomer particularly promising for construction of designed polymer architectures. Further possible modifications through reactions specific to the furan chemistry provide an additional advantage.

9.3.1.3 Cationic Polymerization

As expected, of all the furfural-derived monomers shown in Scheme 9.5, 2-vinyl ethers and 2-alkenyl derivatives display the most pronounced reactivity toward cationic initiation [4]. These polymerizations however are accompanied by two important side reactions, well known in cationic systems, but much more pronounced with these monomers because of the dienic character of the furan heterocycle [4c, 4e]. The first is associated with the already mentioned tendency of the furan ring to undergo electrophilic substitutions as shown in Scheme 9.2, which competes here with regular propagation and gives rise to branched polymers. This inconvenience can be readily eliminated by using monomers bearing a substituent (e.g., a methyl group at C5). The second mechanistic interference has to do with the lability of the hydrogen atom borne by the tertiary carbon of the formed polymer, which is easily abstracted as a hydride ion by a cationic active species. This reaction leaves a positive charge on the polymer, with the subsequent loss of a proton from the neighboring CH_2, thus generating an unsaturation, which in turn facilitates the loss of a second hydride ion from the next monomer unit, and so on. This self-catalytic cyclic mechanism is similar to that depicted below in Scheme 9.10 for the formation of conjugated sequences during the polycondensation of FA. This sequence of events is readily eliminated by working with isopropenyl monomers (i.e., replacing the mobile hydrogen atom by a methyl group).

In the specific instance of 2-vinyl furan, the polymer fragment 7 illustrates the result of alkylation reactions, whereas Scheme 9.8 shows a hydride-ion abstraction from an unsaturated end group, leaving a highly delocalized, and hence stabilized, cationic moiety.

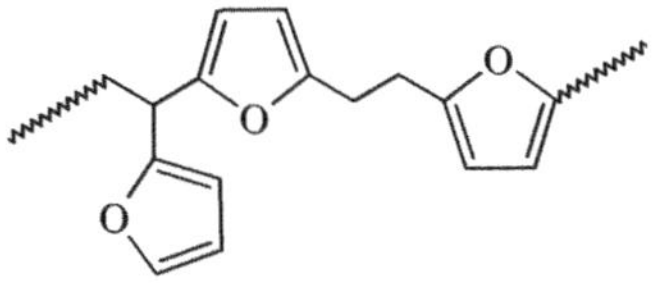

7

The double methyl substitution on this monomer resulted in a clean-cut propagation for the cationic polymerization of 5-methyl-2-isopropenylfuran, leading to the expected linear and saturated structure 8, bearing high DPs when prepared at low temperature [4c,4e].

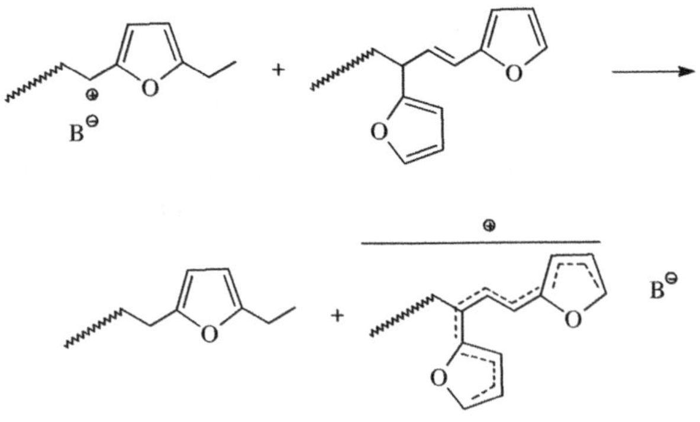

Scheme 9.8

8

The two side reactions can however be exploited to prepare materials with interesting structures and properties. In the case of the electrophilic substitution, the use of a furan compound (e.g., 2-methylfuran) as transfer agent in the cationic polymerization of an aliphatic or aromatic monomer, provides a means to prepare oligomers bearing a furan end group, which can in turn be used as a reactive site for further modifications such as the DA reaction [4c, 7]. In another vein, the use of a difuran derivative bearing free C5 positions such as 2,2-difurylpropane, allows the synthesis of block copolymers by the sequential cationic polymerization of two monomers, each propagating species statistically alkylating one of the free C5 positions of the difuran transfer agent [4c, 7]. These mechanistic tricks have been used in several studies following their original inception [8, 9]. Of course, grafting-onto reactions can also be applied using this concept, for example by conducting the cationic polymerization of a given monomer in the presence of a copolymer incorporating pendant furan moieties which will act as branching sites through their free C5 positions.

The positive exploitation of the H^-/H^+ loss cycle leading to polyunsaturations consists in pushing the mechanism as far as possible, rather than minimizing its impact, to generate very highly conjugated materials. When 2-vinylfuran was polymerized in bulk with triflic acid, the ensuing material was a black cross-linked powder possessing an extremely high proton affinity, which could be put to good use in such operations as heterogeneous acid removal from solvents, reaction media and effluents [10]. These solid proton sponges are easily regenerated after use by suspension in aqueous alkaline solutions.

In conclusion, as in the case of radical polymerization, some furan monomers give interesting materials by cationic polymerization, whereas the drawbacks encountered with others can be turned into useful issues, thanks to the peculiar chemical behavior of the furan heterocycle.

9.3.2   Step-Growth Systems

9.3.2.1   The Polycondensation of Furfuryl Alcohol

The acid-catalyzed polymerization of FA is paradoxically the most widely exploited reaction for preparing furan macromolecular materials with different well-established applications, while remaining for decades a very little understood system in terms of its chemical subtleties [4, 11].

When FA is treated with a Lewis or a Brønsted acid, no matter in what specific conditions of acid strength, homogeneous or heterogeneous medium, concentrations and temperature, the final product, which might take seconds or days to form, is always a black cross-linked solid material. When carried out *in situ*, this polymerization has been applied very successfully to metal casting cores and moulds, corrosion-resistant coatings, polymer concretes, wood adhesives and binders, sand consolidation and well plugging, low flammability and low smoke-release materials and carbonaceous products including graphitic electrodes and carbon micro-particles, without mentioning the more recent important applications discussed below. This state of affairs explains why FA has for decades been by far the most important industrial furan commodity, considering moreover that for many of the applications described above the FA 'resins' are irreplaceable. Although a lack of understanding of the chemical mechanisms underlying this polymerization did not hinder those practical developments, their unravelling has played a key role for the more technical and sophisticated materials that have been investigated and perfected in the last decade.

If FA were a well-behaved beast, it would *only* polymerize by successive acid-catalyzed condensation steps involving the hydroxyl function of one of its molecules and the hydrogen atom at the C5 position of another, with a minor contribution from the alternative condensation involving two OH groups, which however gives the same net result after the loss of a formaldehyde molecule, as shown in Scheme 9.9 for the first step.

This polycondensation does indeed take place [11] but, as soon as oligomers are formed, a first important side reaction starts operating and begins to seriously alter the simple –CH₂–Fu- linear sequence by abstracting hydride ions and protons from two successive methylene moieties. These cycles, already discussed in the case of the cationic polymerization of 2-vinylfuran and furfuryl vinyl ethers, generate conjugated sequences along the oligomer chains, consisting of alternating furan and exo-unsaturated 2,5-dihydrofuran motifs, as illustrated in Scheme 9.10.

The net result of this repeating mechanism is the formation of coloured products, which go from light brown to black as a function of the length of the unsaturated sequences along the polymer chains [11].

If this disturbance was not enough to upset the regular polymer structure, the progressive construction of the conjugated moieties creates a chemical environment favourable to interchain couplings through DA reactions between a furan ring (diene) and an exo-unsaturated 2,5-dihydrofuran moiety (dienophiles), as sketched in Scheme 9.11.

The system thus develops its macromolecular architecture through multiple chain couplings which yield a cross-linked material already at its *early* polymerization

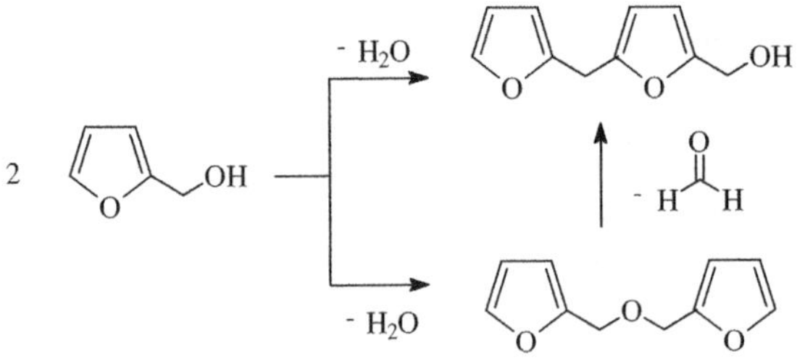

Scheme 9.9

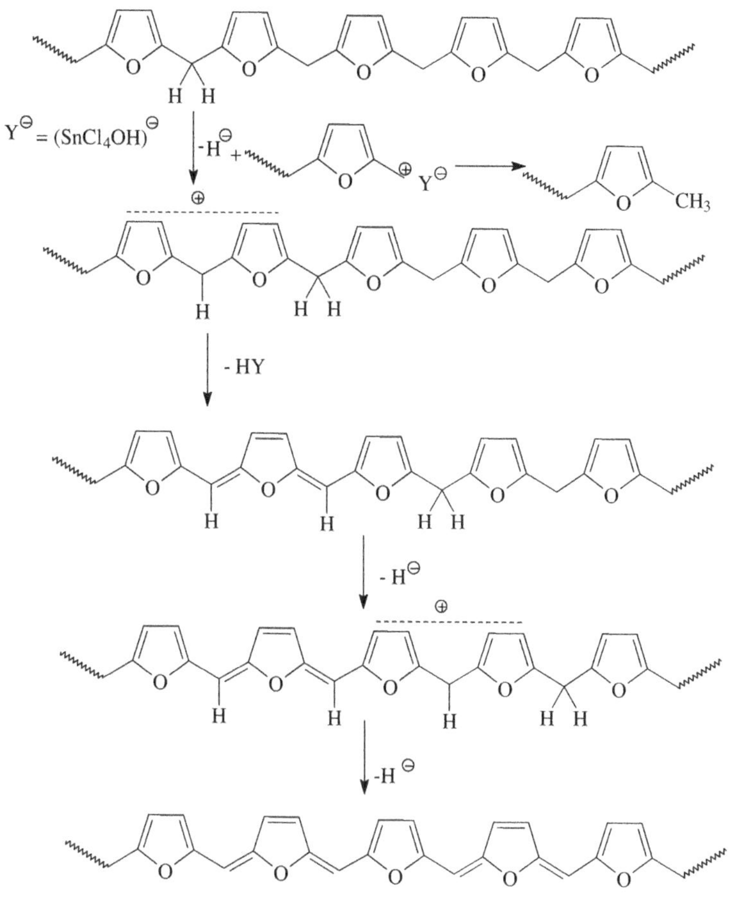

Scheme 9.10

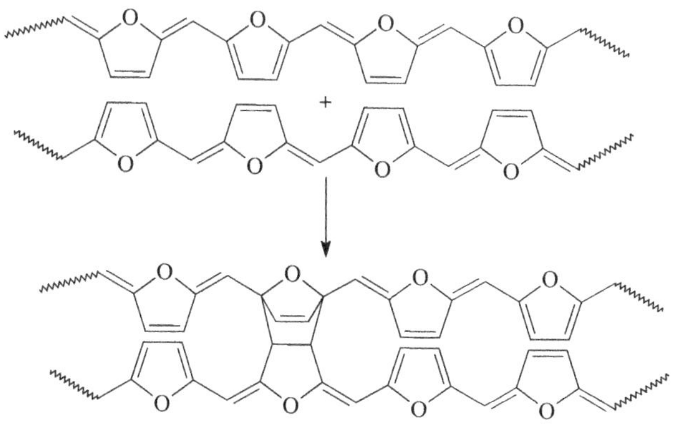

Scheme 9.11

stages. This peculiarity with respect to conventional non-linear polycondensations, stems from the fact that the unsaturations which appear prematurely on the oligomers possess several complementary sites for the formation of the DA adducts and hence the rapid buildup of the network.

In conclusion, what should be a linear colourless thermoplastic polymer bearing the straightforward structure **9**, turns out in fact to be a black cross-linked material because of the extreme reactivity of the methylene groups bridging the furan rings. Despite the understanding of the mechanisms involved in these dramatic structural alterations [11], further research has not provided the means to control or inhibit the two side reactions and the potentially very interesting polymer **9** remains a chimera.

9

Notwithstanding this situation, FA continues to stimulate both fundamental and technologically valuable research [4*f*]. On the one hand, spectroscopic and chemo-rheological studies of its oligomerization under controlled or restricted conditions have provided useful information about the structure of intermediate species [12], confirming the findings and arguments of the pioneering mechanistic study [11]. On the other hand, a variety of novel materials based on polyFA have been prepared and characterized, including carbonaceous foams, nanocomposites and hybrids of different compositions bearing controlled morphologies [13]. Another area in which FA polymerization is gaining momentum is in the synthesis of organic-inorganic hybrids [14], including nanoscopic morphologies [15] and bio-based nanomaterials [16]. An original aspect associated with some of these systems is the formation of furfuryl-alkoxide intermediates (e.g., siloxanes incorporating one or several furfuryl moieties) through alkoxide exchanges, followed by the acid-promoted polymerization of the furfuryl moieties and/or the sol–gel mechanism applied to the hydrolyzed siloxanes, as shown in two approaches given in Scheme 9.12. The interest in these materials resides primarily in their nanomorphology associating the furan resin and the silica particle, rather than in the resin structure, which is always the same as that discussed above.

A further promising field of application for FA, which has recently been revived with a good measure of success, is wood preservation and modification through impregnation with FA and its subsequent *in situ* polymerization promoted by acidic catalysts [17]. The process has been optimized to provide remarkable improvements in such properties as dimensional stability, mechanical and chemical strengthening, excellent resistance to microbial decay and insect attack, as well as ecological soundness, which have led to its commercialization [17]. Little work on the chemistry associated with this impregnation has accompanied the thorough technical development as yet. Recent attempts to unravel the basic query of whether the polymerizing FA structures react with any of the wood components has thus far been limited to the study of lignin model compounds [18], but even such a specific investigation only provided unconvincing conclusions. A recent contribution to this topic dealing with

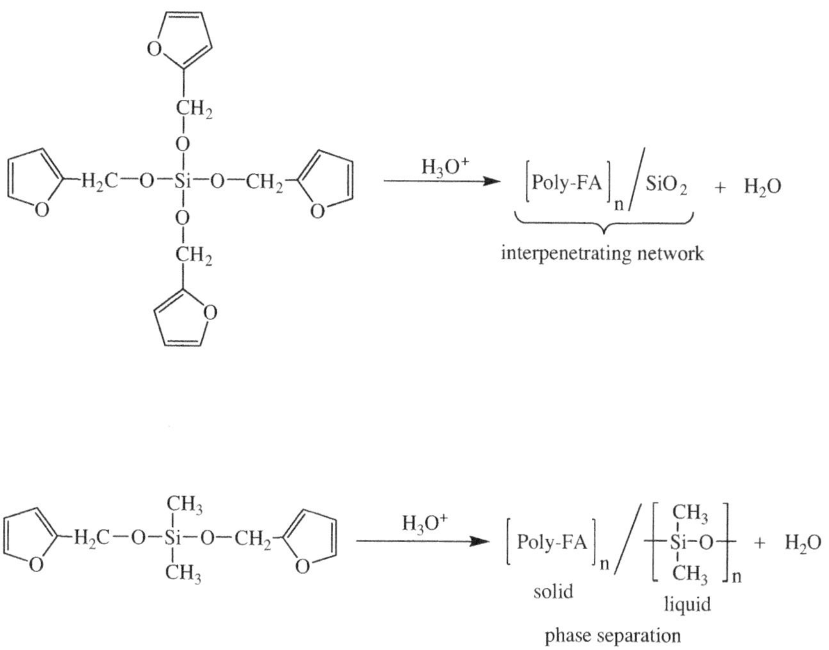

Scheme 9.12

the fluorescence of furfurylated wood [19] provided some insight into the distribution of the polymer within the wood morphology, but again with no new chemical evidence. More fundamental work would certainly be welcomed here.

It is encouraging to give a positive conclusion to this section, by emphasizing that on the whole the potential of FA to generate original and viable materials has not ceased to widen and that many academic and industrial groups are involved in this challenge.

9.3.2.2   The Polycondensation of 5-Methylfurfural

Conjugated polymers are a relatively young family of materials which find numerous applications in high-tech domains such as optoelectronics and which now cover a large array of structures incorporating aliphatic, aromatic and heterocyclic monomer units, often in original combinations. The presence of furan rings in these macromolecules has attracted some attention [4c–4e], but only one structure, poly(2,5-furylene vinylene) (PFV), stands out as promising in terms of practical development given its interesting properties and facile synthesis from a current chemical precursor.

As mentioned earlier, 5MF is a secondary product in the fabrication of F and therefore also a cheap industrial commodity. A comprehensive study of its aldol-crotonic-like polycondensation in the 1990s [4c–4e, 20] led to the preparation of well-defined individual oligomers and polymers bearing the 2,5-furylene-vinylene units and an aldehyde function as one of the end groups (structure 10).

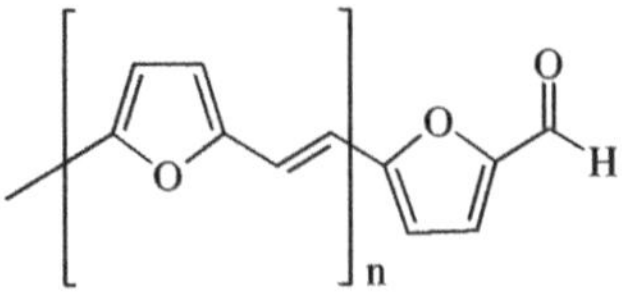

10

Various applications were envisaged for these structures, including, among others: (i) mixed electronic and ionic conductance after coupling the PFV–aldehyde block with a poly(ethylene oxide)–NH$_2$ block through the formation of a Schiff base [4c–e, 20, 21]; (ii) displaying devices based on the photoluminescence and electroluminescence of a mixture of oligomers to cover the whole of the visible spectrum [4c–e, 20]; and (iii) thermoreversible photochemical coupling and cross-linking of natural and synthetic polymers bearing pendant photoactive dimeric structures such as **10** with $n = 1$ [4c–e, 20, 22]. This comprehensive study suggested that polymers and oligomers based on structure **10**, with DPs from that of the dimer to values exceeding 100, readily prepared from 5MF, offered numerous interesting potential applications.

9.3.2.3 Polyamides

The synthesis and characterization of the high-DP furan-aromatic polyamide **11** [4c, 23] simulating the structure and properties of entirely aromatic counterparts such as Kevlar, is a striking example of the viability of furan polymers as materials capable of replacing fossil-based counterparts, here in the specific case of a high-tech product. The expected availability of FDCA as a chemical commodity in the very near future, resulting from the enormous interest in developing an efficient technology for the production of HMF [4f], should make it possible to produce this polyamide at a competitive price.

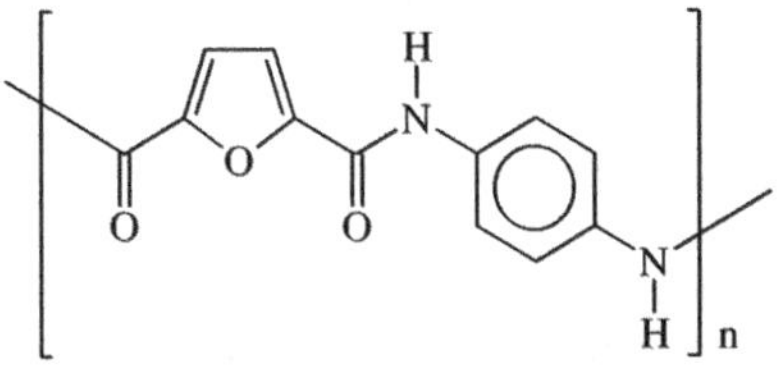

11

A large selection of polyamides based on the difuran-type diacid chloride shown in Scheme 9.6 and a large variety of diamines was synthesized by different procedures and thoroughly characterized [4c–d, 24]. Scheme 9.13 shows the mechanism, applied here to the phase-transfer technique, together with the diamine structures.

The changes in such polymer properties as extent of crystallization, melting temperature, T_g, thermal stability and surface energy as a function of the diamine

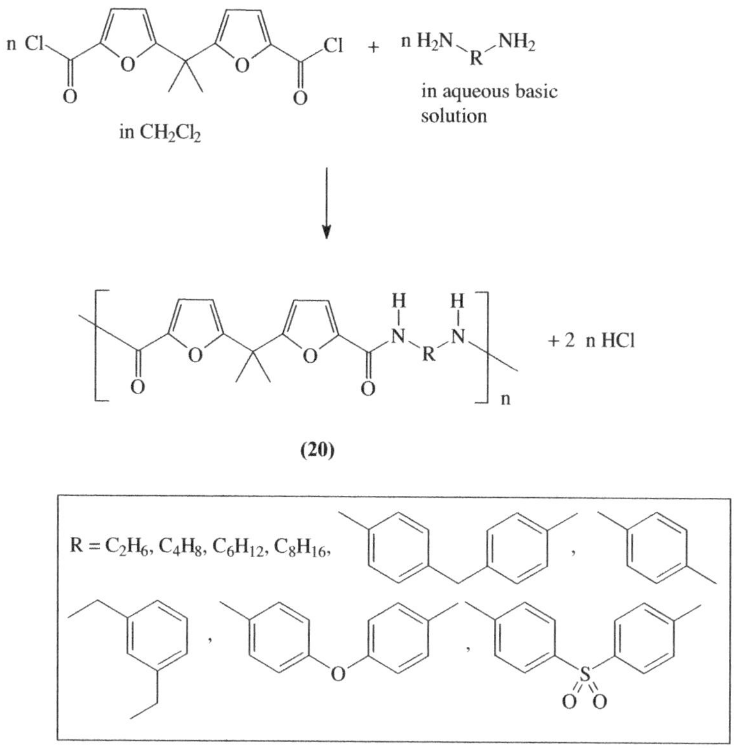

Scheme 9.13

structure allowed a systematic structure–property relationship to be drawn and hence for the possible applications of these polyamides to be assessed.

It is important to point out that furan amines in which the NH_2 group is directly bound to the heterocycle are thermodynamically unstable and tautomerize to give the corresponding dihydrofuran imines. This explains why furan diamines cannot be envisaged as comonomers in the synthesis of polyamides. Furan diamines bearing a methylene group between the ring and the NH_2, such as those shown in Schemes 9.6 and 9.7, are of course thermodynamically stable and were used in the investigations mentioned above [23, 24], but the ensuing polyamide displayed poor thermal stability, associated with the fragility of the methylene groups bridging the furan and amide moieties.

9.3.2.4 Polyesters

The first important contributions to furan polyesters were published by Moore's group some thirty years ago [25]. Interest in these polymers was revived at the end of the last century and is now in full swing. Two families of furan polyesters are relevant here, namely: (i) those in which the diacid unit incorporates two bridged heterocycles (Schemes 9.6 and 9.13); and (ii) those which bear instead the FDCA-derived moiety.

As with the polyamide investigation mentioned above, a comprehensive study of polyesters belonging to the first family was undertaken using different synthetic approaches and a multitude of diols, as well as a variety of diacids in terms of the nature of the bridging moiety separating the two furan rings [4c–d, 26]. The generic polyester structure 12 illustrates the basic backbone in which R was varied through the use of a rich spectrum of diols. In this case, R' and R" represent hydrogen atoms, aliphatic, perfluoroaliphatic, or aromatic groups, appearing in symmetrical (R' = R") or asymmetrical (R' ≠ R") forms. All these polymers were fully characterized in terms of structure, molecular weight and physical properties.

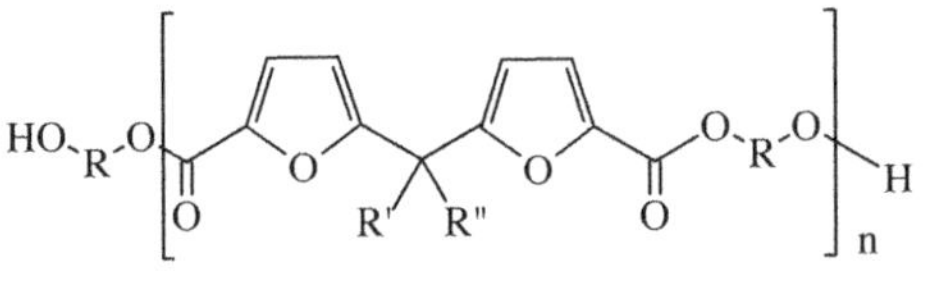

12

Once again, therefore, a rigorous set of structure–property relationships was produced by this investigation with the associated conclusions about the potential applications of the most promising polyesters.

As for the second family of polyesters, the working project, still in progress, started with the member curiously neglected in previous studies, namely poly(2,5-ethylene furandicarboxylate) (PEF), which is the heterocycle homologue of the most important commercial polyester, poly(ethylene terephthalate) (PET). The best results were obtained by the transesterification process illustrated in Scheme 9.14, which produced a regular structure and a DPn of about 250 [27, 28].

This polymer displayed a high crystallinity, a melting temperature of 215 °C, a T_g of 80 °C and was thermally stable up to 300 °C. These properties are very similar to those of PET and were further corroborated by the quality of films cast from solution.

Clearly, this is yet another excellent example of a material which could become a serious alternative to a major polymer prepared from fossil resources, once FDCA is on the market. Interestingly, at that point, PEF will be considered a polyester entirely derived from renewable resources, since ethylene glycol can be prepared from glycerol, the major and very cheap by-product of the biodiesel industry.

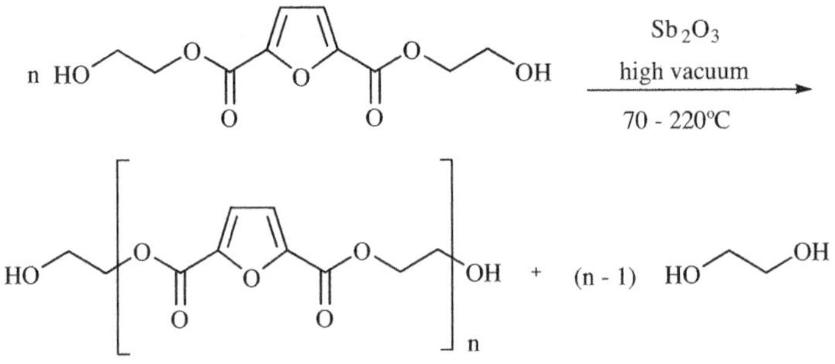

Scheme 9.14

Among the other polyesters prepared from FDCA-based monomers, interesting materials include the structure involving the sugar-based diol isosorbide [28] (**14**) and that involving bis(hydroxymethyl) furan (i.e., the fully furanic polymer **15**).

14

15

An interesting addition to these two polyester families is an unsaturated counterpart arising from the transpolycondensation of monomer **16** [29], readily synthesized from HMF.

16

In fact, the material that provided a viable application resulted from the copolymerization of **16** with an aliphatic homologue, which produced the random structure **17**, in which the presence of the furan units only amounted to about 10% [29].

17

The conjugated furan monomer units were responsible for photoreactivity which allowed the rapid cross-linking of the polymer to be achieved under standard UV irradiation. This material therefore proved highly suited for applications such as negative resists for printing plates and electronic circuit boards.

9.3.2.5   Polyurethanes

Whereas furan polyamides and polyesters have been the objects of investigations, albeit sometimes of poor quality, for many decades, their first polyurethane homologues were reported in the 1980s [30] and a thorough study of their synthesis and properties only took place ten years later [4c, 31]. This very systematic approach called upon: (i) the synthesis of new model compounds and furan diisocyanate and furan diol monomers; (ii) a kinetic appraisal of the corresponding urethane formation, combined with the search for possible side reactions; (iii) the synthesis of a large spectrum of polymers; and (iv) their exhaustive characterization.

The most important findings of this work were, on the one hand, that it is possible to prepare high-DP furan polyurethanes incorporating the heterocycle in all possible modes without any complications and, on the other, that furan isocyanates are very highly reactive toward alcohols, to the point that 2,5-furyl diisocyanate could not be handled reasonably for these macromolecular syntheses because of its extreme reactivity. Even furfuryl isocyanate was found to be more reactive than its phenyl counterpart, despite the fact that a methylene group had been inserted between the heterocycle and the NCO moiety.

A number of representative polymer structures are shown below (18–21), covering some of the most relevant monomer combinations and furan ring number and positioning with respect to the linear backbone.

18

19

20

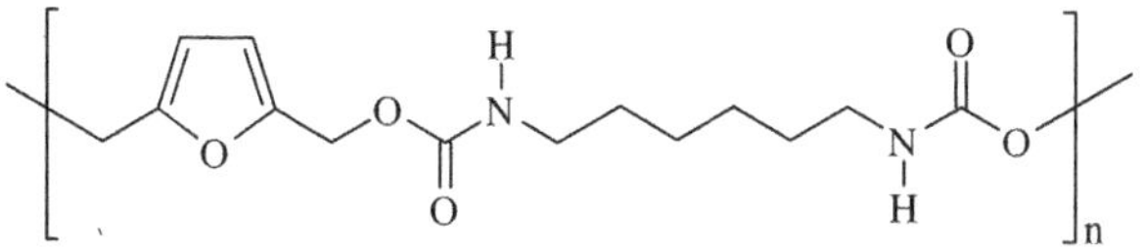

21

Given their regular structure and high molecular weights, most of these polyurethanes readily crystallized and, given the large selection of monomers used, a very abundant choice of linear macromolecular structures was available at the end of the study to provide systematic correlations with thermal, mechanical and surface properties, among others.

Thermoplastic elastomers such as **22** were also prepared and their dynamic mechanical properties assessed.

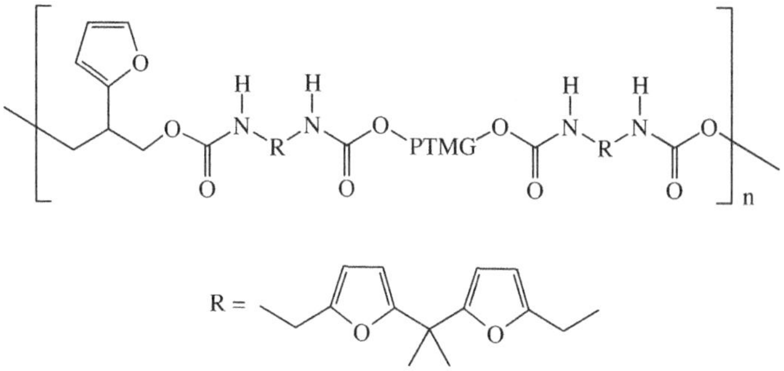

PTMG = Polytetramethylene Glycol

22

The obvious extension from all these linear thermoplastic materials to cross-linked homologues formed with polyfunctional monomers was also tackled, and rigid foams prepared. These foams displayed thermal conductivities comparable with those of commercial fossil-based counterparts and better dimensional stability after accelerated ageing. An important additional advantage was a flame self-extinguishing property when the structures bore a high percentage of furan motifs [32], such as the foam prepared using a five-furan macrodiol formed by the spontaneous reaction of 2-furyl oxirane with bis(hydroxymethyl)furan in a molar proportion of 4:1 [6d] and an aromatic diisocyanate.

9.3.2.6 Miscellaneous Polymers

Other recent additions to the field of furan polycondensates include polyhydrazides (Scheme 9.15) and the corresponding poly-1, 3, 4-oxadiazoles **23** [33], polyamide-imides **24** [34], polyureas **25** and the corresponding polyparabanic acids **26** [35].

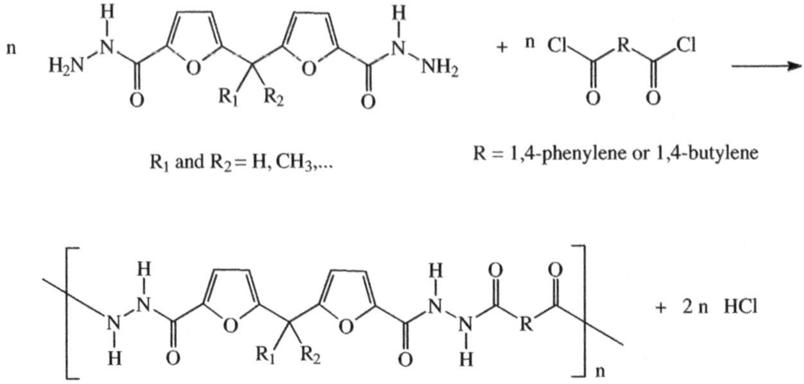

R_1 and $R_2 = H, CH_3,...$

$R = 1,4$-phenylene or $1,4$-butylene

$+ 2n$ HCl

Scheme 9.15

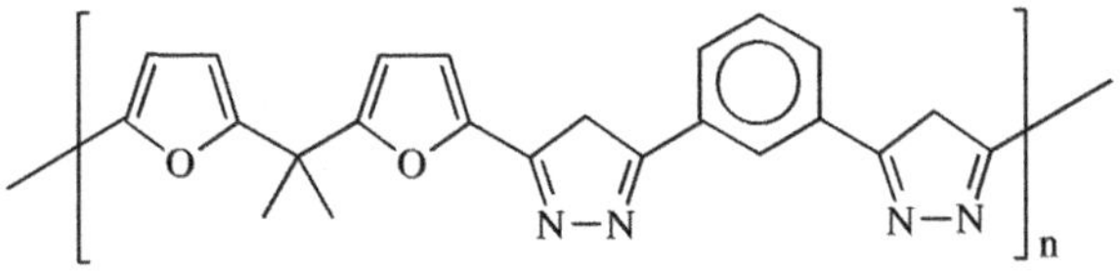

23

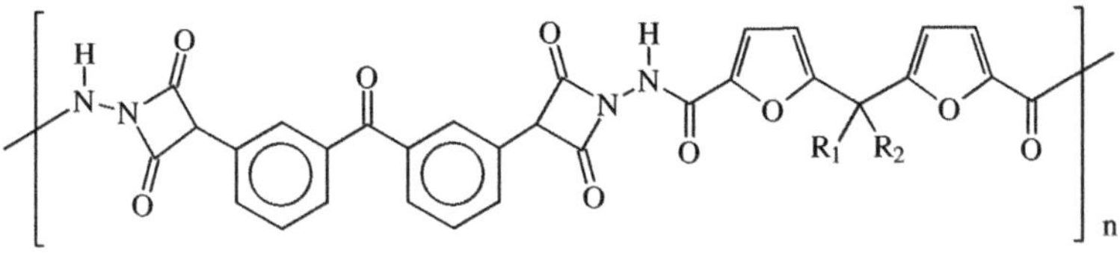

$R_1/R_2 = CH_3/CH_3; CH_3/C_6H_5; CH_3/C_2H_5; CH_3/C_5H_{11}; CH_3/CF_3$

24

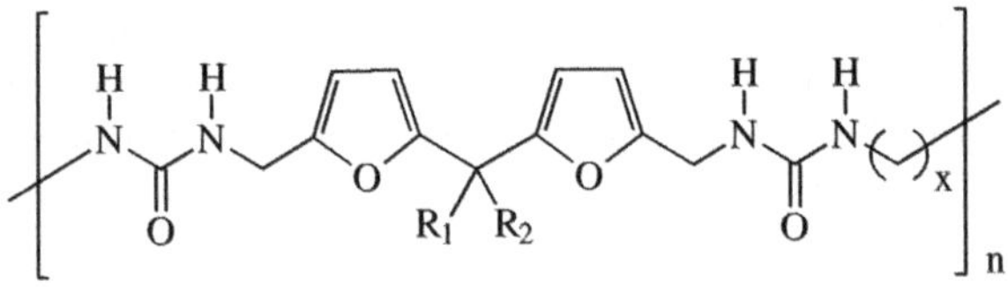

$R_1/R_2 = CH_3/CH_3; C_2H_5/C_2H_5; CH_3/H; CH_{13}/H$

$x = 4, 6$ and 8

25

$$R_1/R_2 = CH_3/CH_3;C_2H_5/C_2H_5;CH_3/H;CH_{13}/H$$
$$x = 4, 6 \text{ and } 8$$

26

All these investigations were carried out with the main aim of increasing the thermal stability of the ensuing materials, which was indeed attained, but sometimes at the expense of a loss of solubility.

9.3.3   The Application of the Diels–Alder Reaction to Furan Polymers

The purpose of this section is to review succinctly the rapidly increasing interest in the application of the DA reaction in polymer chemistry [36], limiting its scope here to the coupling of furans with maleimides. As already mentioned in the introduction, of the common five-membered ring unsaturated heterocycles, furan displays the most pronounced dienic character, when compared with thiophene and pyrrole [4], which makes it highly suitable to intervene as a diene in that beautiful example of click chemistry known as the Diels–Alder reaction. In addition to its clean-cut character as regards the usual absence of side events perturbing its course, the DA reaction is also remarkable for its temperature sensitivity.

The most important recent contribution to the realm of furan polymers is arguably the lively interest in the use of the DA reaction to synthesize a wide variety of macromolecular architectures possessing, among other original properties, the common and peculiar feature of thermal reversibility within a readily accessible domain of temperatures [4, 36].

Scheme 9.3 portrays that reaction showing the *endo* and *exo* isomeric adducts, whose proportion varies as a function of the specific structure of the reagents and the system conditions. In the present context, this feature does not play any significant role, since the object of the exercise is just to produce adducts in any configuration and exploit their thermal reversion. In other words, the relevant mechanism can now be rewritten as in Scheme 9.16.

The key parameters related to the thermal reversibility of this polymerization/ depolymerization equilibrium are of course the most adequate temperatures to be applied to optimize the forward and the reverse reactions (i.e., how to achieve, on the one hand, the highest yield, DP in the case of a polymerization, and, on the other hand, the complete return to the monomers or macromonomers without the intervention of

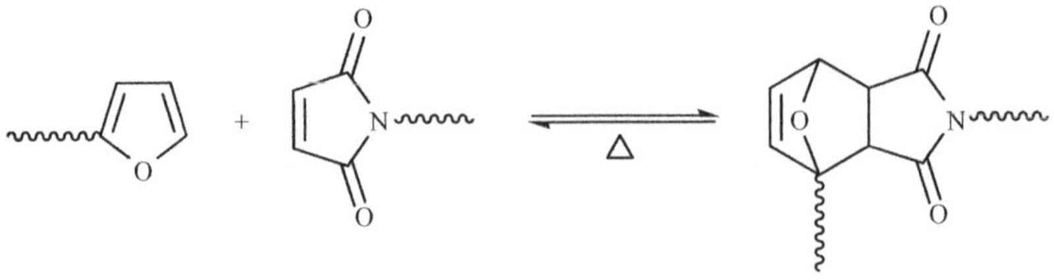

Scheme 9.16

a thermally induced irreversible degradation). For the forward reaction, $\sim 65\,°C$ is an appropriate temperature, because it provides a reasonably high rate of adduct formation, while still avoiding any detectable incidence of the retro-DA reaction. For the backward reaction, $\sim 110\,°C$ is adequate to ensure a major shift of the equilibrium to the left, while avoiding unwanted side reactions. Experience has shown that in these conditions the reproducibility of forward/backward cycles is indeed ensured.

9.3.3.1 Linear Polymerizations

The combination of difuran and bismaleimide monomers to promote the formation of the corresponding linear polyadducts was the first application of the DA reaction to macromolecular chemistry [36] and in the last few years the number of studies has increased vigorously, although with a certain repetitiveness and lack of depth, since the scope tends to be limited to qualitative polymerization–depolymerization features [37].

The first thorough investigation of this type of system called upon a very recent study of the reaction kinetics, first with model compounds, then with a variety of A–A/ B–B monomer combinations, using UV and ^{1}H-NMR spectroscopy to follow both the forward (DA polymerization at 65 °C) and the backward (retro-DA depolymerization at 110 °C) reactions [28a, 38]. This approach provided additional clear-cut evidence of the intermediate structures involved and of the absence of significant side reactions, thanks to the care taken in purifying the monomers. Scheme 9.17 depicts the monomers employed in one of these polymerizations.

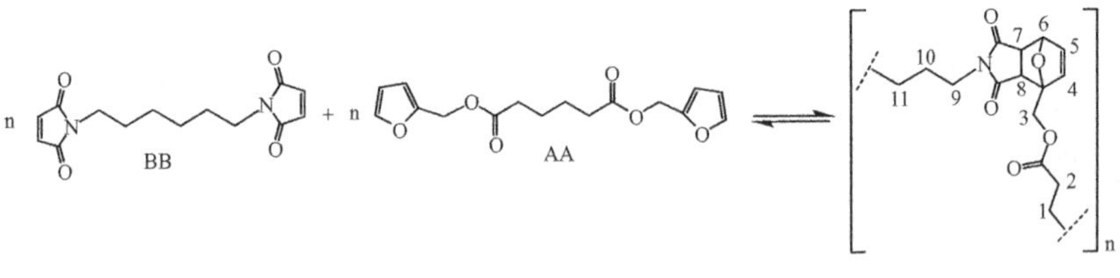

Scheme 9.17

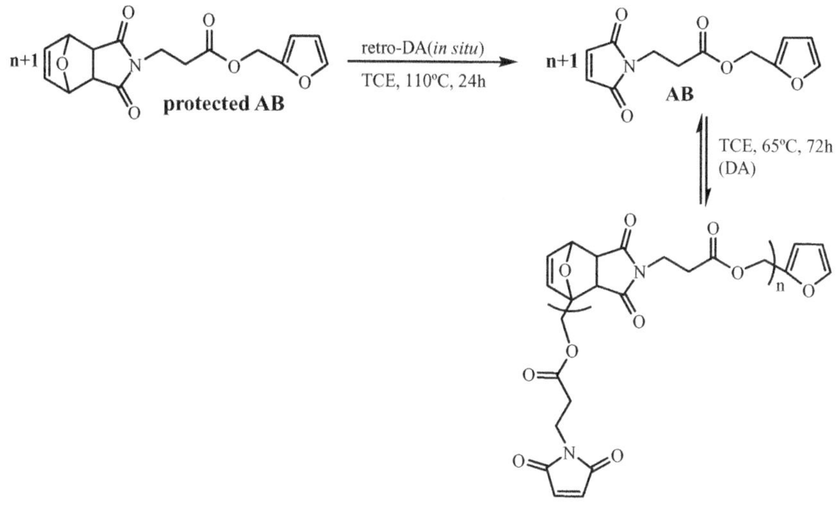

Scheme 9.18

In all these systems, the classical problem of ensuring the exact monomer stoichiometry inevitably cropped up, as suggested by the sometimes incomplete moiety consumption at the end of the polymerizations, and hence molecular weights that were not optimized. In order to circumvent this difficulty, a different strategy was applied, based on the use of an A–B monomer, namely a molecule incorporating both a furan and a maleimide moiety. However, for obvious reasons, only monomers of this type where one of the reactive groups is kept in a masked configuration, can be prepared and characterized before activation through appropriate deprotection. Scheme 9.18 illustrates this approach through a specific example [28a, 39].

This approach seems promising and has the advantage of the added feature related to the synthesis of these monomers, which calls upon the use of other renewable resources in the form of aminoacids, apart from the furan reagent.

9.3.3.2 Non-linear Polymerizations

Monomer combinations comprising one or more reagents with functionality higher than two have been exploited in the context of the fabrication of mendable cross-linked materials [36]. A recent incursion into these systems, which is still in progress, purports to study them in a more quantitative fashion in terms of both kinetics and mechanisms, as well as in view of exploiting the properties of the ensuing materials [28a, 40].

The systems under investigation are typically of the type A3 + B–B and B3 + A–A, where the difunctional monomers are those previously utilized in the study of linear polymerizations [38] and the trifunctional counterparts have structures like 27 and 28 for a furan and a maleimide exponent respectively.

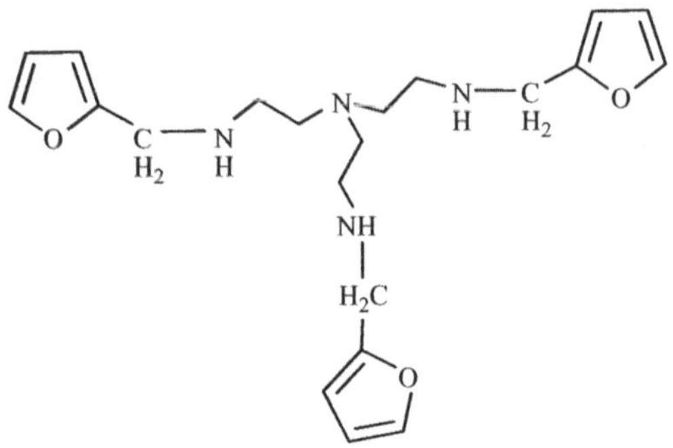

27

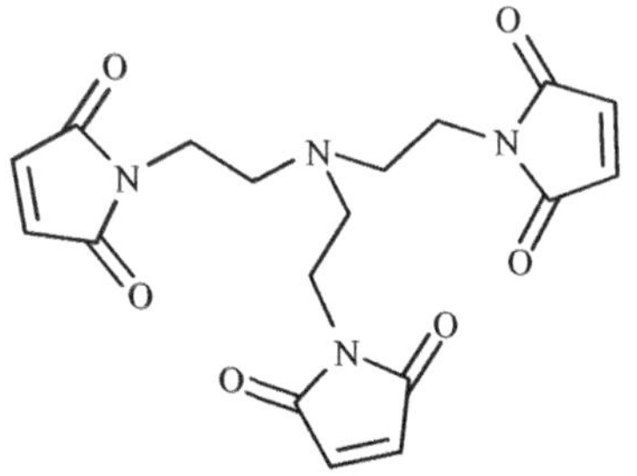

28

An example of such a non-linear polymerization is given in Scheme 9.19.

The use of UV and ^{1}H-NMR spectroscopy again proved valuable here to follow the polymerization up to the gel point, with the rest of the reaction being followed by solvent-extracting the sol fraction as the polycondensation proceeded, albeit in a diffusion-limited fashion. The de-cross-linking occurred readily at 110 °C and thereafter the retro-DA reaction continued all the way to the regeneration of the monomers. These cycles could be repeated several times.

Mendable polymers based on epoxy [41] and poly(ethylene adipate) [42] structures bearing polyfunctional furan and maleimide moieties were recently reported, but these contributions did not add any substantial novelty to this topic [36].

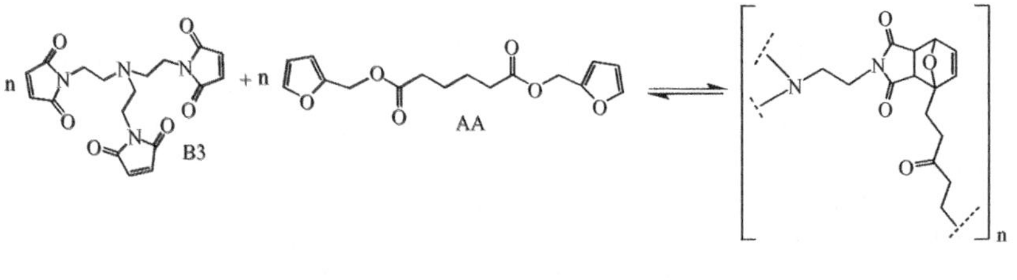

Scheme 9.19

9.3.3.3 Reversible Polymer Cross-Linking

The idea of thermoreversible networks based on the application of the DA reaction to linear polymers bearing furan or maleimide moieties was put forward in the early 1990s and investigated more systematically in the first years of the new millennium [36].

Since then, a number of additional studies have been published on the reversible coupling of furan and maleimide heterocycles, aimed at preparing thermally reversible networks. The strategies vary somewhat, but the overall scenario is essentially the same (i.e., to build a cross-linked material, which can be readily reversed to the starting monomers or polymers). In other words, the original ideas developed for this general purpose [4c–f, 36] are maintained, and only specific issues are in fact altered. Thus: (i) copolymers bearing furfuryl methacrylate units were cross-linked with a bis-maleimide [43]; (ii) shape-memory materials were prepared thanks to this reversible DA reaction [44]; and (iii) thermally reversible cross-linked polyamide [45], epoxy [46–48], hydrogels [49] and bio-based polymers [50], including self-healing structures [51], were described. As already pointed out in the context of DA polymerization reports previous to the contribution from the Aveiro group, these studies focus essentially on the syntheses of monomers and polymers and on the retro-DA reaction applied to the ensuing networks, with little or no emphasis on kinetic aspects and material properties. Plaisted and Nemat-Nasser provided a substantial contribution to this field by a quantitative study of the phenomena associated with multiple healing and reversible cross-linking using a tetrafuran/bismaleimide system [52].

An original approach to strippable imaging materials was recently described [53] based on the polymerization of a DA-adduct diacrylate crosslinker, which can be thermally broken down to regenerate linear polymers.

9.3.3.4 Miscellaneous Systems

Other ingenious applications of DA/retro-DA systems to polymer chemistry include: (i) the synthesis of reversible dendritic structures and hyperbranched macromolecules [4c–f, 36, 54]; (ii) a furan/maleimide star-shaped polymer which was reversibly dismembered through the DA/retro-DA reaction [55]; (iii) block dendrimers that were joined/disjoined through the same mechanism [56]; (iv) the study of the retro-DA reaction of the N-phenylmaleimide–FA adduct in different polymer matrices kept in a viscous state in order to assess the role of diffusion limitations on its decoupling [57]; (v) multi-walled carbon nanotubes (MWCNTs) decorated with both furan and maleimide moieties and the subsequent inter-MWCNT DA reactions [58]; and (vi) gold nanoparticles reversibly coupled with conjugated polymers [59].

Finally, in yet another novel application, drug-delivery vehicles constructed through the furan-DA coupling of antibodies to polymer nanoparticles were recently described [60].

9.4   BIODEGRADABILITY OF FURAN POLYMERS

Regrettably, no thorough study has been reported on the susceptibility of polymers incorporating furan heterocycles to biodegradation. The only investigation in this context deals with copolyesters based on different proportions of difuran and aliphatic dicarboxylic moieties and isosorbide units [61]. The authors found that the higher the furan content, the lower the enzymatic and soil-burial degradability (i. e., the aliphatic structures were more prone to be depolymerized than the furan counterparts). This important topic obviously requires a more systematic and comprehensive investigation.

9.5   CONCLUDING REMARKS

Two different situations characterize the use of furans in the construction of macromolecular materials. These are the classical polymerization systems which make use of one or more furan monomers and the strategies that use the peculiar chemical features of the heterocycle to build specific polymer structures. Whereas in the former approach, the actual quantitative participation of the furan reagent is conspicuous, in some of the examples related to the latter, this participation can be very modest (e.g., a furan compound promoting the synthesis of a block copolymer based on non-furan monomers). Notwithstanding this variable extent in the exploitation of a renewable resource, both working hypotheses can provide materials with a future, thanks to their properties and economy. These materials could contribute to the establishment of a novel wide-ranging family of polymers capable of progressively substituting the present panoply of fossil-based counterparts [1]. Hopefully, this chapter has succeeded in conveying this message.

REFERENCES

[1]   M. N. Belgacem, A. Gandini (eds), *Monomers, Polymers and Composites from Renewable Resources*, Elsevier, Amsterdam, 2008. A. Gandini, *Macromolecules*, 2008, **41**, 9491. A. Gandini, *Green Chem.*, 2011, DOI: 10.1039/C0GC00789G.

[2]   For selected reviews, see D. M. X. Donnely, M. J. Meegan, in *Comprehensive Heterocyclic Chemistry* (eds A. R. Katritzky, C. W. Rees) Vol. 4, Pergamon, Oxford, 1984, p. 657.B. A. Keay, P. W. Dibble, in *Comprehensive Heterocyclic Chemistry II* (eds A. R. Katritsky, C. W. Rees, and E. F. W. Scriven) Vol. 2, Elsevier, Oxford, 1997, p. 395. X. L. Hou, Z. Yang, and H. N. C. Wong, *Progr. Heterocycl. Chem.* 2003, **15**, 167.

[3]   For selected reviews and recent contributions, see C. W. Bird, G. W. Cheeseman, in *Comprehensive Heterocyclic Chemistry* (eds A. R. Katritzky, C. W. Rees) Vol. 4, Pergamon, Oxford, 1984, Chapters 1–3. F. M. Dean, M. Sargent, in *Comprehensive Heterocyclic Chemistry* (eds A. R. Katritzky, C. W. Rees) Vol. 4, Pergamon, Oxford, 1984, Chapters 10–11. X. L. Hou, H. Y. Cheung, T. Y. Hon, *et al.*, *Tetrahedron* 1998, **54**, 1955. D. L. Wright, *Progr. Heterocycl. Chem.* 2005, **17**, 1. J. B. Sperry, D. L. Wright, *Current Opinion Drug Discov. Develop.* 2005, **8**, 723. R. C. D. Brown, *Angew. Chem., Int. Ed.* 2005, **44**, 850. S. F. Kirsch, *Org. Biomol. Chem.* 2006, **4**, 2076. H. E. Hoydonckx, W. M. Van Rhijn, W. Van Rhijn, *et al.*, *Ulmann's Encyclopedia of Industrial Chemistry*, Wiley-VCH, Weinheim, 2007, Vol. 15, p. 187. T. J. Donohoe, J. F. Bower, *Proc. Natl. Ass. Sci.* 2010, **107**, 3373. M. J. Krische, *Proc. Natl. Ass. Sci.* 2010, **107**, 3379. X. Tong, Y. Ma, and Y. Li, *Appl. Catal. A* 2010, **385**, 1.

[4] *a.* A. Gandini, *Adv. Polym. Sci.*, 1997, **25**, 47. *b.* A. Gandini, *ACS Symp. Ser.*, 1990, **433**, 195. *c.* M. N. Belgacem, A. Gandini, *Progr. Polym. Sci.*, 1997, **22**, 1203. *d.* C. Moreau, A. Gandini, and M. N. Belgacem, *Topics Catal.*, 2004, **27**, 9. *e.* A. Gandini, M. N. Belgacem, *Furan Derivatives and Furan Chemistry at the Service of Macromolecular Materials*, in reference [1], Chapter 6. *f.* A. Gandini, *Polym. Chem.*, 2010, **1**, 245.

[5] F. Romani, R. Corrieri, V. Braga, *et al.*, *Polymer* 2002, **43**, 1115. S. Coiai, E. Passaglia, M. Aglietto, *et al.*, *Macromolecules* 2004, **37**, 8414.

[6] *a.* M. C. Salon, H. Amri, and A. Gandini, *Polymer Comm.* 1990, **31**, 209. *b.* H. Amri, M. N. Belgacem, and A. Gandini, *Polymer* 1996, **37**, 4815. *c.* H. Amri, M. N. Belgacem, C. Signoret, *et al.*, *Polym. Intern.* 1996, **41**, 427. *d.* S. Boufi, M. N. Belgacem, and A. Gandini, *Polym. J.* 1997, **29**, 479. *e.* R. Su, Y. Qin, L. Qiao, *et al.*, *J. Polym. Sci. Part A: Polym. Chem.*, 2011, **49**, 1434.

[7] A. Gandini, M. C. Salon, Dutch Patent 8900137, 1989.

[8] S. Hadjikyriacou, R. Faust, *Macromolecules* 1999, **32**, 6393. R. Faust, *Macromol. Symp.* 2000, **157**, 101. Y. Kwon, R. Faust, *Adv. Polym. Sci.* 2004, **167**, 107.

[9] A. Lange, H. P. Rath, and G. Lang, *Macromol. Symp.* 2004, **215**, 209.

[10] A. Gandini, M. C. Salon, unpublished results. M. C. Salon, Doctorate Thesis, Grenoble National Polytechnic Institute, France, 1984.

[11] M. Choura, N. M. Belgacem, and A. Gandini, *Macromolecules*, 1996, **29**, 3839.

[12] N. Guigo, A. Mija, L. Vincent, *et al.*, *Phys. Chem Chem. Phys.*, 2007, **9**, 5359. S. Barsberg and L. G. Thygesen, *Vibrat. Spectrosc.*, 2009, **49**, 52. S. Bertarione, F. Bonino, F. Cesano, *et al.*, *J. Phys. Chem. B*, 2008, **112**, 2580.

[13] D. Kawashima, T. Aihara, Y. Kobayashi, *et al.*, *Chem. Mater.*, 2000, **12**, 3397. J. Yao, H. Wang, J. Liu, *et al.*, *Carbon*, 2005 **43**, 1709. A. J. G. Zarbin, R. Bertholdo, and M. A. F. C. Oliveira, *Carbon*, 2002, **40**, 2413. H. Wang and J. Yao, *Ind. Eng. Chem. Res.* 2006, **45**, 6393. B. Yo, R. Rajagopalan, H. C. Foley, *et al.*, *J. Am. Chem. Soc.*, 2006, **128**, 11307. T. Hirasaki, T. Meguro, T. Wakihara, *et al.*, *J. Mater. Sci.*, 2007, **42**, 7604. F. Casano, D. Scarano, S. Bertarione, *et al.*, *J. Photochem. Photobiol. A Chem.*, 2008, **196**, 143. S. Bertarione, F. Bonino, F. Cesano, *et al.*, *J. Phys. Chem. B*, 2009, **113**, 10571. G. Tondi, A. Pizzi, H. Pasch, *et al.*, *Europ. Polym. J.*, 2008, **44**, 2938. G. Tondi, A. Pizzi, H. Pasch, *et al.*, *Polym. Degrad. Stabil.*, 2008, **93**, 968. A. Pizzi, G. Tondi, H. Pasch, *et al.*, *J. Appl. Polym. Sci.*, 2008, **110**, 1451. G. Tondi, A. Pizzi, L. Delmonte, *et al.*, *Ind. Crops Prods.* 2010, **31**, 327. J. Yao and H. Wang, *Ind. Eng. Chem. Res.*, 2007, **46**, 6264.

[14] Y. Zhai, B. Tu, and D. Zhao, *J. Mater. Chem.*, 2009, **19**, 131.

[15] S. Grund, P. Kempe, G. Baumann, *et al.*, *Angew. Chem. Int. Ed.*, 2007, **46**, 628. S. Spange and S. Grund, *Adv. Mater.*, 2009, **21**, 2111.

[16] L. Pranger and R. Tannenbaum, *Macromolecules*, 2008, **41**, 8682.

[17] S. Lande, M. Westin, and M. Schneider, *Mol. Cryst. Liq. Cryst.*, 2008, **484**, 367.

[18] L. Nordstierna, S. Lande, M. Westin. *et al.*, *Holzforsch.*, 2008, **62**, 709.

[19] L. G. Thygesen, S. Barsberg, and T. M. Venås, *Wood Sci. Technol.* 2010, **44**, 51.

[20] A. Gandini, C. Coutterez, C. Goussé, *et al.*, *ACS Symposium Series*, 2001, **784**, 98. C. Coutterez, Doctorate Thesis, Grenoble Polytechnic Institute, France, 1998. C. Coutterez, A. Gandini, *Recent Advances in Environmentally Compatible Polymers*, (ed. J. F. Kennedy) Woodhead Publ., Cambridge, 2001, p. 17, and references therein.

[21] C. Méleares, A. Gandini, *Eur. Polym. J.*, 1996, **32**, 1269.

[22] V. Baret, A. Gandini, and E. Rousset, *J. Photochem. Photobiol.*, 1997, **A103**, 171. L. Albertin, P. Stagnaro, C. Coutterez, *et al.*, *Polymer*, 1998, **39**, 6187. S. Waig Fang, H. J. Timpe, and A. Gandini, *Polymer*, 2002, **43**, 3505. A. Gandini, S. Hariri, and J. F. Le

Nest, *Polymer*, 2003, **44**, 7565. A. Gandini, S. Waig Fang, H. Timpe, *et al.*, USA Patent USA 6 270 938, 2001.

[23] A. Mitiakoudis, A. Gandini, and H. Cheradame, *Polym. Commun.*, 1985, **26**, 246. A. Mitiakoudis, A. Gandini, *Macromoleçules*, 1991, **24**, 830. A. Gandini, A. Mitiakoudis, USA Patent 4 806 623, 1987.

[24] S. Abid, R. El Gharbi, and A. Gandini, *Polymer*, 2004, **45**, 5793. S. Gharbi, A. Gandini, *J. Soc. Chim. Tunis.* 2004, **6**, 17.

[25] J. A. Moore and J. E. Kelly, *Macromolecules*, 1978, **11**, 568; *J. Polym. Sci., Polym. Chem. Edn.*, 1978, **16**, 2407; *Polymer*, 1979, **20**, 627; *J. Polym. Sci., Polym. Chem. Edn.*, 1984, **22**, 863.

[26] A, Gandini, A. Khrouf, S. Boufi, *et al.*, *Macromol. Chem. Phys.*, 1998, **199**, 2755. A. Chaabouni, S. Gharbi, M. Abid, *et al.*, *J. Soc. Chim. Tunis.*, 2002, **4**, 547. M. Abid, W. Kamoun, R. El Gharbi, *et al.*, *Macromol. Mater. Eng.*, 2008, **293**, 39.

[27] A. Gandini, A. J. D. Silvestre, C. Pascoal Neto, *et al.*, *J. Polym. Sci. Part A: Polym. Chem.*, 2009, **47**, 295.

[28] *a.* A. Gandini, D. Coelho, M. Gomes, *et al.*, *J. Mater. Chem.*, 2009, **19**, 8656. *b.* M. Gomes, M. Sc. Thesis, University of Aveiro, Portugal, 2009.

[29] E. Lasseuguette, A. Gandini H. J. Timpe, *J. Photochem. Photobiol.*, 2005, A174, 222. E. Lasseuguette, A. Gandini M. N. Belgacem, and H. J. Timpe, *Polymer*, 2005, **46**, 5476.

[30] J. L. Cawse, J. L. Stanford, and R. H. Still, *Makromol. Chem.*, 1984, **185**, 708.

[31] A. Gandini, J. Quillerou, M. N. Belgacem, *et al.*, *Polym. Bull.*, 1989, **21**, 555. M. N. Belgacem, J. Quillerou, A. Gandini, *et al.*, *Eur. Polym. J.*, 1989, **25**, 1125. M. N. Belgacem, J. Quillerou, and A. Gandini, *Eur. Polym. J.*, 1993, **29**, 1217. S. Boufi, M. N. Belgacem, J. Quillerou, *et al.*, *Macromoleules*, 1993, **26**, 6706. S. Boufi, M. N. Belgacem, and A. Gandini, *Polymer*, 1995, **36**, 1689.

[32] M. N. Belgacem, J. Quillerou, A. Gandini, *et al.*, unpublished results.

[33] A. Afli, S. Gharbi, R. El Gharbi, *et al.*, *J. Soc. Chim Tunis.* 2003, **5**, 3.

[34] S. Abid, R. El Gharbi, and A. Gandini, *Polymer* 2004, **45**, 6469.

[35] S. Abid, S. Matoussi, R. El Gharbi, *et al.*, *Polym. Bull.* 2006, **57**, 43.

[36] M. N. Belgacem, A. Gandini, *ACS Symp. Ser.*, 2007, **954**, 280. S. D. Bergman, F. Wudl, *J. Mater. Chem.*, 2008, **18**, 41.

[37] N. Teramoto, Y. Arai, and M. Shibata, *Carbohydr. Polym.*, 2006, **64**, 78. M. Watanabe, N. Yoshie, *Polymer*, 2006, **47**, 4946. J. Ahmad, W. A. A. Ddamba, and P. K. Mathokgwane, *Asian J. Chem.*, 2006, **18**, 1267.

[38] A. Gandini, D. Coelho, and A. J. D. Silvestre, *Europ. Polym. J.*, 2008, **44**, 4029.

[39] A. Gandini, A. J. D. Silvestre, and D. Coelho, *J. Polym. Sci. Part A: Polym. Chem.*, 2010, **48**, 2053.

[40] A. Gandini, D. Coelho, and A. J. D. Silvestre, unpublished results.

[41] Q. Tian, M. Z. Rong, M. Q. Zhang, *et al.*, *Polymer*, 2010 **51**, 1779.

[42] N. Yoshie, M. Watanabe, H. Araki, *et al.*, *Polym. Degrad. Stabil.* 2010, **95**, 826.

[43] A. A. Kavitha and N. K. Singha, *J. Polym. Sci. Part A: Polym. Chem.*, 2007, **45**, 4441; *Macromol. Chem. Phys.*, 2007, **208**, 2569; *ACS, Appl., Mater., Interfaces*, 2009, **1**, 1427; *Macromolecules*, 2010, **43**, 3193.

[44] M. Yamashiro, K. Inoue, and M. Iji, *Polym. J.*, 2008, **40**, 657.

[45] Y.-L. Liu, C.-Y. Hsieh, and Y.-W. Chen, *Polymer*, 2006, **47**, 2581. Y.-L. Liu Y.-W. Chen, *Macromol. Chem. Phys.*, 2007, **208**, 224.

[46] Y.-L. Liu and C.-Y. Hsieh, *J. Polym. Sci. Part A: Polym. Chem.*, 2006, **44**, 905.

[47] A. M. Peterson, R. E. Jensen, and G. R. Palmese, *ACS Appl. Mater. Interfaces*, 2009, **1**, 992 and 2010, **2**, 1141.

[48] Q. Tian, Y. C. Yuan, M. Z. Rong, *et al.*, *J. Mater. Chem.*, 2009, **19**, 1289.
[49] H.-L. Wei, Z. Yang, L.-M. Zheng, *et al.*, *Polymer*, 2009, **50**, 2836.
[50] K. Ishida and N. Yoshie, *Macromol. Biosci.*, 2008, **8**, 916.
[51] Y. Zhang, A. A. Broekhuis, and F. Picchioni, *Macromolecules*, 2009, **42**, 1906.
[52] T. A. Plaisted and S. Nemat-Nasser, *Acta Mater.*, 2007, **55**, 5684.
[53] W. H. Heath, F. Palmieri, J. R. Adams, *et al.*, *Macromolecules*, 2008, **41**, 719.
[54] M. L. Szalai, D. V. McGrath, D. R. Wheeler, *et al.*, *Macromolecules*, 2007, **40**, 818. N. W. Polaske, D. V. McGrath, and J. R. McElhanon, *Macromolecules*, 2010, **43**, 1270.
[55] N. Aumsuwan and M. Urban, *Polymer*, 2009, **50**, 33.
[56] M. M. Kose, G. Yesilbag, and A. Sanyal, *Org. Lett.*, 2008, **10**, 2353.
[57] C. Jegat and N. Mignard, *Polym. Bull.*, 2008, **60**, 799.
[58] C.-M. Chang and Y.-L. Liu, *Carbon*, 2009, **47**, 3041.
[59] X. Liu, M. Zhu, S. Chen, *et al.*, *Langmuir*, 2008, **24**, 11967. X. Liu, H. Liu, W. Zhou, *et al.*, *Langmuir*, 2010, **26**, 3179.
[60] M. Shi, J. H. Wosnick, K. Ho, *et al.*, *Angew. Che. Int. Ed.*, 2007, **46**, 6126. M. Shi and M. S. Shoichet, *J. Biomater. Sci. Polym. Edn.*, 2008, **19**, 1143.
[61] M. Okada, K, Tachikawa, and K. Aoi, *J. Appl. Polym. Sci.* 1999, **74**, 3342.

Part II

Part II

10

Food Packaging Applications of Biopolymer-Based Films

N. Gontard*, H. Angellier-Coussy, P. Chalier, E. Gastaldi, V. Guillard, C. Guillaume and S. Peyron

Joint Research Unit Agro-polymers Engineering and Emerging Technologies, Université Montpellier II, Montpellier, France

10.1 INTRODUCTION

Food packaging research and development efforts have, over the last few decades, been mainly devoted to barrier materials (e.g., new polymers, complex and multilayer materials) or new designs for marketing purposes. These endeavours have been supported by the huge development of versatile petrochemical polymers. New food packaging technologies, such as active and intelligent[1] packaging, are now developing

---

[1] Active packaging deliberately incorporates active components intended to release or to absorb substances into, onto or from the packaged food or the environment surrounding the food with their conditions of use. They act on different reactions of degradation of food or as vector of compounds of interest. Intelligent packaging deliberately incorporates intelligent components able to monitor various food quality indicators during processing, transport and storage by using, for example, tags as miniaturized analytical tools with wireless communication, from the producer to the consumer.

Biopolymers – New Materials for Sustainable Films and Coatings, First Edition.
Edited by David Plackett.
© 2011 John Wiley & Sons, Ltd. Published 2011 by John Wiley & Sons, Ltd.

and have been recently introduced in the European Union Food Contact Materials regulation (1935/2004 framework regulation and related directive) as a response to consumer trends towards mildly preserved, fresh, tasty and conveniently packaged food products [17]. In addition, changes in retailing practices, such as globalized markets resulting in longer distribution distances, present major challenges to the food packaging industry, acting as driving forces for the development of new and improved packaging concepts that extend shelf-life while maintaining and monitoring food safety and quality. Packaging safety and waste management are also key points for globalization, harmonization and trade barriers. Simultaneously, and apart from the already practiced weight reduction of conventional materials, the development of new sustainable, recyclable and/or biodegradable packaging materials is also a major driving force to increase the overall presence of eco-friendly packaging solutions across the food chain. Modern concepts of packaging are expected to combine renewable materials with the latest innovative technologies (e.g., nanotechnologies). Bioplastics present a large spectrum of applications such as collection bags for compost, agricultural foil, horticultural products, nursery products, toys, textiles, technical materials and so on [75]. Food packaging represents two-thirds of the total packaging market with a total annual expected turnover of 420 billion Euros in 2009 [70]; however, many of the currently available bioplastics are not applicable to this market. A number of drawbacks have hampered the growth of the bioplastics market, especially in the field of food applications, because of limited processability, mechanical and barrier properties (e.g., fibers, extractable agropolymers) and/or comparatively high production cost (e.g., biopolyesters) or controversial environmental claims. In addition, material sensitivity to water and to microbial spoilage is often incompatible with food safety and quality preservation requirements.

The present chapter will focus on the ability of biopolymer-based packaging to preserve food quality and safety. It will first give an overview on the main properties expected of a food packaging material (i.e., specifications). Then, examples of current commercial applications of biopolymers for food packaging will be presented for different types of foods. Finally, the main knowledge gaps will be discussed and subsequent research lines of thinking likely to facilitate the use of biopolymers for food packaging will be proposed.

10.2 FOOD PACKAGING MATERIAL SPECIFICATIONS

10.2.1 Functional Properties

The shelf-life of food can be defined in terms of degradation reactions: physico-chemical and microbial reactions for non-living products and also physiological reactions for living products. The quality of food is first of all dependent on the initial quality of raw materials. Subsequently, degradation reactions can occur during the processing of the food and of course during the storage and distribution steps (Figure 10.1). Packaging plays a major role here because the shelf-life of food is dependent on its surrounding environmental conditions and amount of the change in quality that is

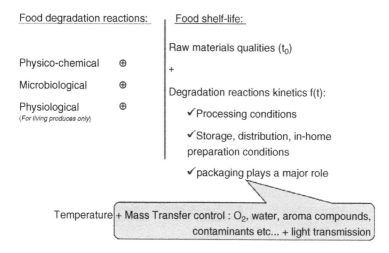

Figure 10.1 Food degradation mechanisms and factors affecting food shelf-life

allowed. It is well known that temperature is an important parameter in degradation reactions. Mass transfer through the packaging is also crucial in controlling food degradation rate.

One of the main roles assigned to packaging materials in terms of food quality and safety preservation is the control of mass transfer between the food, the packaging, and the atmosphere [5]. This is in addition to logistic, cost and environmental issues. As well as protection against light and mechanical aggression, food packaging has to control the transfer of the main gases (oxygen, carbon dioxide, ethylene, etc.) and vapours (moisture, aroma compounds, etc.) involved in food degradation (physical–chemical, biological, microbiological) and also the migration of potentially toxic packaging additives and constituents (especially for complex material formulations; Figure 10.2). The moisture barrier property is an essential basic requirement when packaging many foods, whether dry or moist, in order to preserve texture (such as crispiness, softness, firmness, etc.) and to control the development of aerobic spoilage and pathogen micro-organisms. This property is expressed in terms of the water vapour transmission rate (WVTR), which is the amount of water vapour that permeates per unit of area and time through the packaging materials, or more adequately by the water vapour permeability (WVP), which takes into account the influence of material thickness and water vapour partial pressure gradient. Packaging can also play a crucial role in protecting food from oxygen, a key element of many degradation reactions occurring in food. Oxygen acts as the main factor in organoleptic and nutritional quality degradation of food through the oxidation of vitamins, aroma compounds, pigments, lipids and proteins. Oxygen is also involved in microbial development and the maturation rate of fresh living products such as fresh fruit and vegetables or fermented cheeses. As for a water vapour barrier, the oxygen barrier property is expressed by the oxygen transmission rate (OTR) or more accurately by permeability (OP). Carbon dioxide needs to be preserved in many carbonated drinks. It is involved in the

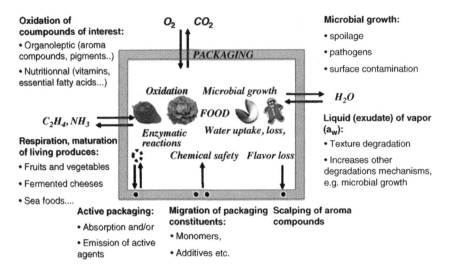

Figure 10.2   Importance of mass transfer control in food packaging

inhibition of respiration rate of living produce and is also used as a bacteriostatic and fungistatic agent. Carbon dioxide transmission rate or permeability is thus one of a number of important selection criteria for food packaging materials. Aroma compound barrier properties are also key elements for preserving flavour and taste of foods such as coffee and juices, for example. Other properties such as ethylene mass transfer (for sensitive produce) or ammonia mass transfer (for cheeses) also need to be considered. Maintaining physical film integrity is essential to guarantee the barrier properties. Mechanical properties are thus essential requirements for food protection. It is important to understand that barrier properties need to be tailored to food requirements and shelf-life.

10.2.2   Safety Issues

Foods are packaged in order to prevent physical, chemical, physiological and microbial spoilage during transport, distribution, handling and storage. When food is in contact with the packaging, a mass transfer process of components from the packaging toward the food and vice-versa, begins. This process is called 'food/packaging interaction' and includes 'chemical migration' (migration of migrants from the packaging toward the food). The Framework Regulation (EC) No 1935/2004 [25] covers all food contact materials and is based on the two general principles of inertness and safety of the material. In the EU, an important effort of harmonization on Food Contact Material (FCM) regulations has been undertaken since 1976, when the Framework Directive was adopted laying down the general principles of FCM safety [73]. Nowadays, the Framework Regulation (EC) 1935/2004 (EC 2004) is the basic European Community legislation on FCM which defines food contact materials as all the articles that one can expect to be brought into contact with food, including packaging, dishes, cutlery, table surfaces in food preparation areas, and conveyor belts. The legislation sets exceptions for active and intelligent materials and for materials that are destined to be consumed with the food.

Regulation 1935/2004 also sets three requirements to ensure safe and good quality food: (1) FCM shall not transfer their components into the food in quantities that could endanger human health; (2) FCM shall not change the composition of the food in an unacceptable way; and (3) FCM shall not cause deterioration in the taste, odour or texture of the food. These requirements constitute the so-called principle of *inertia* for an FCM. The first requirement only deals with migration and food safety, but the other two are related not only to migration, but also to food quality (e.g., scalping of aroma compounds). Regulation 1935/2004 does not include the details for implementation of these principles. However, in Annex I of the text it lays down the list of materials that can be covered by specific legislation. To date, only four (ceramics, regenerated cellulose, plastics and active and intelligent materials) out of 17 categories have been covered by specific legislation. Due to their singular importance, plastic materials were the first to be covered by the Community harmonization. The plastics Directive 2002/72/ EC [22] covers plastic monolayers and multilayers that consist only of plastic. This Directive translates the requirements of Regulation 1935/2004 to plastics in terms of limits that must be respected for migration tests. The requirement of inertia for the material becomes an overall migration limit (OML) (i.e., the total mass released by the packaging material during a migration test). The Directive 2002/72/EC sets this limit to 60 mg/kg of food. Regarding the requirement of safety, Directive 2002/72/EC establishes a specific migration limit (SML) for every substance in a positive list (i.e., authorized in a plastic FCM with the value of the SML based on toxicological data).

The introduction of new materials and technologies has deeply changed the panorama of food packaging and the use of biopolymers is one of the latest technological trends, aimed at improving the performance of currently used materials [68] and/or to produce environmentally friendly packaging. In the absence of specific regulation, these novel materials lie within the scope of European regulation (EC) 1935/2004, which sets the general guidelines for the compliance of any FCM. The procedure to analyze material excluded from the scope of a specific Directive must consequently prove compliance with the general requirements, except if national regulation includes specific *modus operandi*. Very few studies are referenced about the safety assessment of biopolymer-based packaging. The most studied biopolymer is poly(lactic acid) (PLA) and very limited migration was found in foods within the fixed OML [15]. In addition, the potential migrants from PLA are well identified and include lactoyl–lactic acid and other small linear oligomers of lactic acid, and lactide (the cyclic dimer of lactic acid used for the synthesis of PLA). All of these compounds are known to hydrolyze in aqueous systems to lactic acid, which is already authorized as a monomer and additive (Commission directive 2002/72/EC, Annex II) with no restriction or specification. Among the biopolymers produced by bioconversion, poly(3-hydroxybutyrate-co-3-hydroxyvalerate) (PHBV), a biopolyester obtained by means of bacterial fermentation, may be used as a FCM with an SML of 0.05 mg/dm^2 for crotonic acid, a depolymerization product, according to the specification of annex V [22].

Some of the biopolymers derived from biomass, such as starch, pectin or casein, are referred to in EU positive lists of monomers and additives but most (e.g., chitosan, gelatin, keratin, gluten, zein and soybean) are authorized as food additives but not specifically mentioned in FCM legislation. As a consequence, polysaccharide- or protein-based materials are submitted to the OML set by the plastic regulation but no SML is specified. However, the recommended migration tests have been designed mostly for plastics and as such are only applicable to water-resistant materials.

Some bio-derived polymers exhibit low water resistance as well and can only be used to contain dry foodstuffs or short shelf life produce. As a consequence, in these cases, reported migration values tend to be particularly high in common migration tests designed for plastic material [22] and for covering long-term storage in the worst conditions of temperature. Most particularly, some biopolymers such as wheat gluten- and chitosan-based materials proved to be unsuitable in direct food contact with OML values higher than the authorized threshold of 60 mg/kg obtained after contact with aqueous food simulants [32, 56]. The increase of water resistance by combination of biopolymers with natural or synthetic agents (e.g., Starch Mater-Bi®) does not prevent the migration process sufficiently enough that a packaging material can comply with the requirements of the FCM legislation.

These results highlight the inaccuracy of recommended migration tests for such biopolymer-based materials. So far, the only alternative tests that could be recommended are derived from paper testing methods. Paper and board materials, the oldest biodegradable packaging materials, are not yet regulated by a specific directive. A resolution AP2002/1, supported by five technical documents was adopted in 2002 and member states may use it as a basis for their national regulations. However, a fully harmonized EU legislation of paper and board is still needed. For example, the solid fatty simulant Tenax™ (modified polyphenylene oxide, Varian, Netherlands) has been successfully used [56] to determine migration from paper-containing potential organic migrants. In addition, standard conditions for this test have been set in the European Committee for Standardization norm CEN/TC 172 [10]. However, such a limited approach fails to reproduce the variety of foodstuffs packed in these materials and only imitates dry fatty products.

10.2.3   Environmental Aspects

The trends of modern lifestyle, including less time to prepare meals, eating on the move, increased use of convenience food, and decreasing household size have progressively resulted in more packaging being used per unit of product. In developed countries, food packaging represents more than 60% of all packaging [60]. As a result, food packaging waste forms a significant part of municipal solid waste and as such has caused increasing environmental concerns, resulting in strengthening of relevant EU regulations. Concern for packaging waste increased more than any other environmental concern, as indicated in an Internet-based survey in 48 nations by the Nielsen company. Bioplastics have the advantage of being derived from renewable resources and have often proven advantageous when the criteria 'consumption of fossil resources' and 'reduction of CO_2 emissions' are being assessed. If food packaging is also biodegradable, significant advantages could be offered, such as more adequate disposal of expired, spoiled food or food residues directly with the packaging by the end consumer, the food producing companies and retailers or the organizers of large events, such as open air festivals and, sporting events, without further treatments such as opening, sorting, and cleaning.

The Directive 94/62/EC [23] lays down measures for reducing packaging waste and sets recovery and recycling targets updated by Directive 2004/12/EC [24]. This directive set the target that by the end of 2008, 55–80% of packaging waste should be recycled. The Directive 99/31/EC [47] aims to prevent or reduce as far as possible the use of landfills and to develop new approaches for waste treatment, including incineration, recycling and compostability. Among the waste treatment technologies available for food packaging materials based on biopolymers, incineration with recovery of both heat and power is a possible option [63]. However, biopolymers such as starch and natural fiber have relatively low gross calorific values (GCV) and will reduce to some extent the overall GCV if incorporated in wastes with conventional non-biodegradable polymers [44].

For efficient recycling, packaging materials have to be accurately sorted, washed and free from contaminants which is a prerequisite that is not always technically feasible and economically worthwhile due to the costs associated with collection, sorting, logistics and transportation [9, 16, 30].

It should be noted that the disposal of certain biopolymers jeopardizes other efficient plastic waste management systems. For example, PLA addition is known to negatively impact polyethylene terephthalate (PET) recycling systems. The necessary additional sorting systems, the costs of which are not trivial, leave too much PLA as a contaminant in the recycled PET stream, causing expensive downgrading. The concept of recycling is probably not the best option in the case of bio-based packaging materials [27]. Biodegradable and/or compostable polymers broaden the range of waste management treatment options when compared with traditional plastics and this is supported by life cycle assessment. Presently, the most favoured end-of-life disposal options for these materials are domestic and municipal composting in place of landfilling [18]. This implies that bio-based packaging would be collected with organic waste and treated using local composting facilities, in order to reduce the total waste to landfill, the cost of transport and the associated emissions. In the EU directive 2008/98/EC, the member states are forced to take such measures for separate collection of bio-waste and to suitably treat this fraction. European policy is focussing on bio-waste, especially food waste from households (~ 39%), restaurants, caterers and retail premises as well as from food processing plants [42].

In the near future it is assumed that anaerobic digestion will be promoted more and more as it enables energy recovery from bio-waste and, when combined with aerobic digestion in a second step, assists composting. Biodegradability also enables a 'material cycle' to be built up that imitates nature by recycling the carbon. The resulting compost is useful as it can be eco-efficiently used as fertilizer and substrate to improve soil quality [http://www.european-bioplastics.org]. Regarding the greenhouse effect and energy recovery, the combination of anaerobic and aerobic digestion is considered to be very promising. European policy in this area is being implemented in Austria, Belgium, Germany, Switzerland, Luxembourg, the Netherlands, and partly in Italy and Scandinavia. Other countries in Europe are only beginning to do this, or still deposit mixed urban wastes [www.compostnetwork.info].

10.3 EXAMPLES OF BIOPOLYMER APPLICATIONS FOR FOOD PACKAGING MATERIALS

10.3.1 Short Shelf-Life Fresh Food Packaging

Bio-packaging materials currently occupy niches in the food packaging industry; however, more and more such materials are entering the market for fresh and short-shelf-life produce, such as fruit and vegetables. In this type of application, cellulosic materials such as paper bags at retail stores or cardboard boxes in shipping are traditionally used. Even if improved shelf-life of fresh fruit and vegetables using these materials has never been clearly demonstrated, their high porosity leads to higher gas transfer than in most commercial plastic films, thereby allowing the respiration of fresh produce. Most plastic films exhibit an O_2 permeability that is too low, leading when sealed, to a sharp drop in O_2 followed by detrimental anaerobic conditions [80]. As a result, such films need to be macro- or micro-perforated to allow sufficient gas exchange. The density of perforation can be tailored to the O_2 requirement for fresh produce providing a large range of O_2 permeabilities (from 190 to 42 000 mL O_2 m^{-2} 24 h^{-1} [78]).

The drawback of porous or perforated materials is insufficient permselectivity (ratio of CO_2 permeability to O_2 permeability). The consequence of a permselectivity close to one is that CO_2 comes out at the same rate that O_2 goes in (similar diffusion rates), which is not desirable for CO_2-sensitive produce (e.g., mushrooms or asparagus). This might be overcome by use of biopolymer-based plastics [35]. While the permselectivity ratio of common plastic film remains between three and six, bio-based materials generally exhibit higher ratios and values of 16 for pectin film, 17 for chitosan film, 28.4 for wheat gluten film [37, 38], and 17.27 and 10.77 for PLA 4030-D and 4040-D films from NatureWorks LLC have been cited [6]. Due to their hydrophilic nature, another characteristic of bio-based materials is their response to changes in relative humidity (RH) and, to a lesser extent, temperature. For example, the permselectivity of wheat gluten materials sharply increased from 6 in dry conditions to 28.4 at 95% RH [36]. This behaviour appeared compatible with the physiology of fresh fruit and vegetables [31], since the RH rapidly increased to a value higher than 80% within the packaging.

Beyond the fact that the performance properties of bio-packaging might suit fresh produce, it also creates a marketing opportunity in the organic food product sector. Customers who buy fresh and/or organic produce are seen as caring about the environment and sustainability. As a commercial consequence, innovation in biodegradable and bio-based packaging has mainly been directed towards organic fresh produce, as applied by grocery stores such as Sainsbury's (UK), Albert Heijn (Germany), IPER (Italy), Delhaize (Germany), and Auchan (France). In Sainsbury stores, almost 50% of organic fruit and vegetables have been available in compostable packaging since 2008 and this figure rises to 80% if recycled and recyclable materials are included. In the field of commercial flexible films, Natureflex™ a cellulose-based material produced by Innovia Films, MaterBi® (Novamont) and PLA (NatureWorks® LLC) films, converted by companies like Amcor, Triofan, or Vitembal, are commonly used in their perforated or unperforated form for overwrapping or bagging fresh produce. PLA can also be found as transparent punnets and lids and shares the market for rigid bio-based packaging with opaque trays made from sugarcane, palm, or cellulose/lignocellulose fibers.

As an example, in 2006 the Oppenheimer Group, a North American fresh produce company, launched organic kiwifruit packaged in a combination of Earthcycle and Natureflex™. Earthcycle Packaging (Vancouver) provided a renewable and compostable tray sourced from palm fiber and Innovia Films supplied the overwrap Natureflex™ film. The UK grocer Sainsbury's and its potato packer Greenvale chose the Amcor NaturePlus heat-seal film which was co-extruded with Mater-Bi® for bagging baby organic potatoes.

10.3.2   Long Shelf-Life Dry or Liquid Food Packaging

10.3.2.1   Liquid Foods

Using biopolymers for liquid food packing requires that the materials have similar resistance, barrier, inertia, long-term stability and transparency properties to those of a conventional plastic such as PET. The only bioplastic meeting this specification until now is PLA. For example, in 2006 the UK company Belu launched Belu natural water bottled in corn-derived PLA provided by NatureWorks, then a part of the Cargill company. Belu was set up as a commercial enterprise by environmentalists and companies to develop ways to raise funds for clean water projects (Water Aid). Other examples of PLA use for liquid food bottles are the Biota spring water of Colorado or Primo mineral water bottles in the USA. Companies such as US-based Naturally Iowa have been using PLA for packaging such products as organic milk. There is a great interest from the consumer, for eco-friendly water and other 'on the move' drink bottles, in order to reduce landfilling. However, development of these applications has been hampered because of controversy over environmental benefits. PLA needs industrial composting to be degraded and, for recycling purposes, is not easily sorted. In addition, the heat resistance of PLA bottles is not yet satisfactory (e.g., if a water bottle is left in a car during warm weather). It should be added that the products discussed here should not be confused with oxy-degradable PET (such as that produced by Phoenix, Arizona-based ENSO Bottles, LLC.), plant origin PET or mixtures (such as I Lohas mineral water bottles of Coca Cola, Japan.).

10.3.2.2   Dry Foods

Dry food products are especially sensitive to moisture and must be protected against remoistening to avoid, for example, caking and collapse of powder, loss of crispness for dry cereal-based products or stickiness due to absorption of water by sugars and their subsequent migration to product surfaces. Due to their high sensitivity to water, biopolymers (especially agro-polymers such as starch or cellulose) used in monolayers are usually not suitable for packaging dry foods, except if additional strategies are used to compensate for this drawback (i.e., use of humidity scavengers, multi-layers). For instance, paper and board traditionally used for the packaging of dry products such as flour, rice, and pasta are usually co-laminated or coated with an external synthetic layer, which provides moisture resistance. Biopolyesters such as PLA, polyhydroxyalkanoates (PHAs) and derivatives display the lowest water vapour permeability compared with agro-polymers such as wheat gluten, starch and derivatives, obtained from extractable agro-resources, but remain expensive. For example, the WVP of pure polyhydroxybutyrate-co-valerate (PHBV) film is around 1.1×10^{-12} mol m^{-1} s^{-1} Pa and that of PLA is 1.3×10^{-12} mol m^{-1} s^{-1} Pa [72], whereas the corresponding values for wheat gluten- or starch-based films are almost 50 times higher at 6.25×10^{-11} mol m^{-1} s^{-1} Pa [65] and 3.1×10^{-11} mol m^{-1} s^{-1} Pa [13] respectively. A good alternative would be to combine biopolyesters and low-cost agro-polymers (e.g., fibers) in order to obtain bi-layer or composite films with lower sensitivity to water and reduced cost.

Biopolymers are already widely used for packaging dry food goods with mid-term or long shelf-life (e.g., flour, coffee grains, bakery and pastry products). The biopolymers currently used for such applications are mainly based on PLA, paper and other cellulose-based fibers. As examples, the cellulose-based sandwich bag of Biopac (http://www.biopac.co.uk), the bagasse (sugarcane pulp)-based product proposed by Natural-Pack as an alternative to plastic and styrene foam (http://www.natural-pack.com), the punnets and trays made from recycled paper for dry food goods from Green Home, South Africa (http://www.greenhome.co.za), the natural fiber jute bags of W.F. Denny (www.wfdenny.co.uk) or the snack boxes of Napac (http://www.napac.com.au) can be mentioned.

A specific characteristic of biopolymers and especially agro-polymers is their very low O_2 permeability at low relative humidity [36, 58]. For example, the O_2 permeability of wheat gluten-based film at 0% RH is 88×10^{-18} mol m^{-1} s^{-1} Pa [58], whereas that of LDPE is 1078×10^{-18} mol m^{-1} s^{-1} Pa [11]. This specific property could be an advantage in the development of biodegradable packaging for dry or oxidation-sensitive products, which need to be protected against oxygen transfer from the surrounding atmosphere.

10.4 RESEARCH DIRECTIONS AND PERSPECTIVES

10.4.1 Improving/Modulating Functional Properties

The balance of biodegradable packaging properties in terms of combining biodegradability with sufficient durability in service to meet food protection needs is a key issue [39, 45]. In food packaging, a major emphasis is on the development of materials with tailored properties in regard to migration of oxygen, carbon dioxide, flavour compounds, and water vapour. However, to date and without considering cost aspects, the use of biopolymers for food packaging has been limited due to poor barrier properties and/or weak mechanical properties. This is notably so for the agro-polymers. For example, achieving low water vapour permeability is often a critical issue in the development of agro-polymers as sustainable packaging materials. The development of biopolyester-based packaging displaying sufficient permselectivity and high enough O_2 barrier properties for increased shelf-life of fruit and vegetables is another challenge.

In order to improve/modulate/control various properties, biopolymers are frequently blended or otherwise combined with other biopolymers or synthetic polymers to form intimate blends, composites or multilayers. For example, wheat gluten proteins, which display interesting functional properties for food packaging but poor mechanical properties, have been combined with fiber-based materials such as paper and so-called composite materials with superior mechanical properties have been successfully developed [33, 41, 71]. Less frequently, biopolymers are chemically modified by, for example, cross-linking or grafting, addition of functional chemical additives or highly sophisticated bulk or surface chemical modifications [64]. However, the latter strategy is usually focused on the ubiquitous hydroxyl groups in some biopolymers (e.g., starch, cellulose) and generally requires the use of chemical compounds and/or solvents.

The development of biopolymer-based materials for food packaging with tailored barrier properties has driven structural studies of such materials at the nanometric scale. In this context, many studies have been devoted to the use of hybrid organic-inorganic systems and, in particular, to those in which layered silicates are dispersed at a nanometric level in a polymeric matrix [34]. The nanoscale plate morphology of layered silicates and other fillers often provides improved physical properties when incorporated in biopolymers, including improved mechanical properties, thermal stability, and gas barrier properties [1]. Layer-by-layer (LbL) assembly is another new route to develop tailor-made materials and biomaterials since it allows design of functional surfaces in a built-to-order fashion. An LbL assembly treatment can involve a polyanion/polycation multilayer formed by alternate deposition of positively and negatively charged nanolayers on a surface [20]. Recently, this technology was used to develop a material with the lowest oxygen permeability ever reported for a polymer–clay composite (less than 0.002×10^{-6} cc/(m.day atm)) by using alternate layers of branched polyethylenimine (PEI) and sodium montmorillonite (70 pairlayers) resulting in a 231-nm-thick assembly [67]. Additionally, biologically active ingredients can be added to impart desired functional properties to the resulting packaging materials. Considering research and development in this field, in 2001 Yasa-sheet (Plusto, Japan) was awarded the Nikkei Excellent Product Award. This nanostructured multilayer film contains an enzyme (extracted from bamboo) which controls the ethylene concentration inside the packaging and is applied to extend the shelf-life of climacteric fruit and vegetables. Even if such examples of LbL applications are still in some way sourced from petrochemistry, there can be no doubt that tailor-made bio-sourced packaging will soon be developed. This is especially the case considering recent interest in biomolecular architectures with protein assembly [3, 43, 51, 79] and construction of clay–protein ultrathin films [49, 57, 76].

10.4.2   Active Biopolymer Packaging

Active packaging extends the product shelf-life by interacting directly with the food and/or its environment. Research in the area of antimicrobial food packaging materials has significantly increased during the past few years as an alternative method for control of undesirable micro-organisms in foods through incorporation of antimicrobial substances in or coated onto the packaging [50]. The antimicrobial packaging concept allows reduced risk of food-borne microbiological illness and also offers other health-related benefits, for example by providing better nutritional quality with less severe food treatments and lower amounts of additives while ensuring longer shelf life and wider distribution. Antimicrobial activity can be introduced either by using the packaging itself as a vector for active compounds (e.g., acids, bacteriocins, enzymes, fungicides, essential oils) or by using the material's intrinsic antimicrobial properties. As regards the latter, chitosan has received the greatest attention as a *per se* antimicrobial material in this context [28]. Biopolymers such as proteins (e.g., soy proteins, wheat gluten, zein), polysaccharides (e.g., native or modified starches, pectins, celluloses and derivatives, pullulan) and PLA have been intensively explored for physical–chemical structures which would suit controlled release of active ingredients under targeted conditions.

Regarding applications in real food systems, the antimicrobial effectiveness of chitosan coating to preserve fruits has been particularly studied due to complementary properties such as reductions in respiration and loss of water [21]. Chitosan film has been tested on perishable products, for example to obtain inhibition of surface spoilage bacteria on processed meat [61] or to increase shelf life of fish from 5 to 9 days [77]. Investigations on chitosan coatings to control decay of minimally processed fruits and vegetables (strawberry and lettuce) showed that the presence of the film on lettuce induced the development of a bitter taste and a limited antimicrobial effect (4 days), but the effect was maintained on strawberries over 12 days without sensorial changes. Recently, chitosan was proposed as a coating for a Tetra Brik® package to preserve fish soup [29]. In the context of these examples, there is still a strong need to better understand the antimicrobial mechanisms of biopolymers in order to better control their efficacy and safety, and thus applicability. It is, for example, clear from the existing literature that research results on antimicrobial systems are often not reproducible and sometimes contradictory or not yet fully understood. For instance, in the case of chitosan or chitosan-based systems, the mechanisms of biocidal activity under conditions of optimum efficiency have just recently been understood based on the positive migration of glucosamine fractions [46].

10.4.3   Improving Safety and Stability

The structural and physicochemical stability of packaging materials, as well as their microbiological stability throughout the whole life cycle, are essential in ensuring packaged food safety within defined limits and recommended usage conditions. If moisture, temperature and mechanical stress are well-known parameters, others such as food acidity or fat content can also influence biopolymer-based material properties and stability. For the purpose of food-contact materials safety assessment, the key points are: (i) the analytical identification and quantification of harmful substances present in the packaging materials as well as the availability of toxicity data on these chemicals; and (ii) the potential extent of mass transfer of these substances into food. As a result of the complex behaviour of biodegradable packaging, knowledge is lacking regarding the dependency of these two key points on the physicochemical stability of the material (i.e., the potential degradation of the polymeric network, breakdown product formation or unintentionally added substances and enhanced migration). New insights into and understanding of the structural, physicochemical and microbial stability of biopolymer-based materials with respect to targeted food packaging applications are still needed [40].

The use of nanoparticle additives to improve the properties of biopolymer-based packaging materials has been widely studied. As for any new technology, before practical application a sound evaluation, including associated risks and benefits, needs to be carried out. In this respect, the nanobiocomposites discussed earlier might be of less concern health and safety-wise considering that one dimension is in the micrometre size range (i.e., not nano). Further, the high surface area of nanoparticles can potentially bind unwanted substances that might be otherwise released in contact with foods. It must be stressed however that full understanding of the effect on food safety through introducing these and other nanostructures has still to be ascertained. Although potential applications of nanotechnology in food packaging appear promising [4], there is no mention of engineered nanomaterials in European FCM legislation, except in the Regulation EC 450/2009 (articles 5.2 10) [26] related to active and intelligent material which states that ENM must be separated from the food by a functional barrier, which should completely prevent potential migration into the food product. In a recent study [56], it was found that the content of silicon increased for the three aqueous food simulant liquids (FSL) in contact with wheat gluten and montmorillonite materials. In the case of 3% acetic acid, the levels of aluminium also increased. These results were obtained by elemental analysis. Therefore, it is impossible to know if the whole engineered nanomaterial (ENM) has migrated, or only a part, or how the ENM migrated. Such a characterization of ENM in food or food simulants represents a real challenge for current analytical methodologies.

Actually, little is known about the risk that could be linked to the use of ENM in FCM. As pointed out by a recent scientific opinion from the European Food Safety Authority (EFSA, 2009) a suitable risk assessment approach should take into account the 'specific properties of the ENM in addition to those common to the equivalent non-nanoforms'. Given the large variety of ENM, a case-by-case basis seems a sensible approach to the risk assessment. Nevertheless, it is important to note that an individual assessment could involve a disparity of criteria and a high degree of uncertainty until more data become available and the assessment becomes standardized. Additionally, even if nanoparticle incorporation could be expected to limit the process of migration of substances used in manufacturing new biomaterials [19], a first investigation surprisingly demonstrates an increase of OML values of PLA nanocomposites due to nanoclay incorporation [55, 56]. In addition to their inherent potential toxicity, nanoparticles could thus change the stability of biopolymers in unexpected ways.

10.4.4   Towards an Integrated Approach for Biopolymer-Based Food Packaging Development

Nowadays, the development of new biodegradable materials relies essentially on empirical approaches which can have severe economic and safety hazard consequences. For instance, beyond the chilling approach, modified atmosphere is an efficient way to delay senescence of fresh fruits and vegetables. Modified atmosphere packaging (MAP) relies on modification of the atmosphere inside the package in order to extend food shelf-life by reducing the physiological degradation rate (namely respiration) and also physico-chemical (e.g., oxidation) and microbial degradation rates. MAP of fresh fruits and vegetables is achieved by the interplay of two processes: (1) the transfer of gases through the packaging; (2) the respiration of the product; and (3) possibly gas (oxygen) depletion (gas flush or gas scavenger). MAP can be achieved by matching the film permeation rate with the respiration rate of respiring products. Up to now, the suitability of packaging material to extend the shelf-life of a targeted product was only experimentally assessed in real conditions of use (i.e., by packing the product in the film and controlling the partial pressure of O_2 and CO_2 in the packaging headspace and observing, at least, the visual quality of the product). This 'trial and error' methodology is time-consuming and unsatisfying because it does not exploit the wealth of published information on packaging material (i.e., O_2 and CO_2 permeabilities) and respiration rates of fresh fruit and vegetables.

In the last 10 years, some mathematical models have been developed to design passive and/or active MAP [11,12,52–54] for fresh and fresh-cut fruits and vegetables (Figure 10.3). Such numerical tools simplify the package design steps by allowing advance prediction of the required window of packaging permeability in order to maintain packaged food quality and safety. Such tools form part of a real integrated approach, aimed at optimizing the design of biodegradable materials. As illustrated in Figure 10.3, from the required window of packaging

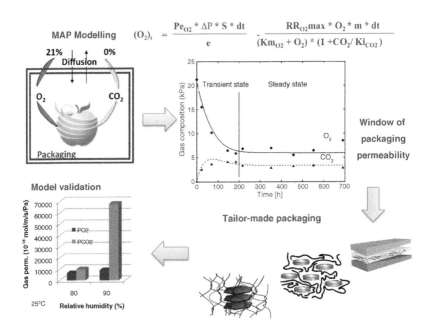

Figure 10.3 Integrated approach based on the modelling of MAP of fresh fruits and vegetables used to design tailor-made biodegradable packaging

permeability, the material formulation (i.e., type of matrix, nanocomposite, composite or multilayer films) and processing (e.g., thermoforming, coating) could be chosen and adapted in order to modulate and control the permeability and match the requirements. Such an approach is based on barrier properties, which is too restrictive considering the numerous other functions of a food packaging. The design of biodegradable packaging for foods needs to take into account numerous other factors such as the cost, availability, potential contaminants of raw materials, processability of constituents, feasibility at industrial scale and the environmental impact, safety and stability of the packaging material throughout the food life cycle, as well as waste management constraints. However these various elements are not always compatible, sometimes contradictory and often pose a dilemma for all stakeholders. The domain of knowledge representation and reasoning could help to solve this dilemma through the development of decision-making tools. On one hand, this approach would allow the formalization of available knowledge for discussion and decision-making, including qualitative expertise as well as quantitative data which are more classically used. On the other hand, decision-making would be possible through the development of methods aimed at resolving conflicts between contradictory interests [8]. Such an approach has not yet been implemented in the food or packaging fields but will be essential in the near future for increasing the chances of success for biodegradable food packaging.

10.5 CONCLUDING REMARKS

It can be concluded from this chapter on biopolymer-based food packaging that there is still an important need for improved knowledge on how 'agro- and bio-polymer science' can facilitate the development of efficient food packaging. An essential step in the application of biodegradable packaging by the food industry is to increase knowledge on the structure/property relationships of agro- and bio-polymers. The properties of these polymers are to some extent subject to the naturally complex behaviour of the plant or microbial cells from which they are engineered. While the field of petrochemical chemistry is mature in this respect, the agro and bio-polymer science field still needs further insights and improvements to facilitate the development of competitive food packaging.

It can reasonably be expected that future projects and studies will be focused on developing integrated studies of bio-materials process–structure–property relationships based on the latest innovative developments in complex materials characterization, composite biomaterial production processes, and mathematical modelling tools to calculate how structure–function relationships at different scales determine end properties.

In order to ensure that the developed packaging materials optimally fulfill the requirements of the food industry, compound producers, packaging converters, food retailers, waste management, and legislation, future projects should adopt a holistic approach aimed at long-term development of biopolymer-based food packaging. This is particularly important for Europe in the competitive and progressive bio-based packaging materials sector, where a significant share of industrial production is moving towards the lower raw materials and processing cost environments found in emerging countries such as Brazil, China or Thailand.

REFERENCES

[1] Alexandre, M., and Dubois, P. 2000. Polymer-layered silicate nanocomposites: preparation, properties and uses of a new class of materials. *Materials Science and Engineering*, 28, 1–63.

[2] Appendini, P., Hotchkiss, J.H. 2002. Review of antimicrobial packaging. *Innovative Food Science Emerging Technology*, 3, 113–126.

[3] Ariga, K., Nakanishi, T., and Michinobu, T. 2006. Immobilization of biomaterials to nano-assembled films (self-assembled monolayers, Langmuir–Blodgett films, and layer-by-layer assemblies) and their related functions. *Journal of Nanoscience and Nanotechnology*, 6: 2278–2301.

[4] Arora, A., Padua, G.W. 2010 Review: Nanocomposites in food packaging. *Journal of Food Science*, 75, 43–49.

[5] Arvanitoyannis, I.S. 1999. Totally and partially biodegradable polymer blends based on natural synthetic macromolecules: preparation, physical properties, and potential as food packaging materials. *Journal of Macromolecular Science, Reviews in Macromolecular Chemistry and Physics*, C39, 205–271.

[6] Auras, R., Harte, B., and Selke, S. 2004. An overview of polylactides as packaging materials. *Macromolecular Bioscience*, 4, 835–864.

[7] Ben Arfa, A., Preziosi-Belloy, L., Chalier, P., *et al.* 2007. Antimicrobial paper based on soy protein isolate or modified starch coatings including carvacrol and cinnamaldehyde. *Journal of Agricultural and Food Chemistry*, 55, 2155–2162.

[8] Bouyssou, D., Dubois, D., Pirlot, M., *et al.*, 2010. Decision-making process (eds Bouyssou, D., Dubois, D., Pirlot, M. & Prade, H). John Wiley & Sons, Inc: Hoboken, NJ 07030, USA, 928 pp.

[9] Brandrup, J. 1998. Ecology and economy of plastic recycling; In: *5th International Scientific Workshop on Biodegradable Plastics and Polymers*, Stockholm, Sweden 9–13 June.

[10] CEN/TC 172 Practical Guide for Users of European Directives, Unit 'Chemical and physical risks; surveillance'of the Health & Consumer Protection Directorate-General of the European Commission.

[11] Charles, F., Sanchez, J., and Gontard, N. 2003. Active modified atmosphere packaging of fresh fruits and vegetables: Modeling with tomatoes and oxygen absorber. *Journal of Food Science*, 68, 1736–1742.

[12] Charles, F., Sanchez, J., and Gontard, N. 2005. Modeling of active modified atmosphere packaging of endives exposed to several postharvest temperatures. *Journal of Food Science*, 70, 443–449.

[13] Chivrac, F., Angellier-Coussy, H., Guillard, V., *et al.* 2010. How does water diffuse in starch/montmorillonite nano-biocomposite materials? *Carbohydrate Polymers*, 82, 128–135.

[14] Coma, V., Sebti, I., Pardon, P., *et al.*, 2001. Antimicrobial edible packaging based on cellulosic ethers, fatty acids, and nisin incorporation to inhibit *Listeria innocua* and *Staphylococcus aureus*. *Journal of Food Protection*, 64, 470–475.

[15] Conn, R.E., Kolstad, J.J., Borzelleca, J.F., *et al.* 1995. Safety Assessment of Polylactide (PLA) for Use as a Food-contact Polymer. *Food and Chemical Toxicology*, 33, 273–283.

[16] Dainelli, D. 2008. Recycling of food packaging materials: an overview. In: Chiellini, E. (Ed.), Environmentally compatible food packaging, Woodhead Publishing Limited. pp 294–324.

[17] Danielli, D., Gontard, N., Spyropoulos, D., *et al.*, 2008. Active and Intelligent Food packaging: legal aspects and safety concerns. *Trends in Food Science and Technology*, 19, 99–108.

[18] Davis, G., Song, J.H. 2006. Biodegradable packaging based on raw materials from crops and their impact on waste management. *Industrial Crops and Products*, 23, 147–161.

[19] De Abreu, D.A.P., Cruz, J.M., Angulo, I., *et al.*, 2010. Mass transport studies of different additives in polyamide and exfoliated nanocomposite polyamide films for food industry. *Packaging Technology and Science*, 23, 59–68.

[20] Decher, G., Hong, J.D., and Schmitt, J. 1992. Buildup of ultrathin multilayer films by a self-assembly proces. Consecutively alternating adsorption of anionic and cationic polyelectrolytes on charged surfaces. *Thin Solid Films*, 210: 831–835.

[21] Devlieghere, F., Vermeulen, A., and Debevere, J. 2004. Chitosan: antimicrobial activity, interactions with food components and applicability as a coating on fruit and vegetables. *Food Microbiology*, 21, 703–714.

[22] Directive 2002/72/EC of 6 August 2002 relating to plastic materials and articles intended to come into contact with foodstuffs.

[23] Directive 1994/62/EC Packaging and Packaging Waste Directive 94/62/EC.

[24] Directive 2004/12/EC amending Directive 94/62/EC on packaging and packaging waste.

[25] EC Framework Regulation (EC) No 1935/2004 of the European Parliament and of the Council of 27 October 2004 on materials and articles intended to come into contact with food.

[26] EC Commission Regulation (EC) No 450/2009 of 29 May 2009 on active and intelligent materials and articles intended to come into contact with food.

[27] European Plastics News, 2001. *Nat. Select.* 28(5), 37–38.

[28] Fimbeau, S., Grelier, S., Copinet, A., *et al.*, 2006. Novel biodegradable films made from chitosan and poly(lactic acid) with antifungal properties against mycotoxinogen strains. *Carbohydrate Polymers*, 65, 185–193.

[29] Fernandez-Saiz, P., Soler, C., Lagaron, J.M., *et al.*, 2010. Effects of chitosan films on the growth of Listeria monocytogenes, Staphylococcus aureus and Salmonella spp. in laboratory media and in fish soup. *International Journal of Food Microbiology*, 137, 287–294.

[30] Brandrup, J. 1998. Ecology and economy of plastic recycling. In: *5th International Scientific Workshop on Biodegradable Plastics and Polymers*, Stockholm, Sweden, 9–13 June.

[31] Flanders (OVAM) and the German Federal Ministry for the Environment, Nature Conservation and Nuclear Safety, Brussels, 9–10 June 2009.

[32] Gällstedt, M., and Hedenqvist, M.S. (2004) Packaging-related properties of alkyd-coated, wax-coated, and buffered chitosan and whey protein films. *Journal of Applied Polymer Science*, 91, 60–67.

[33] Gastaldi, E., Chalier, P., Guillemin, A., *et al.*, 2007. Microstructure of protein-coated paper as affected by physico-chemical properties of coating solutions. *Colloids and Surfaces, A* 301, 301–310.

[34] Giannelis, E.P. Polymer layered silicate nanocomposites, *Advanced Materials*, 1996, 8, 29.

[35] Gontard, N., Guillaume, C. 2009. Packaging and the shelf life of fruits and vegetables. In: *Food Packaging and Shelf Life: A Practical Guide*. Gordon L. Robertson (ed.), CRC Press.

[36] Gontard, N., Thibault, R., Cuq, B., et al., 1996. Influence of relative humidity and film composition on oxygen and carbon dioxide permeabilities of edible films. *Journal of Agricultural and Food Chemistry*, 44, 1064–1069.

[37] Gontard, N., Ring, S. 1996. Edible wheat gluten film: influence of water content on glass transition temperature. *Journal of Agricultural and Food Chemistry*, 44, 3474–3478.

[38] Guilbert, S., Gontard, N., and Morris, L. 1996. Prolongation of the shelf-life of perishable food products using biodegradable films and coatings. *Lebensmittel-Wissenchaft und Technologie*, 29, 10–17.

[39] Guilbert, S., Guillaume, C., and Gontard, N. 2010. New packaging materials based on renewable resources. In: *Food Engineering at Interfaces*. ed. Aguilera, J.M., Barbosa-Canovas, G., Simpson, R., et al., Springer.

[40] Guillard, V., Mauricio-Iglesias, M., and Gontard, N. 2010. Effect of novel food processing methods on packaging: structure, composition and migration properties. *Critical Reviews in Food Science and Nutrition*, 49, 474–499.

[41] Han, J.H., Krochta, J.M. 1999. Wetting properties and water vapor permeability of whey-protein-coated paper. *Transactions of the ASAE*, 42, 1375–1382.

[42] Hannequart, J.P. 2009. *Proceedings of the Conference Bio-Waste – Need for EU-Legislation?* Conference organised by the German Federal Ministry for the Environment, the Public Waste Agency of Flanders, the Environment Ministry of the Czech Republic and the European Commission DG Environment.

[43] He, P.L., Hu, N.F. 2004. Interactions between heme proteins and dextran sulfate in layer-by-layer assembly films. *Journal of Physical Chemistry B*, 108, 13144–13152.

[44] Jacquinet, P. 1985. Calorific value of composites. *Composites*, 2, 14–16.

[45] Kaplan, D.J., Mayer, J.M., Ball, D., et al., 1993. Fundamentals of biodegradable polymers. In: *Biodegradable Polymers and Packaging*, Ching, C., Kaplan, D.L., Thomas, E.L. (eds.). Technomic publication, Basel, pp. 1–42.

[46] Lagaron, J.M. Fernandez-Saiz P., and Ocio J.M. Using ATR-FTIR spectroscopy to design active antimicrobial food packaging structures based on high molecular weight chitosan polysaccharide. *Journal of Agricultural and Food Chemistry*, 2007, 55, 2554–2562.

[47] Landfill Directive (1999/3I/EC) 1999. European Commission. *Official J. Eur. Communities* 1182/1-19, 16 July.

[48] Lee, C.H., Park, H.J., and Lee, S.C. 2004. Influence of antimicrobial packaging on kinetics of spoilage microbial growth in milk and orange juice. *Journal of Food Engineering*, 65, 527–531.

[49] Li, Z., and Hu, N.F. 2003. Direct electrochemistry of heme proteins in their layer-by-layer films with clay nanoparticles. *Journal of Electroanalytical Chemistry*, 558, 155–165.

[50] Lopez-Rubio, A., Almenar, E., Hernandez-Munoz, P., et al., 2004. Overview of active polymer-based packaging technologies for food applications. *Food Review International*, 20, 357–387.

[51] Lu, Z.S., Li, C.M., Zhou, Q., et al., 2007. Covalently linked DNA/protein multi-layered film for controlled DNA release. *Journal of Colloid and Interface Science*, 314, 80–88.

[52] Mahajan, P.V., Oliveira, F.A.R., and Sousa-Gallagher, M.J. 2009. Packaging design system for fresh produce: An engineering approach. In: *The Wiley Encyclopaedia of Packaging Technology*, 3rd edn, Kit, L. Yam,(ed.), Wiley., US.

[53] Mahajan, P.V., Oliveira, F.A.R., Fonseca, S.C., *et al.*, 2006. An interactive design of MA-packaging for fresh produce, In: *Handbook of Food Science, Technology, and Engineering*, Volume 3, Chapter 119, pp 1–16, Y.H. Hui (ed.), CRC Taylor & Francis, New York.

[54] Mahajan, P.V., Oliveira, F.A.R., Montañez, J.C. *et al.*, 2007. Development of user-friendly software for design of modified atmosphere packaging for fresh and fresh-cut produce. *Innovative Food Science and Emerging Technologies*, 8, 84–92.

[55] Martino, V.P., Ruseckaite, R.A., Jimenez, A., *et al.*, (2010). Evaluation of overall migration in nano-biocomposites based on plasticized PLA and organo-modified montmorillonite for food packaging applications. First International *Meeting on Material/Bioproduct Interactions*, 3–5 March 2010, AgroParisTech, Paris

[56] Mauricio, M., Guillard, V., Peyron, S., *et al.*, 2010. Wheat gluten nanocomposite films as food contact materials: migration tests and impact of a novel food stabilization technology (high pressure). *Journal of Applied Polymer Science*, 116, 2526–2535.

[57] Miao, S.D., Bergaya, F., and Schoonheydt, R.A. 2010. Ultrathin films of clay-protein composites. *Philosophical Magazine*, 90, 2529–2541.

[58] MujicaPaz, H., and Gontard, N. 1997. Oxygen and carbon dioxide permeability of wheat gluten film: Effect of relative humidity and temperature. *Journal of Agricultural and Food Chemistry*, 45, 4101.

[59] Nerin, C., Asensio, E. 2007. Migration of organic compounds from a multilayer plastic–paper material intended for food packaging. *Analytical and Bioanalytical Chemistry*, 389, 589–596.

[60] Northwood, T., Oakley-Hill, D. 1999. *Wastebook*. Luton Friends of the Earth, Environment Agency and the Building Research Establishment.

[61] Ouattara, B., Simard, R.E., Piette, G., *et al.*, 2000. Inhibition of surface spoilage bacteria in processed meats by application of antimicrobial films prepared with chitosan. *International Journal of Food Microbiology*, 62, 139–148.

[62] Oussalah, M., Caillet, S., Salmieri, S., *et al.*, 2004. Antimicrobial and antioxidant effects of milk protein-based film conaining essential oils for the preservation of whole beef muscle. *Journal of Agricutural and Food Chemistry*, 52, 5598–5605.

[63] Patel, M., Bastioli, C., Marini, L., *et al.*, 2003. Life-cycle assessment of bio-based polymers and natural fibre composites. In: Steinbüchel, A. (ed.), *Biopolymers*, vol. 10 John Wiley.

[64] Petersen, K., Nielsen, P.V., Bertelsen, G., *et al.*, 1999. Potential of biobased materials for food packaging. *Trends in Food Science and Technology*, 10, 52–68.

[65] Pommet, M., Redl, A., Morel, M.H., *et al.*, 2003. Study of wheat gluten plasticization with fatty acids. *Polymer*, 44, 115–122.

[66] Pranoto, Y., Rakshit, S.K., and Salokhe, V.M. 2005. Enhancing antimicrobial activity of chitosan films by incorporating garlic oil, potassium sorbate and nisin. *LWT-Food Science and Technology*, 38, 859–865.

[67] Priolo, M.A., Gamboa, D., and Grunlan, J.C. 2010. Transparent Clay-Polymer Nano Brick Wall Assemblies with Tailorable Oxygen Barrier. *ACS Applied Materials and Interfaces*, 2, 312–320.

[68] Ray, S.S., and Bousmina, M. 2005. Biodegradable polymers and their layered silicate nanocomposites: In greening the 21st century materials world. *Progress in Materials Science*, 50, 962–1079.

[69] Report on the food contact application of composites and cushioning materials, EC project of the Sixth Framework Programme: Innovation and sustainable Development in the Fibre Based Packaging Value Chain (SustainPack NMP3 CT-2004-500311).

[70] Rexam, 2006. Consumer packaging report. Future Innovation Today. http://www.rexam.com.

[71] Rhim, J.W., Lee, J.H., and Hong, S.I. 2006. Water resistance and mechanical properties of biopolymer (alginate and soy protein) coated paperboards. *LWT-Food Science and Technology* 39, 806–813.

[72] Sanchez-Garcia, M.D., Gimenez, E., and Lagaron, J.M. 2008. Morphology and barrier properties of solvent cast composites of thermoplastic biopolymers and purified cellulose fibers. *Carbohydrate Polymers*, 71, 235–244.

[73] Schäfer, A. 2007. Regulation of food contact materials in the EU. In: *Chemical migration and food contact materials* (eds) Barnes, K.A. Sinclair, R. and Watson, D.H. CRC Press.

[74] Sebti *et al.* 2003. Controlled diffusion of an antimicrobial peptide from a biopolymer film. *J. Chem. Eng. Res. Design, Trans IchemE*, 81, 1099–1104.

[75] Siracusa, V., Rocculi, P., Romani, S., *et al.*, 2008. Biodegradable polymers for food packaging: a review. *Trends in Food Science and Technology*, 19, 634–643.

[76] Szabo, T.S., Szekeres, M., Dekany, I., *et al.*, 2007. Layer-by-layer construction of ultrathin hybrid films with proteins and clay minerals. *Journal of Physical Chemistry*, 111, 12730–12740.

[77] Tsai, G.J., Su, W.H., Chen, H.C., *et al.*, 2002. Antimicrobial activity of shrimp chitin and chitosan from different treatments and applications of fish preservation. *Fish Science*, 68, 170–177.

[78] Varoquaux, P., Ozdemir, I. 2005. Packaging and produce degradation. In: Produce Degradation: Pathways and Prevention, Lamikanra, O., Imam, S.H., and Ukuku, D. O. (eds.) Boca Raton, Florida: CRC Press, pp 117–153.

[79] Wang, C.Y., Ye, S.Q., Sun, Q.L., *et al.*, 2008. Microcapsules for controlled release fabricated via layer-by-layer self-assembly of polyelectrolytes. *Journal of Experimental Nanoscience*, 3, 133–145.

[80] Zagory, D., Kader, A.A. 1988. Modified atmosphere packaging of fresh produce. *Food Technology*, 42, 70–77.

11

Biopolymers for Edible Films and Coatings in Food Applications

Idoya Fernández-Pan and Juan Ignacio Maté Caballero

Department of Food Technology, College of Agricultural Engineering, Universidad Pública de Navarra, Pamplona, Spain

11.1 INTRODUCTION

An edible film or coating could be defined simply as a thin and continuous layer of edible material that is usually placed on food surfaces for preservation or quality purposes. The terms edible film and edible coating are often used interchangeably; however, strictly speaking, edible films are regarded as stand-alone films, as they are formed before use. Such films can be used as covers, wraps or separation layers [1]. On the other hand, edible coatings are formed directly on the food surface. In this case, coatings are regarded typically as a part of the final product. Materials used to form edible films and coatings include biopolymers such as proteins or carbohydrates as well as lipids. However, since lipids are not polymers they do not generally form cohesive stand-alone films for practical applications.

Biopolymers – New Materials for Sustainable Films and Coatings, First Edition.
Edited by David Plackett.
© 2011 John Wiley & Sons, Ltd. Published 2011 by John Wiley & Sons, Ltd.

The use of edible films and coatings constitutes one of the most important applications of biopolymers in the area of food science and technology and has been the focus of great research efforts in the last 20 years [2–8]. However, the idea of covering foodstuffs has been with us for a long time. Fruit waxing has been applied for centuries to retard dehydration and to impart brightness [9]. Collagen and cellulose-based casings are commonly used to enrobe meat emulsions to provide structural integrity. Zein coatings have been developed for confectionery items such as jellies, liquorice and butter creams to prevent sticking and provide a shiny surface [7].

Edible films or coatings are used on foodstuffs to improve food quality and to increase shelf-life. Specific reasons for this procedure could include the following:

Controlling mass transfer. Undesirable mass transfer phenomena between the food surface and the environment can lead to a loss of food quality. These mass transfer problems include surface dehydration, moisture absorption, aroma and flavour loss or gain, oxygen uptake and fat migration. Edible films and coatings on the food surface can prevent these problems by acting as a mass transfer barrier between the food and the environment [10]. In this case, edible films and coatings are not intended to substitute the conventional protective packaging, but to complement it. Since one or several functions of the conventional package can be performed by the edible coating, a simpler, less expensive, possibly lighter and more recyclable package could potentially be used if an edible coating is applied. An illustrative example is shown in Figure 11.1. Furthermore, edible films and coatings continue to protect the product after the package has been opened [1]. In addition, they can also act as a mass transfer barrier between components within a heterogeneous food system, which could obviously not be achieved by means of a conventional synthetic non-edible material.

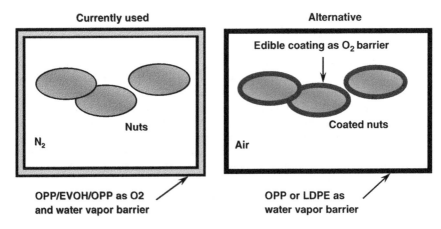

Figure 11.1 Nuts need to be protected against oxygen and moisture absorption and a laminated plastic of OPP/EVOH/OPP (nitrogen flushed) is an expensive currently used package. A less expensive and more environmental friendly alternative is the use of an edible coating as oxygen barrier on the nut surface together within a bag of LDPE or OPP acting as water vapour barrier. In this case there is no need for a nitrogen atmosphere

Carrier of food ingredients or additives. Edible films and coatings can be used as a means of adding functional ingredients/additives to a food system in order to enhance food quality, stability and safety [11]. These ingredients include antioxidants, antimicrobials, texture enhancers, nutraceuticals, probiotics, colorants and spices [12]. Edible coatings can also be used for flavouring encapsulation [13]. In addition, standalone edible films are used in both food and pharmaceutical industries to prepare premeasured single doses of ingredients [14].

Controlling surface conditions. There are some deteriorating phenomena in foodstuffs that start at and are focused on the food surface. Microbial growth and enzymatic browning in fruits are among them [15]. Adding some active compounds directly to the food surface to control some critical condition could delay this problem. However, since the additive diffuses into the bulk of the product, lowering the concentration at the food surface [16], the effective time span of this method is limited. Incorporating the active compounds into an edible film or coating that is placed directly on the food surface is an alternative, providing that the release rate of the active compound on the food surface is under control (Figure 11.2). Thus, an edible coating could serve as a means to control the time that an active compound is on the food surface above a certain critical concentration. By using this approach, a strong localized functional effect would be imparted, without elevating the overall concentration of the additive in the food [11].

Physical and mechanical protection. An additional layer of edible material on the surface can be used to protect food from physical damage caused by mechanical impact, pressure, vibrations and other mechanical factors. This mechanical protection is especially important in fragile food such as breakfast cereals, freeze-dried food and bleached fruits and vegetables. In addition, there are some foods that do not have a defined solid structure, such as meat emulsions or some extruded foods. The structural integrity and the manageability of these foods can be improved by enrobing them with an edible coating [17].

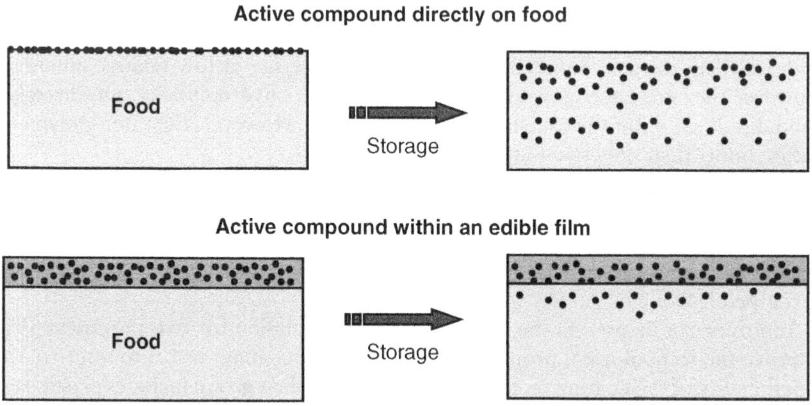

Active compound directly on food

Food

Storage

Active compound within an edible film

Food

Storage

Figure 11.2   Theoretical evolution of concentration of active compounds when added directly to the food surface or through an edible coating

Sensorial improvement. Since the food surface is usually responsible for the consumer's first sensorial impressions, an edible film or coating covering a food product can be used to change the organoleptic properties of the coated food. Films or coatings can be used to impart brightness, to provide a homogeneous stable colour or to provide a non-sticky or non-greasy surface [18–19]. Edible coatings have been frequently used as an adhesive to add sugar, salt, seasonings, colourings, or flavours onto the surface of a large variety of food products, especially snack foods [7].

Other objectives. Edible films and coatings have been tested as processing aids. For example, they have been used to improve the effectiveness of the popping process for popcorn. In addition, Japanese polysaccharide films are used for meat products such as ham before they are smoked or steamed. The films dissolve during the processing and the coated meat shows improved texture and reduced moisture loss [12]. Edible coatings have also been used in agriculture as a means to add fungicides on seed surfaces. The coating helps to maintain the appropriate fungicide concentration on the surface avoiding losses caused by the weather and thus reducing the required fungicide doses [20]. Biopolymers employed for edible coatings have also been used in the pharmaceutical sector as encapsulant materials for drug delivery [21].

11.2 MATERIALS FOR EDIBLE FILMS AND COATINGS

The formulation of films and coatings must include at least one component capable of forming a cohesive structural matrix. The main materials used to form a three-dimensional structural matrix are biopolymers and lipids. The biopolymers typically used for edible films are hydrocolloids that can be further classified into two categories: proteins and carbohydrates (Figure 11.3). The physical and chemical characteristics of the biopolymers and lipids greatly determine the properties of resulting films and coatings.

Both carbohydrates and proteins can generally be considered good film formers with excellent oxygen, aroma and lipid barrier properties at low relative humidity. However they are poor moisture barriers due to their hydrophilicity. On the other hand, lipids are naturally excellent moisture barriers. However, their non-polymeric nature limits their cohesive film-forming capacity [6].

Composite films based on the combination of hydrocolloids and lipids have been successfully employed to improve the overall characteristics of edible coatings. In addition, with the purpose of obtaining a film with unique properties, biopolymer composites, based on a mixture of different biopolymers, have been developed [11].

Additives can be used in the film or coating formulation for two purposes: (i) to improve the technological properties of the film and coating, or (ii) to improve the functionality of the coating on the food product. The first group includes plasticizers to improve the mechanical properties and surfactants to improve both the wettability of the coating solution on the food surface and the stability of the emulsion of a

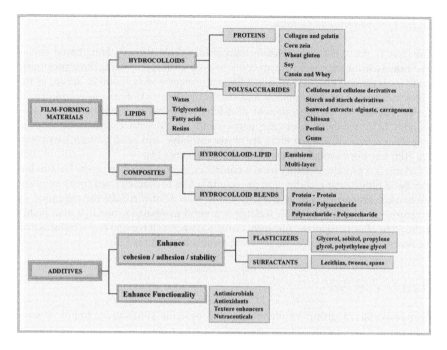

Figure 11.3   Main components for formulation of edible films and coatings for food applications

composite film-forming solution. The second group includes antimicrobial, antioxidant and other functional ingredients/additives (Figure 11.3).

It is important to emphasize that all film components, including any functional additives in the film-forming materials should be food-grade. That is, ingredients acceptable for edible films and coatings should be used within any limitations specified by the particular legal system. In addition, if a coating formulation includes a known allergen (e.g., milk protein, gluten) it must also be clearly labelled [22].

11.2.1   Protein–Based Films and Coatings

Protein film-forming materials are derived from many different animal and plant sources. Proteins are macromolecules, heteropolymers with specific amino acids as their monomer units. A large number of sequences and molecular structures with a wide range of potential interactions and chemical reactions can be found in the corresponding literature [11]. The structure of a protein matrix can be modified by heat, pressure or other actions in order to further control the physical and mechanical properties of the resulting films and coatings [23].

11.2.1.1 Collagen and Gelatin

Collagen is the most commercially successful edible protein film, based on its traditional use for the production of edible sausage casings, replacing those previously made from natural gut. Collagen is a fibrous protein widely found in nature as the major constituent of skin, tendon and connective tissue. Collagen is a hydrophilic protein because of the high content of acidic, basic and hydroxylated amino acid residues [25]. Collagen films are particularly strong with excellent mechanical properties overall. Collagen films are water insoluble and good oxygen barriers at 0% RH, but not particularly good moisture barriers [25].

Gelatin is a protein resulting from a controlled hydrolysis of collagen. Gelatin has very good film-forming ability and the resulting film is characterized by its excellent mechanical properties. Edible coatings made with gelatin reduce the migration of moisture, oxygen and oil [23]. Gelatin is a good gas barrier, but it is also highly hydrophilic. Both origin and film-processing parameters have significant influence on the functional properties of the resulting gelatin-based films [26].

11.2.1.2 Corn Zein

Zein comprises a group of alcohol-soluble proteins (prolamins) found in corn endosperm. Commercial zein is essentially a by-product of the corn wet-milling industry. Zein has a hydrophobic nature related to its high content of non-polar amino acids [27].

Zein films are easily cast from alcohol solutions with added plasticizers such as glycerol, polyethylene glycol or mixtures. Addition of plasticizers and cross-linking agents are employed to affect both mechanical and barrier properties. Besides, zein films may be formed by extrusion of dry resin pellets. The tensile strength (TS) of zein films varies substantially with RH and temperature (T), but is similar of that of wheat gluten films and two- to threefold lower than that of methylcellulose and hydroxypropylcellulose films. Zein films are insoluble in water although they are not good moisture barriers, with WVP values lower or similar to those of other protein films, cellulose ethers and cellophane [28], but much higher than those of LDPE. Zein coatings are traditionally used as oxygen and lipid barriers for nuts, candies, confectionery products and other foods. Zein-based coatings have been applied to fresh and dried fruits as a substitute for shellac coatings [29].

11.2.1.3 Wheat Gluten

Wheat gluten, a general term for water-insoluble proteins of wheat flour, is composed of a mixture of polypeptide molecules, considered to be globular proteins. Wheat-gluten-based films are homogeneous, transparent, mechanically strong, water insoluble and semi-permeable to oxygen and carbon dioxide [31]. Cohesion of wheat gluten-based films depends mainly on the type and density of intra- and intermolecular interactions. In general, wheat gluten films show low TS, but high elongation (E) compared with other films; however, their mechanical properties depend greatly on processing conditions. Both moisture and gas permeabilities are highly dependent on

RH, T and on the nature and amount of plasticizer added [32]. The addition of plasticizers is essential to avoid film brittleness as a consequence of extensive intermolecular associations [14].

11.2.1.4 Soy Protein

The protein content of soybeans is much higher than the protein content of cereal grains. Most of the protein in soybeans is insoluble in water, but soluble in dilute neutral salt solutions. Soy protein isolate (SPI) has been used to develop edible and biodegradable films. SPI, with higher protein content than other soy protein products, has superior film-forming ability [33]. SPI films are usually obtained by the casting method and are characterized by having high barrier properties against both oxygen and oil at low relative humidity. However, their mechanical properties and relatively high moisture sensitivity limit the usage of pure SPI films [34]. Consequently, addition of other molecules to the protein is essential for improving the SPI film for practical purposes.

11.2.1.5 Milk Proteins

Milk proteins can be divided into casein and whey protein. Casein represents approximately 80% of the total milk protein. Sodium and potassium caseinates are used in preference to calcium and magnesium caseinate as they are more soluble and have better functional characteristics [10]. Casein forms films from aqueous solutions without further treatment due to its random-coil nature and ability to hydrogen bond extensively [23]. Casein films, being transparent, flavourless, and flexible are attractive for food applications. Pure caseinate films are highly water soluble. The ability to function as a surfactant makes casein a very promising material for the formation of emulsion films [29].

Whey protein is the protein that remains soluble after casein has been precipitated at pH 4.6 and 20 °C. Whey proteins are globular and heat labile in nature [6]. Whey protein isolate (WPI)-based edible films are water insoluble, have good mechanical properties and are excellent oxygen, lipid and aroma barriers. Formulation of WPI-based films must include low-molecular weight plasticizers to enhance their flexibility. Due to their inherent emulsifier capacity these films have been used in a variety of lipid-based composite films [29].

11.2.2 Polysaccharide-Based Films and Coatings

Polysaccharides used for edible films and coatings include cellulose derivatives, starch and starch derivatives, chitosan, seaweed extracts, pectin and gums. The sequence of polysaccharides is simple compared with that of proteins, since only a few monomers are involved in the range of existing polysaccharides. However, the conformation of polysaccharide structures presents much larger molecular weights than proteins [11]. Because of the make up of the polymer chains, polysaccharide films exhibit good gas permeability, but their hydrophilic nature makes them poor barriers for water vapour [35–36].

11.2.2.1   Cellulose and Cellulose Derivatives

Cellulose is the most abundant natural biopolymer on earth and is essentially a linear natural high-molecular weight polymer of D-glucose units linked through β-1,4 glycosidic bonds. As a consequence of its chemical structure, it is highly crystalline, fibrous, and water-insoluble [24]. Cellulose ethers are polymer substances obtained by partial substitution of hydroxyl groups in cellulose by ether functions. Methyl-cellulose (MC), hydroxypropylcellulose (HPC) and hydroxypropylmethylcellulose (HPMC) are nonionic cellulose ethers which are soluble in cold water and form gels upon heating. Carboxymethylcellulose (CMC) is an ionic, cold and hot water-soluble cellulose ether, usually available in its sodium salt form [37]. MC, HPMC, HPC and CMC possess good film-forming characteristics. The films are generally odourless and tasteless, flexible, of moderate strength, transparent, resistant to oil and fats, water-soluble, and exhibit moderate moisture and oxygen transmission properties [6, 30].

11.2.2.2   Starch and Starch Derivatives

Starch, the main polysaccharide energy storage material in the plant kingdom, is primarily derived from cereal grains (corn, wheat, rice), potatoes and tapioca. Starch is a mixture of amylose and amylopectin polymers whose content varies depending on its botanic origin. Amylose is a nearly linear polymer composed of α-1,4 anhydro-glucose units. In contrast, amylopectin is a highly branched polymer consisting of short α-1,4 chains linked by α-1,6 glucosidic branching points occurring every 25–30 glucose units [38]. Amylose is known to form coherent and relatively strong, freestanding films, but amylopectin films can be brittle and non-continuous [24, 37].

The potential of starch as a biomaterial for edible films has been widely recognized [6]. Starch provides a good barrier to oxygen and carbon dioxide transmission, but is also very hydrophilic [39–40]. Different studies have been conducted to improve the barrier properties of starch-based films by adding lipids [41]. Starches can be chemically modified to enhance their properties or to create starch products with unique properties for various applications. It is also possible to find cross-linked, substituted, oxidized, acid-hydrolyzed starches, and high-amylose corn starch.

11.2.2.3   Chitosan

Chitosan is obtained by alkaline deacetylation of chitin. This biopolymer, obtained from crustacean wastes, is the second most abundant naturally occurring biopolymer after cellulose. Chitosan is a cationic copolymer [30]. Chitosan exhibits antibacterial and antifungal activity and it forms films without the addition of plasticizers. Chitosan films are characterized by having low oxygen permeability, and by pre-senting excellent mechanical properties [42]. However, several researchers confirmed significant variations in the mechanical and barrier properties of the films which are mainly due to the type and concentration of chitosan and the type of solvent used [43–44]. The main disadvantage of chitosan films is their high sensitivity to moisture. The functional properties of chitosan films have been improved when combined with other film-forming compounds [43, 45–47].

11.2.2.4 Seaweed Extracts

Seaweed extracts are classified in three main groups: alginates, carrageenans and agar. Alginates are linear, unbranched polymers that contain D-mannuronic and L-guluronic acid units and are therefore highly anionic under certain pH conditions [48]. The ability of alginates to react with divalent and trivalent cations has been utilized to induce ionic interactions followed by inter-chain hydrogen bonding during alginate film formation [24, 36–37]. Alginates possess good film-forming properties, producing uniform, transparent and water-soluble films. Alginate-based films are good oxygen barriers, impervious to oils and fats but have high water vapour permeability [35]. Carrageenan is a water-soluble galactose polymer and ι-, κ-, and λ-carrageenans are used in food applications. Gelation of ι- and κ-carrageenan occurs with both monovalent and divalent cations, whereas λ-carrageenan is prevalent as a non-gelling and thickening agent [24]. Agar, also a galactose polymer, forms strong gels characterized by melting points far above the initial gelation temperature [24, 36–37].

Other polysaccharides used to form edible films and coatings are pectins and gums. Pectin is a complex anionic polysaccharide composed of D-galacturonic acid residues, where the uronic acid carboxyls are to different degrees methyl esterified. Gums, include exudate (arabic, karaya), microbial fermentation (pullulan, xanthan gum, dextran and gellan gum) and seed gums (guar gum) [35].

11.2.3 Lipid-Based Films and Coatings

Many lipid compounds have been used to make edible films and coatings. Suitable lipids include waxes, neutral lipids of glycerides, fatty acids and resins.

Waxes are esters of long-chain aliphatic acids with long-chain aliphatic alcohols. Since waxes are highly hydrophobic, they are insoluble in bulk water, but soluble in typical organic solvents. Wax coatings are substantially more resistant to moisture transport than other lipid or non-lipid edible coatings and therefore they are used worldwide as the most efficient barriers to water vapour transfer [29]. Edible waxes can be divided into synthetic and natural forms. Paraffin is a synthetic wax permitted for use in a restricted list of fresh fruit and on cheese. The most important natural waxes employed for edible coating applications are carnauba, beeswax and candelilla.

Triglycerides or neutral lipids are esters of fatty acids with glycerol. They have increased polarity relative to waxes, are insoluble in bulk water, but spread to form a stable monolayer [24].

Fatty acids are also considered to be polar lipids and are used primarily as emulsifiers and dispersing agents. The most important fatty acids used for preparation of edible films and coatings include lauric, palmitic, stearic, oleic and linoleic acids. Their properties, including the water vapour permeability of films, are markedly dependent on their physical state, chain length and degree of saturation [49].

Resins are represented by shellac and are mainly used to impart gloss to products. Shellac is composed of a complex mixture of aliphatic alicyclic hydroxyacid polymers.

It is soluble in alcohols and in alkaline solutions and is compatible with most waxes, resulting in improved moisture-barrier properties and increased gloss for coated products [49]. Shellac resin has been used as a varnish and in edible coatings for pharmaceuticals, confectionery, fruit and vegetables.

Lipid-based coatings are characterized by high water vapour barrier properties in comparison with those of polysaccharide- and protein-based coatings [29]. On the other hand, the non-polymeric nature of lipids limits their ability to form cohesive films. Therefore, pure lipids are used as coatings in most cases. In order to be employed as free-standing films their fragility requires their use in conjunction with a hydrocolloid as a supporting matrix, forming composite edible films [49].

11.2.4   Composite/Multilayer Films

Composite films are based on more than one component able to form cohesive structural matrices and are made to create desirable film structures for specific applications (see Figure 11.3). Traditionally, composite films and coatings have been developed to combine the advantages of both lipid and hydrocolloid components. The lipid component in the formulation would serve as a good barrier to water vapour, whereas the polysaccharide or protein would provide the film integrity, acting as a supporting matrix [50–51]. According to their preparation, emulsion or bilayer films can be obtained. Bilayer films can potentially provide better moisture barriers. However, they tend to delaminate and exhibit poor mechanical properties compared to emulsion films. The moisture barrier properties of emulsion films can be improved by using viscoelastic lipids, increasing lipid content, reducing lipid particle size and improving film-drying conditions [52].

More recently, numerous multi-component films based on different biopolymer combinations have been developed to form films with properties that combine the most desirable attributes of each component [52].

Osés et al. [53] developed composite edible films made of mixtures of WPI and mesquite gum (MG) and suggested the use of MG to improve mechanical properties of WPI films as an alternative to using larger amounts of low-molecular weight plasticizers such as glycerol or sorbitol. Jia et al. [54] developed composite edible films based on konjac glucomannan, chitosan and SPI. WVP and mechanical properties depended on glucomannan, chitosan and SPI concentration. Fabra et al. [55] developed composite edible films from blends of sodium caseinate, alginate or λ-carrageenan as polysaccharides and oleic acid and beeswax as lipids to improve the mechanical and barrier properties of films. Composite films obtained from different ratios of SPI and cod gelatin were prepared by Denavi et al. [56]. The WVP of the composite films diminished significantly as compared to pure-gelatin films. Incorporation of caseinates in SPI-based films containing lipids (oleic acid or oleic acid-beeswax blends) resulted in increased elastic modulus and TS and contributed to increasing the WVP [57]. Ghanbarzadeh et al. [58] prepared laminated WPI and zein films. Laminated films had higher TS than the single whey protein films and the single zein films plasticized by olive oil. In order to improve water resistance and mechanical properties of films made from chitosan, Xu et al. [59] developed chitosan-starch composite films. Flores et al. [60] developed tapioca starch and xanthan gum-based composite films. The resulting composite films were flexible and homogeneous. The xanthan gum increased film solubility and produced a reinforcing effect.

11.2.5   Additives

The functional, organoleptic, nutritional, and technological properties of edible films can be modified by the addition of various chemicals in minor amounts. Due to the fact that cohesive forces between film polymer molecules can result in undesirable brittleness, the most frequently used additives in edible films and coatings are plasticizers. Food-grade plasticizers are small low-molecular weight, non-volatile compounds added to the film or coating formulation to interact with the polymer chains, decreasing intermolecular forces resulting from chain-to-chain interaction. By reducing intermolecular forces and thus increasing the mobility of the polymer chains, plasticizers lower the glass transition temperature of films and improve film flexibility, elongation and toughness [61]. The main drawback of plasticizers is the increase in gas, water vapour and solute permeability that results from decreased film cohesion [62]. Common plasticizers for edible films and coatings are saccharides (glucose, fructose-glucose syrups, sucrose and honey) and polyols (glycerol, sorbitol, glyceryl derivatives and polyethylene glycols). Glycerol is a low-molecular weight, hydrophilic plasticizer that has been widely used both in casting and thermoplastic processing of films [27].

Film functionality depends not only on cohesion (attractive forces between film polymer molecules), but also on adhesion (attractive forces between film and substrate). Emulsifiers are added to improve the latter. Edible emulsifiers such as lecithin, Tweens or Spans are surface-active agents of amphiphilic nature able to reduce the surface tension at the water–lipid or the water–air interface. As a consequence, once they have been added to a coating formulation they improve the wettability and adhesion of the film on the food surface [1, 11]. Emulsifiers are also essential for the formation and stability of protein or polysaccharide films containing lipid particles [61].

11.3   EDIBLE FILMS AND COATINGS FOR FOOD APPLICATIONS

11.3.1   Edible Coatings on Fruit and Vegetables

Edible films and coatings may contribute to maintaining the quality and extending the shelf-life of whole and minimally processed fruits and vegetables, mainly by reducing dehydration [63–64] and decreasing respiration rate. A hydrophobic edible coating (e.g., lipid-based) could reduce water loss by acting as water vapour barrier. Waxes have been used for centuries on whole citrus fruits and are now used on a large variety of fruit and vegetables. Waxes are primarily used to prevent moisture migration and damage to the surface of the produce and are also used to impart gloss. In addition, protein- and polysaccharide-based coatings with selective permeability are able to produce an internal modified atmosphere (with high CO_2 and low O_2 levels) by isolating the coated product from the environment. As a consequence, the respiration rate could be reduced, which would slow down the post-harvest metabolism of fruits and vegetables, therefore increasing their shelf-life [64].

Many different biopolymers have been found to be effective in decreasing respiration rates. These biopolymers include carbohydrates such as starch [29], chitosan [65], and pullulan [66] and proteins such as zein, wheat gluten, soy protein, caseinate and WPI [63]. Emulsion coatings have also been used to reduce both dehydration and respiration rate [67].

White blush on the surface of processed raw carrots (baby carrots) is a phenomenon that occurs during product storage, resulting in loss of quality. White blush consists of a physical phenomenon resulting essentially from superficial dehydration associated with the collapse of surface cells [50, 68–69]. Emulsion coatings based on sodium caseinate with stearic acid have proven effective against this disorder [70]. Mei et al. [71] reported similar results when they coated carrots with a xanthan gum solution. Vargas et al. [72] developed chitosan coatings which were able to diminish the white blush during storage, and reported that coatings applied with a vacuum pulse significantly improved the WVP of the samples.

Edible coatings for minimally processed fruit can contain antioxidant agents to protect the produce against enzymatic browning [18, 64, 73–74]. McHugh and Senesi [18] delayed browning of fresh-cut apples coated with a mixture of apple purée, pectin and vegetable oils containing ascorbic acid and citric acid. Olivas et al. [75] preserved fresh-cut pear from surface browning by using an MC-based coating containing ascorbic acid and citric acid. Edible films and coatings have also been developed to enhance the texture of fruit and vegetables [76]. Lee et al. [73] indicated that incorporating calcium chloride in WPC coatings improved the firmness of fresh-cut apple pieces. Rojas-Graü et al. [77] observed that apple wedges coated with alginate and gellan edible coatings enriched with calcium salts maintained their initial firmness during refrigerated storage. Similar results were obtained by Oms-Oliu, et al. [78] on fresh-cut melon coated with alginate, gellan and pectin.

Improvement of the food safety of fruit and vegetables has also been a focus of various researchers. Thus, Eswaranandam et al. [79], incorporated malic and lactic acid as antimicrobial agents in SPI coatings to extend the shelf-life of fresh-cut cantaloupe melon. Rojas-Graü et al. [80]combined alginate and gellan edible coatings with lemongrass, oregano and vanillin essential oils (EOs) to prolong the shelf-life of fresh-cut apples. Raybaudi-Massilia et al. [81] applied alginate-based coatings with cinnamon, clove or lemongrass EOs on apples, reducing the growth of E. coli. On the other hand, the inherent antifungal activity of chitosan-based coatings was used to protect highly perishable fruits such as strawberries, raspberries, and grapes from fungal decay [68]. Nutraceuticals have also been incorporated into the formulation of edible coatings. For example, Park and Zhao [82] developed a chitosan film with calcium, zinc and Vitamin E incorporated. Han et al. [83] improved both the nutritional and physicochemical quality of strawberries and raspberries by employing chitosan-based coatings enriched with calcium and vitamin E.

11.3.2 Edible Films and Coatings on Meat and Poultry

Edible films and protective coatings have been used for a long time to prevent quality loss in meat and poultry products [84]. According to Gennadios et al. [37], Cutter [36] and Coma [85], edible coatings can improve the quality of fresh, frozen and processed meat and poultry products by delaying moisture loss, reducing lipid oxidation and discoloration, enhancing product appearance and functioning as a carrier of food additives. Collagen films acting as casing for sausages is probably the most successful commercial application of edible film technology [7].

Meat, poultry and most of their fresh and ready-to-eat products are highly perishable food commodities, mainly due to their enriched nutrient composition, high pH and high water activity. These characteristics make fresh meats an ideal substrate for the growth of several pathogenic and spoilage bacteria [86]. Active edible films and coatings with antimicrobial properties have been developed. The most important antimicrobial agents added to edible coating formulations are bacteriocins, organic acids and EOs [87–88]. In addition, chitosan-based coatings have been used to protect meat products because of the inherent antimicrobial properties of this biopolymer [89].

A total inhibition of L. monocytogenes growth on ham, turkey meat and beef was achieved by Ming et al. [90] when bacteriocins were incorporated on a cellulose-based matrix. When nisin was incorporated in agar coatings and applied on a fresh poultry product, reductions in S. typhimurium growth were recorded during refrigerated storage [91–92]. Rossi-Márquez et al. [93] developed an active film based on WPI and nisin and obtained more than four log cycles reduction of Brochotrix thermosphacta from the surface of a ham sample after 8 days of incubation at 4 °C. Grape seed extracts combined with nisin and organic acids in WPI-based coating inhibited the growth of L. monocytogenes, E. coli and S. typhimorium in a turkey frankfurter system stored at refrigeration temperature [88]. Siragusa et al. [89] obtained a significant reduction of L. monocytogenes growth on raw beef by the application of organic acids immobilized in calcium alginate gels.

Other antimicrobial compounds have been incorporated in edible coatings. For instance, Ravishankar et al. [94] successfully developed apple-based edible films containing cinnamaldehyde or carvacrol against S. enterica or E. coli inoculated on chicken breasts and L. monocytogenes on ham. The effectiveness of antimicrobial films incorporating different levels of sodium lactate and ε-polylysine into WPI films against beef spoilage flora during storage at 5 °C was determined by Zinoviadou et al. [95]. These authors obtained significant inhibition in growth of the total viable count of lactic acid bacteria and pseudomona when different film compositions were tested on fresh beef cut portions.

Lipid oxidation is another main spoilage mechanism for quality degradation during storage in meat and poultry products [96]. Recently, Haque et al. [97] successfully employed sour whey powder-based edible coatings with and without CMC to reduce oxidative degradation of cut beef steaks and extend quality. Chidanandaiah et al. [98] reported that alginate coating on buffalo meat patties followed by dipping in calcium chloride provided protection against oxidative and microbiological deterioration. Furthermore, the alginate coat improved the sensorial quality of the product.

Edible films and coatings offer a number of benefits to meat and poultry products other than fresh and ready-to-eat. On grilled pork, Yingyuad et al. [98] reported a significant reduction in the total viable count of samples when a chitosan coating was associated with vacuum packaging during refrigerated storage. Yu et al. [99] reported that sodium alginate coating improved the quality of pork during frozen storage.

Sodium alginate coating decreased thawing loss of meat without significant differences between controls and treatments in terms of structure.

11.3.3 Edible Films and Coatings on Foods with Low Water Content

Food products with reduced water content include, among others, nuts, cereal grains, cereal-based products, bakery products and sweets. Some of these products such as dry nuts can be utilized not only for direct retail consumption but also as ingredients in other foods such as ice creams or confectionery.

Nuts are typically low-moisture, high-fat foods. The most common forms of spoilage of nuts and nut-containing products are rancidity due to lipid oxidation, sogginess due to moisture uptake, lipid migration in food systems, and loss of flavour [100].

There are two different strategies to protect nuts from oxidative rancidity with an edible coating. The first is to reduce the oxygen level within the nut by a high oxygen barrier coating. The second is to use the coating as a carrier of antioxidants which can act on the nut surface. To prevent oxidation, almonds and hazelnuts were protected with low methoxyl pectin or cellulose derivative coatings [101]. Zein coatings were used as oxygen, lipid, and moisture barriers for nuts, candies, and confectionery products [28]. Whey protein coating has been proved to be a very effective oxygen barrier on dry roasted peanuts, delaying oxidative rancidity in terms of hexanal content and peroxide value [102]. Maté and Krochta [103] found that acetylated monoglyceride coatings carrying tocopherol as antioxidant significantly reduced hexanal production in walnuts. More recently, Javanmard [104] coated dried pistachio kernels with WPC plasticized with glycerol. The results indicated that the coated pistachios were oxidized at a significantly slower rate than the uncoated ones. The WPC coatings, regardless of composition or thickness, provided a glossy appearance to the coated nuts.

Edible coatings have been used as a moisture barrier on dried food products. In fact, the main use of edible films on bakery products is to slow down moisture migration [7]. For this purpose, lipid-based coatings have been found to be more efficient than hydrocolloid coatings. Chocolate coatings are used as a moisture barrier in ice cream to maintain the crispness of the cone [105]. Blends of beeswax and vegetable oils or acetylated monoglycerides have been used to reduce moisture loss in raisins. As an alternative, films based on composite edible films (polysaccharides and lipids) were shown to be effective in controlling moisture transfer in crackers [106] and in ice cream cones [107].

11.3.4 Edible Coatings on Deep Fat Frying Foods

Fried foods contain up to 45% oil [108]. Nowadays, there is a trend towards healthy eating which demands a reduction of the amount of fat in fried products. In addition, the fact that several components of the fried foods are released into the oil accelerates its degradation and limits its operating life. Controlling the mass transfer between the fried food and the oil would be beneficial for both the final quality of the fried food and the number of times the oil can be used. Thus, an application of an edible coating prior to frying, acting as a lipid and moisture barrier could imply a reduction of fat intake and reduction of the transfer of food components into the oil [7].

Based on their hydrophilicity, hydrocolloid-based coatings have been used in the past to reduce oil absorption in fried foods. Among them, many different protein- or carbohydrate-based coatings have been studied as oil barriers. Thus, Mallikarjunan *et al.* [109] coated a fat-free starchy product with HPMC and MC film-forming solutions. The MC-based coating proved to be the most efficient one. García *et al.* [110–111] and Quasem *et al.* [112] also reported that MC coatings were very effective in reducing the oil uptake of fried foods. Khalil [113] reported that pectin or sodium alginate coatings that were fixed with calcium ions significantly reduced the oil uptake of French-fried potatoes. The best treatment included a secondary layer of CMC. Most authors claimed significant oil uptake reduction in fried food, regardless of formulation. Most of the authors tried a coating independent from the batter. However, Dogan *et al.* [114] mixed the protein coating solution with the batter before applying it on chicken nuggets. WPI-added batters not only reduced the oil content of the fried nuggets, but also improved the crunchiness of the product. Albert and Mittal [115], compared the behaviour of different polysaccharides and proteins as oil barriers in fried products and concluded that SPI/MC and SPI/WPI composite coatings provided the highest index value (reduction in fat uptake/decrease of water loss).

11.4   CONCLUDING REMARKS

The use of edible films and coatings to improve food quality and to extend the shelf-life of food products has become an attractive technique for the food industry. This is caused by the existence of a large variety of problems that can be addressed and solved with formulations that are already available as shown in Table 11.1. In addition, the use of an edible coating can mean a reduction in packaging waste and cost, and an improvement in the recyclability of the package is an extra plus for the industry and for our society.

Research lines that are now going on and have the interest of the industry include: *Hydrocolloid-based edible films with outstanding moisture barrier properties.* There has been a very important industrial demand and an equally important research response to obtain formulations and procedures for improving the water vapour barrier of hydrocolloid-based films, while maintaining their mechanical properties. With this objective, investigations on bilayer and microemulsion films will continue. In addition, research and development on bio-nanocomposite materials for edible films is expected to grow in the next decade, with the development of new bio-based materials and composites that are expected to improve the physical properties of biopolymeric matrices.

Active edible coatings. Incorporating active components, especially antimicrobials and antioxidants, in edible films for food application has been the focus of many researchers. This research line is very likely to continue because of the increasing concerns of our society for food safety.

Film stability. The properties of edible coatings generally vary in accordance with storage time, especially if compared with synthetic films, due to the intrinsic

Table 11.1 Examples of the use of biopolymers as edible films or coatings in food applications

Problem	Product	Biopolymer	Main function	Reference
FRUITS AND VEGETABLES				
Senescence	Strawberry	CMC, WPI, casein	Gases, moisture barrier	[117]
	Mango	Wax, shellac, zein	Gases, moisture barrier	[118]
	Peach	Wax, CMC	Moisture barrier	[119]
	Avocado	MC	Gases, moisture barrier	[120]
Enzymatic browning	Potato	Caseinate, WPI	Oxygen barrier	[75]
	Fresh-cut pear	MC	Antioxidant carrier	[76]
	Fresh-cut pear	Alginate, gellan	Antioxidant carrier	[121]
	Fresh-cut apple	Pectin, apple puree	Antioxidant carrier	[18]
	Fresh-cut apple	WPC, beeswax	Antioxidant carrier	[74]
	Fresh-cut apple	Carrageenan, WPC	Antioxidant carrier	[73]
	Fresh-cut melon	SPI	Antioxidant carrier	[79]
Microbial spoilage	Fresh-cut apple	Alginate	Antimicrobial carrier	[15]
	Strawberry	Starch	Antimicrobial carrier	[122]
	Fresh-cut melon	Alginate	Antimicrobial carrier	[81]
White blushing	Baby carrot	Chitosan	Dehydration prevention	[123]
	Baby carrot	Caseinate	Dehydration prevention	[70]
Firmness loss	Fresh-cut apple	Alginate, gellan	Texture enhancer carrier	[77]
	Melon	Alginate, pectin	Texture enhancer carrier	[78]
Internal liquids loss	Pineapple	Alginate	Texture enhancer carrier	[124]
Nutritional value loss	Strawberry	Chitosan	Nutraceutical carrier	[65]
	Strawberry	Chitosan	Nutraceutical carrier	[97]
MEAT AND POULTRY				
Lack of structure	Sausages	Collagen	Structural integrity	[125]
Microbial spoilage	Beef	Cellulose	Antimicrobial carrier	[89]
	Turkey frankfurter	WPI	Antimicrobial carrier	[88]

	Food	Polymer	Function	Reference
	Beef	WPI	Antimicrobial carrier	[95]
	Cooked ham	Chitosan	Antimicrobial	[126]
	Cooked ham	Chitosan	Antimicrobial carrier	[126]
	Cooked meats	Cellulose	Antimicrobial carrier	[90]
	Chicken breast	Apple puree	Antimicrobial carrier	[94]
	Ham	Apple puree	Antimicrobial carrier	[94]
	Grilled pork	Chitosan	Antimicrobial	[98]
Rancidity	Beef	WPI	Antioxidant carrier	[127]
	Beef steaks	Whey protein	Oxygen barrier	[96]
	Buffalo patties	Alginate	Oxygen barrier	[97]
DRIED PRODUCTS				
Lipid migration	Peanuts	SPI, CC, CMC	Oil barrier	[128]
	Peanuts	WPI	Oxygen barrier	[129]
	Peanuts	WPI	Increases surface hydrophilicity	[130]
	Peanuts	WPI	Antioxidant carrier	[131]
	Peanuts	HPC, CMC	Oxygen barrier	[132]
	Peanuts	WPI	Oxygen, moisture barrier	[102–133–135]
	Walnuts	WPI	Oxygen, moisture barrier	[103–133]
DRIED PRODUCTS				
Moisture absorption	Crackers	CS, MC	Oxygen, moisture barrier	[106]
	Ice cream cones	MC	Oxygen, moisture barrier	[107]
FRIED PRODUCTS				
Oil absorption	Fried chicken	WPI	Oil barrier	[114]
	Fish nuggets	CMC or HPMC	Oil, moisture barrier	[136]
	Potato chips	CMC	Oil barrier	[137]
	Banana chips	MC, pectin, alginate, CMC	Oil barrier	[138]
	Chicken	CMC or HPMC	Oil, moisture barrier	[139]
Moisture loss	Fish nuggets	CMC or HPMC	Oil, moisture barrier	[136]
	Chicken	CMC or HPMC	Oil, moisture barrier	[139]

instability of their raw materials. The effect of storage time and conditions on some physical changes, such as polymer reorganization (e.g., starch retrogradation), or the migration of low-molecular weight components (e.g., plasticizers), as well as chemical changes (oxidation) is sure to be the focus of upcoming projects, which will undoubtedly increase the range of applications for edible films and coatings.

Sensorial evaluation. A fair amount of laboratory research related to edible films and coatings cannot be implemented because of sensorial limitations. New ways to overcome taste difficulties associated with certain additives (e.g., surfactants) will be at the centre of future research.

REFERENCES

[1] Krochta, J.M., Proteins as raw materials for films and coatings: definitions, current status, and opportunities, in A. Gennadios (ed.), *Protein-Based Films and Coatings*, CRC Press, Boca Ratón, Florida, USA (2002).

[2] Krochta, J.M., Baldwin, E.A., and Nisperos-Carriedo, M.O. (1994) *Edible Coatings and Films to Improve Food Quality*, Technomic Publishing Company, Inc., Lancaster.

[3] Gennadios, A. (2002) *Protein-Based Films and Coatings*, CRC Press LLC, Boca Ratón, Florida.

[4] Han, J.H. (2005) *Innovations in Food Packaging*, Elsevier Academic Press, San Diego.

[5] Embuscado M.E. and Huber, K.C. (2009) *Edible Films and Coatings for Food Applications*, Springer, New York.

[6] J.M. Krochta, and DeMulder-Johnston, *Food Technol*. 1997, 51, 61–74.

[7] S. Chapman and L. Potter, *Edible Films and Coating: a Review*. 2004, Review N° 45, Campden & Chorleywood Food Research Association Group.

[8] T. Bourtoom and M.S. Chinnan, *Food Sci. Technol. International*. 2009, 15, 149–158.

[9] R.E. Hardenburg, *Agr Research Bulletin*, 1967, 51–15.

[10] T.H. McHugh and J.M. Krochta, *J AmOil Chemists' Society*, 1994, 71, 307–312.

[11] Han, J.H. and Gennadios, A., Edible Films and Coatings: a Review, in J.H. Han (ed.), *Innovations in Food Packaging*, Elsevier Academic Press, San diego, USA (2005).

[12] Martin-Belloso, O., Rojas-Graü, M.A., and Soliva-Fortuny, R., Delivery of Flavor and Active Ingredients Using Edible Films and Coatings, in M.E. Embuscado and K.C. Huber, (eds), *Edible Films and Coatings for Food Applications*, Springer: New York, USA, pp 295–314 (2009).

[13] Reineccius, G.A., Flavor Encapsulation, in J.M. Krochta, E.A. Baldwin, and M.O. Nisperos-Carriedo (eds), *Edible Coatings and Films to Improve Food Quality*, Technomic Publishing Company, Inc. Lancaster, pp 105–120. (1994).

[14] Buffo, R.A. and Han, J.H., Edible Films and Coatings from Plant Origin Proteins, in J.H. Han, (ed.), *Innovations in Food Packaging*, Elsevier Academic Press, San Diego, USA, pp 277–300 (2005).

[15] M.A. Rojas-Graü *et al.*, *Postharvest Biol. Technol*. 2007, 45, 254–264.

[16] Baker, R.A., Baldwin, E.A., and Nisperos-Carriedo, M.O., Edible Coatings for Processed Foods, in J.M. Krochta, E.A. Baldwin, and M.O. Nisperos-Carriedo

(eds), *Edible Coatings and Films to Improve food Quality*, Technomic Publishing Company, Inc., Lancaster, pp 89–104 (1994).
[17] J.J. Kester and O. Fennema, *Edible films and coatings: a review. Food Technol.*, 1986, **40**, 47–49.
[18] T.H. McHugh and E. Senesi, *J. Food Sci.* 2000, **65**, 480–490.
[19] Zhao, Y. and McDaniel, M., Sensory Quality of Foods Associated with Edible Film and Coating Systems and Shelf-life Extension, in J.H. Han (ed.), *Innovations in Food Packaging*, Elsevier Academic Press, San Diego, USA, pp 434–454 (2005).
[20] K. Ziani *et al.*, *Food Hydrocolloids*, 2009, **23**, 2309–2314.
[21] Radwick, A.E., and Burgess, D.J., Proteins as Microencapsulating Agents and Microencapsulate Active Compounds in Pharmaceutical Applications, in A. Gennadios (ed.), *Protein-Based Films and Coatings*, CRC Press LLC, Boca Ratón, Florida, pp 341–366 (2002).
[22] L.R. Franssen, T.R. Rumsey, and J.M. Krochta, *J. Food Sci.* 2004, 69.
[23] Lacroix, M. and Cooksey, K., Edible Films and Coatings from Animal-Origin Proteins, in J.H. Han (ed.), *Innovations in Food Packaging*, Elsevier Academic Press: San Diego, USA, (2005).
[24] D.S. Cha and M.S. Chinnan, *Critical Rev Food Sci Nutrition*, 2004, **44**, 223–237.
[25] Arvanitoyannis, I., Formations and Properties of Collagen and Gelatin Films and Coatings, in A. Gennadios (ed.), *Protein-Based Films and Coatings*, CRC Press LLC, Boca Ratón, Florida, pp 275–304 (2002).
[26] J. Gómez-Estaca *et al.*, *J. Food Eng.* 2009, **90**, 480–486.
[27] V.M. Hernandez-Izquierdo and J.M. Krochta, *J. Food Sci.* 2008, 73.
[28] Padua, G.W. and Wang, Q., Formation and Properties of Corn Zein Films and Coatings, in A. Gennadios (ed.), *Protein-Based Films and Coatings*, CRC Press LLC, Boca Raton, Florida, pp 43–68 (2002).
[29] D. Lin and Y. Zhao, *Comprehensive Reviews Food Sci. Food Safety*, 2007, **6**, 60–75.
[30] T. Bourtoom, *International Food Research J.* 2008, **15**, 237–248.
[31] N. Gontard *et al.*, *J. Agr. Food Chem.* 1996, **44**, 1064–1069.
[32] H. Mujica-Paz and N. Gontard, *J. Agr. Food Chem.* 1997, **45**, 4101–4105.
[33] J.W. Rhim, *Food Sci. Biotechnol.* 2007, **16**, 691–709.
[34] J.W. Rhim, *Lebensm-Wiss Technol.* 2004, **37**, 323–330.
[35] Lacroix, M. and Le Tien, C., Edible Films and Coatings from non-starch Polysaccharides, in J.H. Han (ed.), Elsevier Academic Press, San Diego, pp 338–361 (2005).
[36] C.N. Cutter, *Meat Sci.* 2006, **74**, 131–142.
[37] A. Gennadios, M.A. Hanna, and L.B. Kurth, *Lebensm-Wiss Technol.* 1997, **30**, 337–350.
[38] Z. Liu, Edible Films and Coatings from starches, in J.H. Han (ed.), *Innovations in Food Packaging*, Elsevier Academic Press, San Diego, pp 318–337 (2005).
[39] I. Arvanitoyannis, A. Nakayama, and S.I. Aiba, *Carbohyd. Polym.* 1998, **36**, 105–119.
[40] C. Pagella, G. Spigno, and D.M. De Faveri, Characterization of starch based edible coatings. *Food and Bioproducts Processing: Transactions of the Institution of Chemical Engineers, Part C*, 2002. 80(3), 193–198.
[41] M.A. García, M.N. Martino, and N.E. Zaritzky, *J. Food Sci.* 2000, **65**, 941–947.
[42] N.E. Suyatma *et al.*, *J. Agr. Food Chem.* 2005, **53**, 3950–3957.
[43] C. Caner, P.J. Vergano, and J.L. Wiles, *J. Food Sci.* 1998, **63**, 1049–1053.
[44] S.Y. Park, K.S. Marsh, and J.W. Rhim, *J. Food Sci.* 2002, **67**, 194–197.

[45] P. Mayachiew and S. Devahastin, *Drying Technol.* 2008, **26**, 176–185.
[46] K. Ziani *et al.*, *Lebensm-Wiss Technol.* 2008, **41**, 2159–2165.
[47] N. Niamsa and Y. Baimark, *Am. J. Food Technol.* 2009, **4**, 162–169.
[48] Nieto, M.B., Structure and Function of Polysaccharide Gum-Based Edible Films and Coatings, in M.E. Embuscado and K.C. Huber (eds), *Edible Films and Coatings for Food Applications*, Springer, New York, pp 57–112 (2009).
[49] Rhim, J.W. and Shellhammer, T.H., Lipid-Based Edible films and Coatings, in J.H. Han (ed.), *Innovations in Food Packaging*, Elsevier Academic Press, San Diego (2005).
[50] R.J. Avena-Bustillos *et al.*, *Postharvest Biol. Technol.* 1994, **4**, 319–329.
[51] J.W. Park *et al.*, *J. Food Sci.*, 1994, **59**, 916–919.
[52] Pérez-Gago, M.B. and Krochta, J.M. *Emulsions and Bi-layer Edible Films*, in J.H. Han (ed.), *Innovations in Food Packaging*, Elsevier Academic Press, San Diego, pp 384–402 (2005).
[53] J. Osés *et al.*, *Food Hydrocolloids*, 2009, **23**, 125–131.
[54] D. Jia, Y. Fang, and K. Yao, *Food Bioproducts Processing*, 2009, **87**, 7–10.
[55] M.J. Fabra, P. Talens, and A. Chiralt, *Carbohyd. Polym.* 2008, **74**, 419–426.
[56] G.A. Denavi *et al.*, *Food Hydrocolloids*, 2009, **23**, 2094–2101.
[57] F.M. Monedero *et al.*, *J. Food Eng.* 2010, **97**, 228–234.
[58] B. Ghanbarzadeh and A.R. Oromiehi, *J. Food Eng.* 2009, **90**, 517–524.
[59] Y.X. Xu *et al.*, *Industrial Crops Products*, 2005, **21**, 185–192.
[60] S.K. Flores *et al.*, *Materials Sci. Eng.* 2010, **30**, 196–202.
[61] Sothornvit, R. and Krochta, J.M., Plasticizers in Edible Films and Coatings, in J.H. Han (ed.), *Innovations in Food Packaging*, Elsevier Academic Press, San Diego, pp 403–433 (2005).
[62] N. Gontard, S. Guilbert, and J.L. Cuq, *J. Food Sci.* 1993, **58**, 206–211.
[63] Olivas, G.I. and Barbosa-Cánovas, G., Edible Films and Coatings for Fruits and Vegetables, in M.E. Embuscado and K.C. Huber (eds), *Innovations in Food Packaging*, Springer, New York. 2009.
[64] E.A. Baldwin, M.O. Nisperos-Carriedo, and R.A. Baker, *Critical Reviews Food Sci. Nutrition*, 1995, **35**, 509–524.
[65] M. Vargas *et al.*, *Postharvest Biol. Technol.* 2006, **41**, 164–171.
[66] T. Diab *et al.*, *J. Sci. Food Agr.* 2001, **81**, 988–1000.
[67] Baldwin, E.A. and Baker, R.A., Use of Properties in Edible Coatings for Whole and Minimally Processed Fruits and Vegetables in A. Gennadios (Ed.), *Protein-Based Films and Coatings*. CRC Press LLC, Boca Raton, Florida, pp 501–516 (2002).
[68] A. Simões *et al.*, *Postharvest Biol. Technol.* 2010, **55**, 45–52.
[69] L. Cisneros-Zevallos, M.E. Saltveit, and J.M. Krochta, *J. Food Sci.* 1997, **62**, 363–398.
[70] R.J. Avena-Bustillos *et al.*, *J. Food Eng.* 1994, **21**, 197–214.
[71] Y. Mei *et al.*, *J. Food Sci.* 2002, **67**, 1964–1968.
[72] M. Vargas *et al.*, *Food Hydrocolloids*, 2009, **23**, 536–547.
[73] J.Y. Lee *et al.*, *Lebensm-Wiss Technol.* 2003, **36**, 323–329.
[74] M.B. Pérez-Gago, M. Serra, and M.A.D. Río, *Postharvest Biol. Technol.* 2006, **39**, 84–92.
[75] G.I. Olivas, J.J. Rodriguez, and G.V. Barbosa-Cánovas, *J. Food Processing Preservation*, 2003, **27**, 299–320.
[76] C. Le-Tien *et al.*, *Biotechnol. App. Biochem.* 2004, **39**, 189–198.
[77] M.A. Rojas-Graü, M.S. Tapia, and O. Martín-Belloso, *Lebensm-Wiss Technol.* 2008, **41**, 139–147.

[78] G. Oms-Oliu, R. Soliva-Fortuny, and O. Martín-Belloso, *Lebensm-Wiss Technol.* 2008, **41**, 1862–1870.
[79] S. Eswaranandam, N.S. Hettiarachchy, and J.F. Meullenet, *J. Food Sci.* 2006, **71**, 307–313.
[80] M.A. Rojas-Graü *et al.*, *J. Food Eng.* 2007, **81**, 634–641.
[81] R.M. Raybaudi-Massilia, J. Mosqueda-Melgar, and O. Martín-Belloso, *International J. Food Microbiol.* 2008, **121**, 313–327.
[82] S.I. Park and Y. Zhao, *J. Agr. Food Chem.* 2004, **52**, 1933–1939.
[83] C. Han *et al.*, *Postharvest Biol. Technol.* 2004, **33**, 67–78.
[84] Ustunol, Z., Edible Films and Coatings for Meat and Poultry, in M.E. Embuscado and K.C. Huber (eds), *Edible Films and Coating for Food Applications*, Springer, New York (2009).
[85] V. Coma, *Meat Sci.* 2008, **78**, 90–103.
[86] Samelis, J Managing Microbial Spoilage in the Meat Industry, in C.W. Blackburn (ed.), *Food Spoilage Microorganisms*, CRC Press LLC, Boca Raton, Florida (2006).
[87] A. Cagri, Z. Ustunol and E.T. Ryser, *J. Food Protection*, 2004, **67**, 833–848.
[88] V.P. Gadang *et al.*, *J. Food Sci.* 2008, **73**, 389–394.
[89] P.K. Dutta *et al.*, *Food Chem.* 2009, **114**, 1173–1182.
[90] X. Ming *et al.*, *J. Food Sci.* 1997, **62**, 413–415.
[91] N. Natrajan and B.W. Sheldon, *J. Food Protection*, 2000, **63**, 1189–1196.
[92] N. Natrajan and B.W. Sheldon, *J. Food Protection*, 2000, **63**, 1268–1272.
[93] G. Rossi-Márquez *et al.*, *J. Sci. Food Agr.* 2009, **89**, 2492–2497.
[94] S. Ravishankar *et al.*, *Journal Food Sci.* 2009, **74**, 440–445.
[95] K.G. Zinoviadou, K.P. Koutsoumanis, and C.G. Biliaderis, *Food Hydrocolloids*, 2010, **24**, 49–59.
[96] Z.U. Haque, J. Shon, and J.B. Williams, *J. Food Quality*, 2009, **32**, 381–397.
[97] Chidanandaiah, R.C. Keshri, and M.K. Sanyal, *J. Muscle Foods*, 2009, **20**, 275–292.
[98] S. Yingyuad *et al.*, *Packaging Technol. Sci.* 2006, **19**, 149–157.
[99] X.L. Yu *et al.*, *J. Muscle Foods*, 2008, **19**, 333–351.
[100] Trezza, T.A. and Krochta, J.M., Application of Edible Protein Coatings to Nuts and Nut-Containing Food Products, in A. Gennadios (ed.), *Protein-Based Films and Coatings*, CRC Press LLC, Boca Raton, Florida, pp 527–550 (2002).
[101] F. Debeaufort, J.A. Quezada-Gallo, and A. Voilley, *Critical Reviews Food Sci. Nutrition*, 1998, **38**, 299–313.
[102] J.I. Maté, E.N. Frankel, and J.M. Krochta, *J. Agr. Food Chem.* 1996, **44**, 1736–1740.
[103] J.I. Maté and J.M. Krochta, *J. Agric. Food Chem.* 1997, **45**, 2509–2513.
[104] M. Javanmard, *J. Food Process Eng.* 2008, **31**, 247–259.
[105] T.P. Labuza and C.R. Hyman, *Trends Food Sci. Technol.* 1998, **9**, 47–55.
[106] B. Bravin, D. Peressini, and A. Sensidoni, *J. Food Eng.* 2006, **76**, 280–290.
[107] D.C. Rico-Peña and J.A. Torres, *J. Food Sci.* 1990, **55**, 1468–1469.
[108] I.S. Saguy and E.J. Pinthus, *Food Technol.* 1995, **49**, 142–145.
[109] P. Mallikarjunan *et al.*, *Lebensm-Wiss Technol.* 1997, **30**, 709–714.
[110] M.A. García *et al.*, *Food Sci. Technol. International*, 2004, **10**, 339–346.
[111] M.A. García *et al.*, *Innovative Food Sci. Emerging Technol.* 2002, **3**, 391–397.
[112] J.M. Quasem *et al.*, *Am. J. Agr. Biol. Sci.* 2009, **4**, 156–166.
[113] A.H. Khalil, *Food Chem.* 1999, **66**, 201–208.
[114] S.F. Dogan, S. Sahin, and G. Sumnu, *Eur. Food Research Technol.* 2005, **220**, 502–508.

[115] S. Albert and G.S. Mittal, *Food Research International*, 2002, **35**, 445–458.
[116] Y.M. Stuchell and J.M. Krochta, *J. Food Sci.* 1995, **60**, 28–31.
[117] C. Ribeiro *et al.*, *Postharvest Biol. Technol.* 2007, **44**, 63–70.
[118] T.T. Hoa *et al.*, *J. Food Quality*, 2002, **25**, 471–486.
[119] H. Togrul and N. Arslan, *Food Hydrocolloids*, 2004, **18**, 215–226.
[120] N. Maftoonazad and H.S. Ramaswamy, *Lebensm-Wiss Technol*, 2005, **38**, 617–624.
[121] G. Oms-Oliu, R. Soliva-Fortuny, and O. Martín-Belloso, *Postharvest Biol. Technol.* 2008, **50**, 87–94.
[122] M.A. García, M.N. Martino, and N.E. Zaritzky, *Nahrung - Food*, 2001, **45**, 267–272.
[123] M. Vargas *et al.*, *Postharvest Biol. Technol.* 2009, **51**, 263–271.
[124] M. Montero-Calderón, M.A. Rojas-Graü, and O. Martín-Belloso, *Postharvest Biol. Technol.* 2008, **50**, 182–189.
[125] Osburn, W.N., Collagen Casings, in A. Gennadios (ed.), *Protein-Based Films and Coatings*, CRC Press LLC, Boca Ratón, Florida, pp 445–466 (2002).
[126] B. Ouattara *et al.*, *J. Food Sci.* 2000, **65**, 768–773.
[127] M. Oussalah *et al.*, *J. Agr. Food Chem.* 2004, **52**, 5598–5605.
[128] J. Han, S. Bourgeois, and M. Lacroix, *Food Chem.* 2009, **115**, 462–468.
[129] X.G. Chen *et al.*, *J. Agr. Food Chem.* 2002, **50**, 5915–5918.
[130] S.Y.D. Lin and J.M. Krochta, *J. Agr. Food Chem.* 2005, **53**, 5018–5023.
[131] J.H. Han, *et al.*, *J. Food Sci.* 2008, **73**, 349–355.
[132] E.A. Baldwin and B. Wood, *HortScience*, 2006, **41**, 188–192.
[133] J.I. Maté, M.E. Saltveit, and J.M. Krochta, *J. Food Sci.* 1996, **61**, 465–472.
[134] J.I. Maté and J.M. Krochta, *J. Food Sci.* 1996, **61**, 1202–1210.
[135] J.I. Maté and J.M. Krochta, *J. Food Eng.* 1998, **35**, 299–312.
[136] S.D. Chen *et al.*, *J. Food Eng.* 2009, **95**, 359–364.
[137] A.D. Garmakhany *et al.*, *Eur. J. Lipid Sci. Technol.* 2008, **110**, 1045–1049.
[138] J. Singthong and C. Thongkaew, *Lebensm-Wiss Technol.* 2009, **42**, 1199–1203.
[139] S.G. Sudhakar *et al.*, *J. Food Sci. Technol.* 2006, **43**, 377–381.

12

Biopolymer Coatings for Paper and Paperboard

Christian Aulin and Tom Lindström

Innventia AB, Stockholm, Sweden, and Wallenberg Wood Science Center, Royal Institute of Technology, Stockholm, Sweden

12.1 INTRODUCTION

Large quantities of paper and board products are surface sized or coated, often in more than one stage. The primary reason for coating paper and board products is to improve the printability [1, 2]. Good printability requires controlled interaction between ink and the surface to be printed, such as surface smoothness and ink absorption. Surface sizing improves the surface strength of paper and modifies the surface chemistry and porosity, often in a stage preceding coating operations. Surface sizing is usually done by, for example, applying modified starches, and common coatings are made from various pigments, latexes and/or starches/proteins, water retention agents and dispersants. None of these operations target the barrier properties (e.g., oxygen and water vapour permeability, grease and oil resistance) of paper and board and coatings are generally porous (pigment concentrations are above the CPVC or Critical Pigment Volume Concentration) in order to receive the printing ink. Hence, general coating formulations used by the industry do not constitute potential barriers, as these require more binders and lower pigment loads beneath the

Biopolymers – New Materials for Sustainable Films and Coatings, First Edition.
Edited by David Plackett.
© 2011 John Wiley & Sons, Ltd. Published 2011 by John Wiley & Sons, Ltd.

CPVC [3–6]. Greaseproof and glassine papers made from heavily beaten pulps can, however, provide grease and fat resistance [7].

Surface treatments with polysaccharides and their derivatives can provide oil and grease barrier and oxygen barrier properties; however, the latter are typically subject to moisture sensitivity due to uptake of water in the coatings [8–11]. Historically, the application of various waxes and internal sizing agents has been used to impart water repellence and water vapour resistance to paper/board products. Paper and paperboard can however acquire barrier properties and functional performance through lamination with plastics such as PE (polyethylene), PP (polypropylene), polyethylene terephthalate (PET) or ethylene vinyl alcohol (EVOH) and aluminium foils in liquid packaging applications [12]. It is only during the last 20 years that there has been a growing interest in using dispersion coating formulations involving high latex additions to create water vapour barriers [2]. The most frequently used lattices in such formulations are based on polymers or copolymers of styrene, acrylate, methacrylate, butadiene, or vinyl acetate (i.e., the same types used in ordinary coating formulations, hence not being bio-based).

The base sheet properties of paper/board materials are of paramount importance in terms of coating quality. A base sheet of paper is a highly porous material with variable smoothness and porosity and these properties are dependent on a multitude of variables, such as fiber type and treatment, filler content, forming conditions, wet pressing, and pre-calendering. Depending on the base-sheet properties, rheological characteristics of the coating colour and the colour application method, the coating colour hold-out properties and the final coating properties will vary. The application of surface sizes was historically made in pond-type size applicators and gate roll-type size presses but these were later replaced by film-type applicators [2] during the 1990s. These are all classified as pre-metering-type applicators.

Coating processes are dominated by post-metering type of equipment. First, the coating colour is applied using roll coaters, short dwell coaters, jet applicators and the like. Post-metering is done by the application of blade, rod or an air-knife. More recently spray-coating, curtain and extrusion coating, and spot/pattern coating applications have also been practised. The central area of application of barrier materials falls primarily in the domain of primary food packaging [8, 13, 14]. Water vapour barriers for tertiary packaging materials, such as linerboard, is an important area, where wax-based coatings are used, but these coatings are considered incompatible with recycling procedures used in this sector.

Biodegradable films and coatings, which are also edible are commonly used to protect various foods in primary packaging applications and can be considered an integral part of a food [10, 17, 21–33]. Many functions of such films are similar to functions provided by synthetic food packaging and must obviously be chosen according to specific food applications and the major mechanisms of quality deterioration. This is a rapidly growing market, particularly for active/intelligent packaging applications, whereas bio-based coatings on paper/board materials are currently only in a research and development stage [8]. This large field has relevance for commercial coatings on paper/paperboard, but does not directly address paper/board applications. The materials used for edible films and coatings can be classified as polysaccharides and their derivatives (e.g., starches, modified cellulosics, alginate, carrageenan, pectin, chitosan, and gums), various proteins (e.g., gelatin, casein, whey protein, corn zein, gluten soy proteins) and lipids (e.g., beeswax, carnauba wax, paraffin wax, shellac resins). Water-resistant bioplastics can also be produced from blends of thermoplasticized starch with biodegradable polymers such as polycapro-lactone using compatibilizing agents, known as Mater-Bi® products, and through use of microbial polymers (e.g., poly(3)-hydroxybutyrate, known as Biopol® and discussed in Chapter 4) or polylactic acid-based polymers discussed in Chapter 3 (e.g., Ecopia®, Lacea®) [20, 40–44].

12.2   BIOPOLYMER FILMS AND COATINGS

The use of renewable resources for the production of packaging materials in particular has recently received increased interest [15–17]. Depending upon the application, low oxygen and water vapour permeability as well as mechanical strength and flexibility are important target properties for such packaging films. Several studies have been undertaken in recent years to investigate the potential of bio-based materials for packaging applications [6]. The basic materials used to form edible films and coatings can be classified in three major categories: polysaccharides, proteins and lipid compounds [18].

Films prepared from polysaccharides (cellulose and derivatives, starches, hemi-celluloses, chitosan) provide efficient barriers against oils and lipids [19–21], but their moisture barrier properties are poor [19, 20]. Although not so extensively studied, protein-based films have highly interesting properties. Many proteins have been tested, including collagen, zein, wheat gluten and cottonseed proteins, and soy-bean [23–28]. Lipid compounds, such as animal and vegetable fats (e.g., natural waxes and derivatives, acetoglycerides) have been used to make edible films and coatings [11, 21–23]. They are generally used for their excellent moisture barrier properties, but can cause textural and organoleptical problems due to oxidation and their waxy tastes. Due to the hydrophilic character of biopolymers, they must generally be combined with hydrophobic materials, such as fatty acids and waxes. Two primary preparation methods exist for lipid based bio-films; emulsion films and bilayer films, the latter being a better barrier for oxygen and water vapour [34–39].

Edible and bio-based packages formed with several compounds (composites) have been developed to take advantage of the complementary functional properties of these different constitutive materials and to overcome their respective drawbacks. Most composite films studied to date combine one or several lipid compounds with one polysaccharide-based structural matrix [18, 29, 31]. Several polysaccharide biopo-lymers have the ability to form films with a relatively high degree of crystallinity and with low permeability to oxygen [3, 19, 32, 33], thus making them interesting for use in packaging applications. The polysaccharide polymers in particular will be high-lighted in this review. Biopolymer films for use as packaging materials based on starch [33–38], chitosan [57–58], cellulose derivatives [63, 77–82], and hemicellu-loses [43–46] are described in detail in the literature. Bio-based materials have also been used, for example, in composite formulations (e.g., starch/low-density polyeth-ylene (LDPE) blends) [24, 25]. Starch–PVOH blends have shown properties competitive with those of conventional LDPE and polystyrene (PS) films [11, 89–91]. The use of bio-based materials with respect to food packaging requirements is discussed by Krochta and de Mulder-Johnston [9], Petersen et al. [26], Sorrentino et al. [27] and Rhim [28]. Starch, cellulose and hemicellulose-based products are commercially available. Examples are Paragon (starch) manufactured by Avebe®, Cellophane (regenerated cellulose) produced by Innovia Films® and Xylophane (xylan) produced by Xylophane AB®.

Bio-based film-forming materials can be used in the surface treatment of cellulose-based substrates, either by coating or by extrusion/lamination. There is, however, a lack of literature in the field concerning the improvement of the functional properties of paperboard via coating with biopolymers [3, 39, 40, 58, 75]. In contrast, the use of biomaterials for the production of edible films or thin coatings applied directly on food (e.g., on fruit for mechanical protection or barrier purposes) has received considerable attention [9, 29, 30]. In the following section, some special features of different types of bio-based films and coatings are discussed. The water vapour and gas permeability of polymer films depend strongly on the temperature and the RH gradient across the barrier film, which is why adequate reference to the specific test conditions is necessary. Strong cohesion within the film reduces the permeability of gases and solutes through the film [20, 31]. The mechanical and barrier properties of amorphous materials are also strongly affected by the glass transition temperature of the material [6]. A direct comparison of reported barrier properties of various biopolymer coatings from literature values is not readily carried out because of the strong influence of the substrate structure, presence of sizing, pre-coatings, etc. The final barrier properties on a surface-treated paperboard not only depend on the thickness of the functional biopolymer layer, but also on the substrate porosity and smoothness. For this reason, a review of past research on material properties of both biopolymer films and coatings is considered appropriate.

12.2.1 Starches

Starch is one of the natural biopolymers most widely used to develop environmentally friendly packaging materials as a substitute for petrochemical-based non-biodegradable plastics. Since starches are inherently biodegradable, renewable and low-cost materials, there is high potential for use in food or non-food packaging applications. However, there are limitations due to their poor water vapour barrier and mechanical properties. As a packaging material, starches alone do not form films with appropriate mechanical properties unless first plasticized or chemically modified. Common plasticizers for hydrophilic polymers, such as starches are glycerol and other low-molecular-weight polyhydroxy compounds. Several companies offer starches from various plant sources. Examples are Lyckeby Stärkelsen Industrial Starches AB® (potato), Roquette® (maize, wheat and potato) and National Starch and Chemical® (maize, potato). Potato is the major source of starch for use in the paper industry in Europe [32], whereas corn starch is more common in the United States. Starch with different amylose/amylopectin ratios and hydroxypropylated starch have been used in food packaging and for edible films [9, 33]. Commercial starches for paper sizing and coating are usually highly oxidized and contain 0.4–1.2 wt% carboxyl groups, corresponding to a degree of substitution (D.S.) of 0.015–0.044. Oxidation of starch is carried out to reduce the chain length and thus to lower the viscosity of the starch solution. Oxidation also reduces the gelatinization temperature of the starch. Incorporation of hydrophilic hydroxypropyl groups improves the low-temperature stability (retrogradation), the stability of the dispersion viscosity and the flexibility of the resulting films. Hydroxypropylated starches have found use in both food and non-food applications, and show excellent film-forming properties along with a high adhesive strength and beneficial rheological properties [34]. The linear amylose fraction in starch is able to form continuous, strong and flexible films through hydrogen bonding. Amylopectin forms, on the other hand, hard and brittle films. [35]. The film-forming behaviour depends on the molecular weight of the amylose and amylopectin fractions, the water-holding properties and the colloidal stability of the starch solution. The amylose fraction strongly influences the adhesive properties when applied to paper, while the amylopectin fraction controls water retention and rheology. The effect on the formation of a closed-packed structure with linear amylose on the barrier properties was illustrated by Jansson and Järnström [34], who showed that the water vapour permeability of a high amylose starch film was about 10 times lower than that of a film from normal potato starch, presumably due to its ability to crystallize.

Oxygen and water vapour permeability data for starch and a number of other biopolymer films are shown in Tables 12.1 and 12.2 respectively. Although it can be difficult to compare permeability data obtained by various research groups using different equipment, procedures and units in which to express permeability results, the data in Tables 12.1 and 12.2 give an indication of the oxygen and water vapour permeability of the materials. Figure 12.1 shows a two-dimensional plot of the oxygen permeability versus the water vapour permeability for various biopolymer and synthetic films. Starch films are moisture sensitive due to their hydrophilic character and are therefore poor barriers towards water vapour [31]. The crystallinity of amylose starch films improves the barrier properties toward gases [36]. Gas permeability and water vapour permeability can also take place through pores present in the films, which may be 20–100 nm in size [36]. High-amylose starch films are moderate to good oxygen barriers with an oxygen permeability that is even lower than that of most synthetic polymers [10]. However, hydroxyproplyated starch films have a lower oxygen permeability [34, 37]. Apparently, the starch structure in these films produces a combination of high cohesive energy density, low free volume and high crystallinity that is not common in many synthetic polymers [10].

Water transport through a coated surface is mainly driven by capillary forces and is affected by surface roughness, surface chemical properties and pore structure [6]. In general, starches increase both the roughness and the hydrophilicity of the coated surface. The water resistance of starches can be enhanced by incorporation of hydrophobic groups or by cross-linking [34, 38].

12.2.2 Chitosan

Chitosan (β-1,4-linked 2-amino-2-deoxy-D-glucose) is a partially deacetylated derivative of chitin, which is one of the most abundant natural biopolymers next to cellulose. Since chitosan is biodegradable, non-toxic, and biocompatible, it has been studied extensively for various industrial and packaging applications. However, like other hydrophilic natural biopolymers, its properties as a packaging material also need to be improved. Chitosan exhibits high cohesive energy density due to its relatively high crystallinity and the hydrogen bonds between the molecular chains [57–63]. A study on composite films consisting of mixtures of chitosan and

Table 12.1 Oxygen permeability of biopolymer films and conventional synthetic films

Film type	Oxygen permeability $(cm^3\ \mu m)/(m^2\ day\ kPa)$	Source and conditions
Cellulose-based		
MFC (carboxymethylated)	0.0006	[20], 0% RH
MFC (carboxymethylated)	0.9	[20], 50% RH
MFC (not pre-treated)	3.5–5.0	[39], 50% RH
TEMPO-oxidized nanocellulose	0.004	[40], 0% RH
Regenerated cellulose (NMMO)	0.4	[41], 0% RH
Cellophane	9.5	[42], 50% RH
Hemicellulose-based		
O-acetylgalactoglucomannan-sorbitol 0.26: 0.14	2	[43], 50% RH
O-acetylgalactoglucomannan-xylitol 0.26:0.14	4.4	[43], 50% RH
O-acetylgalactoglucomannan-alginate (0.26:0.14)	0.55	[43], 50% RH
O-acetylgalactoglucomannan-CMC (0.26:0.14)	1.3	[44], 50% RH
O-acetylgalactoglucomannan (benzylated)	130	[44], 50% RH
O-acetylgalactoglucomannan-alginate (0.26:0.14), laminated with benzylated O-acetylgalactoglucomannan	8	[44], 83% RH
O-acetylgalactoglucomannan-alginate (0.26:0.14), vapour phase grafting with styrene	1.8	[44], 50% RH
65% Glucuronoxylan/35% sorbitol	0.2	[45], 50% RH
Arabinoxylan (barley husks)	2	[46], 50% RH
Protein-based		
Whey protein-glycerol (3:1–0.8:1)	40–330	[47], 50% RH
Whey protein-sorbitol (1.5:1)	1.0	[48], 30% RH
Collagen	1.2	[48], 0% RH
Starch-based		
Amylomaize (16% glycerol)	65	[49], ~ 100% RH

Amylose-glycerol (2.5:1)	7	[36], 50% RH
Amylopectin-glycerol (2.5:1)	14	[36], 50% RH
Chitosan-based		
Chitosan-glycerol (4:1)	0.01–0.04	[50], 0% RH
Chitosan-glycerol (2:1)	0.04–0.08	[50], 0% RH
Wax/lipid-based		
Beeswax	1540	[48], 0% RH
Carnauba wax	160	[48], 0% RH
Synthetic		
Polyvinylidene chloride (PVDC)	0.1–3	[51], 50% RH
Poly (vinyl alcohol) (PVOH)	0.20	[16], 50% RH
Polyamide (PA)	1–10	[51], 0% RH
Poly(ethylene-terephthalate)(PET)	10–50	[51], 50% RH
Poly(vinyl chloride)PVC	20–80	[51], 50% RH
Poly(lactic acid) (PLA)	180	[40], 0% RH
Polypropylene (PP)	500–1000	[51], 50% RH
Polystyrene (PS)	1000–1500	[51], 50% RH
Low-density polyethylene (LDPE)	1900	[48], 50% RH
Ethylene vinyl alcohol (EVOH)	0.01–0.1	[51], 0% RH
Ethylene vinyl alcohol (EVOH)	12	[52], 95% RH

Table 12.2 Water vapour permeability of biopolymer films and conventional synthetic films

Film type	WVP 10^{-11} g/(m s Pa)	Source and conditions
Cellulose-based		
Hydroxypropylmethyl cellulose (HPMC)	9.2	[30], 27 °C, 0–97%
HPMC/stearic acid	0.1	[30], 27 °C, 0–97%
Cellulose acetate	3.2	[23], 25 °C, 0–100%
Cellophane (300P, CP films®)	6.9	[23], 25 °C, 0–90%
Starch-based		
Amylose (potato) (2 wt% glycerol)	120	[36], 23 °C, 50%
Amylopectin (potato) (2 wt% glycerol)	140	[36], 23 °C, 50%
Hydroxypropylated high-amylose (potato) (0–30 pph glycerol)	0.3	[34], 23 °C, 50%
Hydroxypropylated and oxidized native (potato) (0–30 pph glycerol)	5	[34], 23 °C, 50%
Hemicellulose-based		
Arabinoxylan (corn hull) (20 wt % glycerol)	2.3–4.3	[53], 22 °C, 54%
Arabinoxylan (maize bran), hydrophobized with stearic acid (15% glycerol)	12–15	[54], 25 °C, 54%
Chitosan-based		
Chitosan	13	[55], 25 °C, 76.2%
Wax/lipid based		
Beeswax	0.06	[23], 25 °C, 0–100%
Stearic acid	0.2	[23], 25 °C, 12–56%
Palmitic acid	0.7	[23], 25 °C, 12–56%
Synthetic		
Polyvinylidene chloride (PVDC)	0.0007–0.0024	[23], 38 °C, 0–90%
Ethylene vinyl alcohol copolymer	2.1	[23], 38 °C, 0–90%
Polypropylene (PP)	0.05	[23], 38 °C, 0–90%
Polystyrene (PS)	0.5	[23], 25 °C, 0–100%
Low-density polyethylene (LDPE)	0.07–0.1	[23], 38 °C. 0–90%

cellulose fibers showed that the films provided a good oxygen barrier [64]. Free-standing films of chitosan exhibit very low oxygen permeability as was reported by Butler et al. [65].

The possibility of obtaining an oxygen barrier by coating paper with chitosan has also been tested in an experiment in which a copy paper was coated with five layers

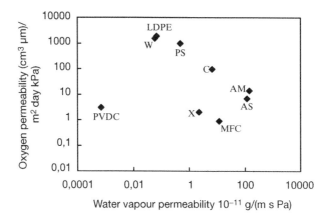

Figure 12.1   Oxygen permeability versus water vapour permeability for biopolymer films and conventional synthetic films. The oxygen permeability was measured at 23 °C and 50% RH. (MFC: carboxymethylated microfibrillated cellulose [20], PS: polystyrene [23, 56], AS: amylose [36], AM: amylopectin [36], LDPE: low-density polyethylene [23, 48], C: cellophane [23], X: arabinoxylan [46, 53], PVDC: polyvinylidene chloride [23, 48] and W: beeswax [23, 48])

using a bench-scale rod coater [60]. The total coat weight obtained was 6.9 g/m² and the oxygen permeability of this material was reported to be 1.1 (cm³ μm)/(m² day kPa) when using a modified curtain coater on a bench scale. Gällstedt *et al.* [57] produced a chitosan-coated paperboard material with a low oxygen permeability value of 1.48 (cm³ μm)/(m² day kPa). Moreover, chitosan exhibits good barrier properties against grease [62] and has natural antibacterial and fungicidal properties [67, 68]. These characteristics make it attractive for use in the coating of paper and paperboard for food or medical applications [69]. In a study by Kjellgren *et al.*, chitosan was used for producing grease-resistant paper and board [61]. Furthermore, chitosan has been incorporated into starch to give antimicrobial functionality to edible coatings [69]. Chitosan is commercially available in a range of different grades. Manufacturers are, for example, France Chitine and Aqua Premier Co., Ltd. Chitosan also possesses good mechanical properties, as has been shown in the work of Arvanitoyannis *et al.* [70]. These researchers studied the mechanical properties of films composed of chitosan/ gelatin and chitosan/poly(vinyl alcohol) and found that the tensile strength increased when the fraction of chitosan in the film increased.

12.2.3   Hemicelluloses

As one of the major constituents of wood and plants, the hemicellulose sugars are amongst the the most naturally abundant biopolymers on Earth. Hemicelluloses are potentially available in large quantities from agricultural wastes and films from hemicelluloses have been the target of research efforts by many groups. These polysaccharides are hydrophilic in nature and films produced from these materials are generally hygroscopic, resulting in poor gas and water vapour barrier properties at high humidity. As recently described in review papers, chemical modification of hemicellulose by either bulk or surface modification is a way to circumvent these problems [71–73]. Hemicelluloses such as arabinoxylans extracted from cereal cell walls (e.g., barley husks) have been investigated for the production of biopolymer films or coatings with good oxygen barrier properties. The films were strong, with a stress at break of more than 50 MPa, elongation at break of 2.5%, and a Young's modulus of 2900 MPa (Table 12.3). Furthermore, the films were shown to be highly hygroscopic with a primarily amorphous structure [46]. An oxygen permeability of 2 $(cm^3 \mu m)/(m^2$ day kPa) at 50% RH was reported. The hydrophobicity of films from arabinoxylan extracted from maize was increased by grafting functional aliphatic acrylates [74]. The study showed a significant improvement in the water vapour

Table 12.3 Mechanical properties of biopolymer films and conventional synthetic films

Film type	Tensile strength (MPa)	Strain at break (%)	Elastic modulus (MPa)	Source
Cellulose-based				
MFC (enzymatic pre-treatment)	129–214	3.3–10.1	10400–13700	[75]
MFC (not pre-treated)	146	8.6	17500	[39]
TEMPO-oxidized nanocellulose, softwood	233	7.6	6900	[40]
TEMPO-oxidized nanocellulose, hardwood	222	7.0	6200	[40]
Chitosan-based				
Chitosan	6	7	18000	[76]
Starch-based				
Amylose (2 wt% glycerol)	20	31	1000	[36]
Amylopectin (2 wt% glycerol)	6	29	500	[36]
Hemicellulose-based				
Arabinoxylan (corn hull) (0–20 wt % glycerol)	9.7–60.7	1.2–12.1	365–1320	[53]
Arabinoxylan (barley husks)	50	2.5	2900	[46]
Xylan (cotton stalk)	1.1–1.4	45.6–56.8	0.11–0.49	[77]
Glucuronoxylan (20% xylitol)	40	2	~ 5000	[45]
Synthetic				
Polyvinylidene chloride (PVDC)	48.4–138	20–40	2900–3300	[78]
Poly(ethylene-terephthalate) PET	175	70–100	2800–3100	[78]
Polystyrene (PS)	72	20	3000–3600	[78]
Low-density polyethylene (LDPE)	13	500	200–400	[78]

barrier properties after grafting. The oxygen barrier properties of glucuronoxylan from aspen wood have been examined by Gröndahl *et al.* [45]. Films from unmodified xylan were brittle (Tg $\sim$ 180 °C) and therefore sorbitol and xylitol were used as plasticizers in concentrations of 20, 35, and 50 wt %. All the films were semi-crystalline (44–47%) regardless of plasticizer content. Tensile testing showed that the addition of 20 wt % plasticizer resulted in a stress at break of 40 MPa. The elongation at break was however only 2%. Increasing the plasticizer content resulted in a reduction of strength and a concurrent increase in the elongation at break. An oxygen permeability of 0.21 $(cm^3 \mu m)/(m^2$ day kPa) at 50% RH was reported for a film plasticized with 35 wt % sorbitol, which was similar to the value found for a poly (vinyl alcohol) (PVOH) film under the same conditions. The hemicellulose O-acetylgalactoglucomannan (AcGGM) originating from wood was studied by Hartman *et al.* as a potential oxygen barrier [43]. Solvent-cast AcGGM films were obtained by concentrating process water from thermomechanical pulping. Films were cast with one of the plasticizers glycerol, sorbitol or xylitol. In addition, mixtures (2:1 on weight basis) of AcGGM with alginate or carboxymethyl cellulose (CMC) were utilized for film formation. Oxygen permeabilities were the lowest for the AcGG-M–alginate and AcGGM–CMC films. The sorbitol-containing AcGGM film had an oxygen permeability of 2.0 $cm^3 cm^3/(\mu m\,m^2$ day kPa), which was markedly lower than the value of 4.4 $cm^3/(\mu m\,m^2$ day kPa) found for the xylitol-plasticized AcGGM film. A second study of AcGGM was performed by Hartman *et al.* [43] and an overview of the oxygen permeability of the films formed in this work is given in Table 12.1. AcGGM was mixed with alginate or CMC at a ratio of 7:3 (weight basis) or hydrophobized using either vapour-phase grafting with styrene or plasma treatment followed by styrene grafting. The blends were cast as films.

Past research in producing barrier films from hemicellulose demonstrates the difficulties of modifying a hygroscopic material to form a moisture barrier; however, modified hemicelluloses have shown major improvements in the barrier properties compared with the parent compounds. Table 12.2 shows the water vapour barrier properties of several synthetic polymers as well as a number of biopolymers. The water vapour permeability (WVP) values for LDPE, polypropylene (PP), polystyrene (PS), and polyvinylidene chloride (PVDC) are several orders of magnitude lower than the corresponding values for the different hemicellulose films. Corn hull arabinoxylan films [53] exhibited similar WVP barrier to cellulose acetate, but slightly better barrier properties than cellophane. Films from maize bran arabinoxylan with various hydrophobic modifications [74] had comparable barrier properties to those of cellophane and hydroxypropylmethylcellulose (HPMC). All the hemicellulose films targeted for coatings were superior to amylose and amylopectin films in terms of WVP values. Although improvements in barrier properties can be significant for modified hemicelluloses, these films are not yet capable of competing with commercial films from non-renewable sources such as PET, LDPE and PS in terms of water vapour barrier properties.

12.2.4 Cellulose Derivatives

The making of films and coatings from water-soluble cellulose/cellulose derivatives has been known for a long time. In fact, cellulose nitrate was the first artificial or synthetic plastic, commercially produced in 1868 [79]. The nature of the high crystallinity renders a modified cellulose with high rigidity, modulus, and hardness and reduced solvent solubility. Since cellulose is not a thermoplastic material, decrystallization and changes in the architecture of cellulose molecules have to be made in order that cellulose can be melted and extruded. These changes can be achieved by chemical modification reactions. Cellulose derivatives are made by either etherification, esterification, or graft-copolymerization, in which cellulose esters (e.g., cellulose nitrate and cellulose acetate) are used so far in the greatest volume. Since cellulose nitrate has a glass transition temperature of 53 °C, plasticizers must be added to obtain films/coatings with suitable flexibility. The temperature sensitivity of cellulose nitrate excludes it from all fabrication methods involving heat, such as moulding and extrusion. Most early work was aimed at utilizing cellulose nitrate in explosives. Later development of solvents and plasticizers for cellulose nitrate led to many new and important non-explosive uses. The development of stable cellulose nitrate with low viscosity resulted in fast-drying lacquer coatings, which have been extensively used in automobile and furniture production. Cellulose nitrate films can be prepared using both solvent-casting and filtration techniques.

Cellulose acetate films can be obtained either by extrusion or solvent coating. The thermoplastic characteristics of cellulose acetate are greatly improved as the acetyl content is increased from ∼ 20% to ∼ 39% [80]. Higher acetyl content gives better moisture resistance; lower acetyl content gives better impact strength. To minimize thermal degradation of cellulose esters during processing, plasticizers are used to lower the melting point. Higher softening grades are preferred for compression moulding and softer grades for extrusion. Cellulose acetate adsorbs moisture and always needs to be dried before being processed. Excessive moisture in cellulose acetate often causes surging problems during extrusion and surface imperfections in finished products. Cellulose acetate film has many uses in the fields of packaging and display. It permits the passage of water vapour and gases while remaining impervious to liquid water, which makes it particularly useful when a 'breathing' film is required as, for example, in the packaging of fresh produce.

Hydroxypropyl cellulose (HPC), ethyl cellulose, methyl cellulose, carboxymethyl cellulose and hydroxymethyl cellulose are examples of cellulose ethers that can be cast as films or coated from water and/or ethanol solutions. These derivatives are, except for HPC, not thermoplastic and hence do not give heat-sealable coatings [6, 9]. All these ethers offer poor moisture barrier properties, but they have found use as oil and fat barriers. A number of groups have investigated composite films composed of solid lipids, such as beeswax and fatty acids [22, 23, 9, 81]. Many of these films have water vapour permeabilites as low as LDPE. These composite films were all polymer–lipid bilayers formed either in one step from aqueous/ethanolic solutions of cellulose ether with fatty acids or in two steps by laminating pre-formed films with wax.

Cellophane, a regenerated form of cellulose (cellulose II) made from the viscose process, was the first transparent, flexible cellulose-based packaging film. Biodegradable cellophane makes a strong package due to its good tensile strength and elongation [76, 82]. Other attributes include excellent printability and good machinability [83]. Like other biopolymers, cellophane is sensitive to moisture and hence is a poor gas barrier at high levels of RH. Because of their inherent hydrophilic nature, sensitivity to moisture is perhaps the greatest problem with all biopolymers.

Cellophane is not thermoplastic and is therefore not heat sealable. Cellophane exhibits good oxygen, oil and fat barriers at low relative humidity (RH) [52, 83], but its barrier properties are compromised at intermediate and high RH. Coating with waxes or polyvinylidene chloride reduces the influence of RH on these barrier properties. Coated cellophane has a water vapour permeability as low as that of high-density polyethylene (Table 12.2). The major factor that has prevented large-scale commercial use of cellulose ethers and other cellulose derivatives as coating materials is the high costs associated with the derivatization processes. The highly crystalline structure of cellulose makes it less accessible for derivatization reactions compared with, for example, starch [6].

12.3   BIO-NANOCOMPOSITE FILMS AND COATINGS

12.3.1   Nano-Sized Clay

Nanotechnology is one of the fastest growing research fields in the world. The concept of using nano-sized materials to enhance the functionality of paper and board covers the whole range of applications from improvement of optical properties through abrasion and scratch resistance to barrier or mechanical reinforcement, and is believed to be one of the most important strategic development lines in the coming years [6]. Nanoparticles are generally defined as having a size below 100 nm and can be either inorganic (silicates, metal oxides) or organic (polymers, dyes).

One potential way to control the material properties of biopolymer films and coatings is to use nano-fillers with specific properties. Traditionally, mineral fillers such as clay, silica and talc are incorporated in film preparations to reduce cost or to improve performance in some way. Recently, polymer–clay nanocomposites [84–88] have received significant attention as an alternative to conventionally filled polymers. As a result of their nanometer-size dispersions, polymer-clay nanocomposites exhibit large improvements in mechanical, physical and barrier properties compared with pure polymers or conventional composites. These enhancements generally include, for example, increased modulus and strength and decreased gas permeability [88]. Extensive research has been performed in the area of nanocomposites and several extensive review articles regarding recent advances in the production of nanocomposites with various polymers and their performance and applications are available [27, 28, 55, 84, 86]. In particular, J.J. De Vlieger [86], J.W. Rhim [28, 55], Sorrentino et al. [84] and Gorrasie et al. [89] have prepared comprehensive review articles focusing on bio-based nanocomposite materials. Montmorillonite, hectorite and saponite are frequently used layered silicates, which are combined with polymeric materials to form nanocomposites. Among the nano-scale silicates, montmorillonite is of particular interest and has been widely studied. Montmorillonite is a clay mineral consisting of stacked silicate sheets with a high aspect ratio and a plate-like morphology. The high aspect ratio plays an important role in enhancing the mechanical and physical properties of composite materials. Dispersions of layered silicates in a polymer matrix can result in a number of different states, leading to intercalated or exfoliated nanocomposites. Exfoliation can be carried out by both chemical and/or mechanical means. In an intercalated nanocomposite, a single polymer chain may often be inserted between the clay silicate layers, but the system still remains quite well ordered in a stacked type of arrangement. In an exfoliated nanocomposite the silicate layers are completely delaminated from each other and are well dispersed. It is believed that complete and homogenous dispersion in which the clay platelets are arranged in a non-parallel manner (exfoliation) will give the highest performance improvements in coatings (Figure 12.2) [90].

Typically, commercial nanoclays have aspect ratios between 50 and 1000, which is much larger than for typical clay pigments used in paperboard coatings with aspect ratios of 10–30. The large aspect ratio of nanoclays makes them effective for barrier improvement even at very low concentrations (< 5%) [6, 90, 92, 93]. The use of nanoclays in paper and paperboard coating is thus advantageous, particularly given that less material is required to reach the desired barrier or mechanical properties. Less material leads to reduced costs and reduced amounts of waste. Southern Clay Products and Nanocor are amongst the largest suppliers of montmorillonite clay for nanocomposite applications.

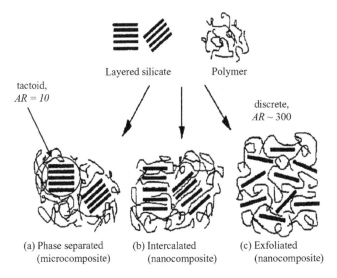

Figure 12.2   When silicates are mixed with a polymer, three different types of composites are obtained: (a) tactoids, (b) intercalates, and (c) exfoliated structures. In tactoids, complete clay particles are dispersed within the clay matrix and the layers do not separate. The mixture of polymer and clay are micro-scale composites and the clay only serves as a conventional filler. In an intercalated nanocomposite, a single polymer chain will be driven between the clay silicate layers, but the system still remains quite well ordered in a stacked type of arrangement. In an exfoliated nanocomposite the silicate layers are completely delaminated from each other and are well-dispersed. Reprinted from S.A. McGlashan and P.J. Halley, Preparation and characterisation of biodegradable starch-based nanocomposite materials, *Polymer International* 2003, 52, 1767–1773. Copyright (2003) with permission from John Wiley and Sons

Although addition of nanofiller is carried out in order to reinforce a polymer (i.e., to increase the strength and toughness of the material), another issue is enhancement of barrier properties at high RH [90]. Polymer nanocomposites can have excellent barrier properties against gases and water vapour. Studies have shown that such reductions in permeability strongly depend on the aspect ratio of clay platelets, where high aspect ratios dramatically enhance gas barrier properties. Yano et al. [94] prepared polyimide/clay nanocomposite films with four different sizes of clay in order to investigate the effect of their aspect ratio on the barrier properties of the hybrids. These authors found that at constant clay content, the relative permeability coefficient decreased exponentially when the length dimension of the clay was increased. Yano et al. [94] also showed that the water vapour and oxygen permeability of polyimide/clay nanocomposite films decreased exponentially with an increase in clay content. Generally, the best gas barrier properties are associated with exfoliated clay minerals having a large aspect ratio. The plate-like structure of clay increases the path for diffusing molecules, thus decreasing the permeability of molecules through the material [27, 55].

De Carvalho et al. [95] prepared biocomposites of thermoplastic corn starch and kaolin clay through mixing and hot pressing and tested their properties. Both modulus and tensile strength of the composites increased up to 50% clay addition, and then decreased, while the elongation at break decreased almost monotonically. Park et al. [96] prepared 'green' nanocomposites from cellulose acetate and organically modified clay. The cellulose acetate was compounded with 5 wt% of Cloisite® 30B (Southern Clay Products) through extrusion and the melted hybrid was injection moulded to obtain the desired nanocomposites. The addition of clay nanoplatelets increased the tensile strength by more than 50%. In addition, several studies have been reported on solvent casting mixtures of chitosan and montomorillointe nanoclay [97, 98]. X-ray diffraction supports the superior properties when clay is present in the polymer matrix in an exfoliated state. The chitosan-based bio-nanocomposites resulted in drastically enhanced mechanical properties in combination with better gas barrier properties.

12.3.2 Nanocellulose

Currently, the isolation, characterization and search for applications of novel cellulosics, variously termed nanofibers, nanocrystals, nanowhiskers, microfibrillated cellulose (MFC) or nanofibrils, is generating much research activity. Such isolated cellulosic materials with at least one dimension in the nanometer range are generally referred to as 'nanocelluloses' [99, 100]. These nanocellulosic materials combine in a unique manner important cellulose properties, such as hydrophilicity, broad chemical modifying capacity, and the formation of versatile semi-crystalline fiber morphologies, with specific features of nanoscale materials, mainly caused by their very large surface area. The properties of nanocellulose differ greatly from the properties of wood pulp fibers, mainly because of the large specific surface area and high aspect ratio of the fibrils. Figure 12.3 shows a transmission electron microscope image of liberated carboxymethylated MFC with a thickness of about 5–10 nm. In aqueous suspensions the fibrils form a highly entangled network which behaves as a pseudoplastic gel [101, 102]. The rheological properties enable the use of the

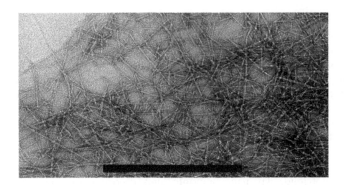

Figure 12.3 A transmission electron micrograph of microfibrillated cellulose (MFC). The scale bar corresponds to 0.5 μm [103]. Reprinted from L. Wågberg *et al.*, The build-up of polyelectrolyte multilayers of microfibrillated cellulose and cationic polyelectrolytes, *Langmuir*, 24, 784–795. Copyright (2008) with permission from American Chemical Society

nanofibrillar cellulose materials as thickeners or stabilizers in suspensions or emulsions in many applications other than coatings, such as foods, paints, cosmetics and pharmaceuticals [102].

The Young's modulus of crystalline cellulose has been reported to be up to 220 GPa [104]. Therefore, strong reinforcement effects of MFC are anticipated owing to high nanofibrillar aspect ratios, the inherently good mechanical properties of the nanofibrils, and its ability to form a network. The extremely good strength properties and good thermal stability of the microfibrils make them suitable for use as reinforcement in bio-nanocomposites and paper products [85, 105].

Homogeneous thin films of MFC can be prepared upon drying and these films can be utilized for example in paper coatings [20, 39]. The first published data on properties of MFC films were presented by Taniguchi and Okamura a decade ago [106], but no absolute values of film properties were given and it was only concluded that the films were stronger than print paper. Since then, many investigators have published data on MFC films and some of these are summarized in Table 1. Despite random-in-plane MFC orientation, MFC films have interesting mechanical properties. Aulin *et al.* reported a Young's modulus of 30 GPa for solvent-cast caroboxymethylated MFC films [20]. Analysis using the method of Cox *et al.* [107] has been used to relate the modulus of highly oriented polymeric model fibers E to the modulus of randomly oriented fibers in a network or film by the following equation:

$$E = 3E_{\text{film}} \tag{12.1}$$

and the modulus of a highly oriented polymeric fiber E to the crystal E_C and amorphous modulus E_A by the relationship:

$$\frac{1}{E} = \frac{v_1}{E_C} + \frac{v_2}{E_A} \tag{12.2}$$

where v_1 and v_2 are the volume fractions of the crystalline and amorphous phases respectively. One obtains a value of $\approx$48 GPa for E_A from equations 1 and 2, using values of 90 and 220 GPa for the highly oriented and crystalline fiber respectively. This result is in large agreement with a previous study by Jeronimidis et al. [108], in which the elastic modulus for amorphous cellulose was measured to be 50 GPa.

In another study by Yano and co-workers [109], the authors found that the water retention value (WRV) correlated with both the modulus of elasticity and the tensile strength. The higher the WRV, the stronger the material, which is a well-known correlation for papermakers. By means of vacuum filtration methodology to dewater MFC, Henriksson et al. formed cellulose nano-papers, using enzymatically pre-treated MFC from sulfite pulp with various degrees of polymerization [75]. The films were found to be very strong, particularly those with a high DP (Table 12.1). The porosity of the films was between 20 and 28%. Results similar to those of Henriksson et al. were also obtained using non-pretreated MFC by Syverud and Stenius [39]. Typically, an MFC film with a basis weight of 35 g/m^2 had a tensile strength of 146 MPa and an elongation of 8.6%. The modulus of elasticity (17.5 GPa) of a film composed of randomly oriented fibrils was comparable to values for cellulose fibers with a fibril angle of 50°. The use of MFC as a surface layer on base paper increased the strength of the paper sheets significantly and reduced the air permeability dramatically. Oxygen permeabilities as low as 3.52–5.03 (cm^3 μm)/(m^2 day kPa) were obtained for films prepared from pure non-pretreated MFC [39]. Furthermore, Fukuzumi et al. prepared transparent and high gas barrier films of cellulose nanofibers by TEMPO-mediated oxidation [40]. The films exhibited very high strength properties together with very low oxygen permeability, unique characteristics which are promising for potential applications in packaging materials. Aulin et al. also reported similar results [20], where free-standing MFC films were prepared by solvent-casting carboxymethylated MFC dispersions. The films showed excellent barrier and optical properties (transparency > 90%). An oxygen permeability of 0.0006 (cm^3 μm)/(m^2 day kPa) was reported for 5 g/m^2 films at 0% RH. Commercial papers were coated with MFC suspensions using a laboratory rod-coater, resulting in a dramatic decrease in air permeability and good oil/fat barrier properties. It is not exactly clear how manufacturing procedures (MFC pre-treatment, extent of delamination, and forming processes including vacuum filtration, casting, drying conditions, etc.) affect the mechanical and barrier properties of the films and coatings.

In light of the more recent shift in interest towards optical and barrier properties of MFC films and coatings, the density issue of films will presumably be much more in focus in future studies. As a consequence of the high degree of crystallinty ($\sim$ 63% [20]), high cohesive energy density and excellent film-forming properties of nanocellulose, this material is believed to have potential in packaging applications where high gas barrier properties are required. Research in the area of oxygen barrier materials from nanocellulose is ongoing [19, 32, 87] and seems to hold great promise for future applications. MFC has also been suggested to work as a reinforcing component in paper coatings [110–112]. Although MFC is a promising barrier material, there is a need to further study the fundamental properties of MFC films and coatings such as the effect of moisture uptake and film density on the barrier and mechanical properties.

12.4 CONCLUDING REMARKS

Packaging materials derived from sustainable resources are now receiving considerable interest, not least because of growing public awareness of environmental issues and the advantages in product marketing. This is particularly important in the food packaging sector, where there is a strong need for efficient barrier materials in order to preserve food and minimize food waste. Bio-based polymers should be a first choice in the development of new functionalities for paper/paperboard. High gas barriers are key requirements for these packaging materials, in addition to good mechanical properties in terms of high strength and flexibility. In this respect, biopolymer films and coatings have been produced with oxygen permeability values that are comparable to those of synthetic polymers such as EVOH and superior to polyethylene.

However, moisture barriers in the form of edible films and coatings have only been produced with moderate success from biopolymers. The main obstacle in this field is the hygroscopic nature of the initial material. Increased hydrophobicity may be achieved by either chemical modification or addition of hydrophobic compounds (e.g., stearic acid and nanoclays). In some cases, the water vapour barrier properties have been improved significantly, but the performance has generally not been comparable with that observed for non-edible commercial synthetic films such as LDPE. Significant developments are however expected in starch-based plastics and polylactides or other biopolyesters in conjunction with nanoclays and nanocellulosic materials. Hence, the application of nanotechnology may provide one possible avenue for progress. Several examples of biopolymers for the preparation of nanocomposites with nano-sized clays have been discussed. Even though a limited number of research studies have been conducted on bio-nanocomposites, materials prepared with biopolymers such as starch and cellulose are expected to have improved mechanical properties and decreased water sensitivity without sacrificing biodegradability. Moreover, research has shown that improved properties are generally attained at low nanoclay content when compared to conventional fillers.

There is today virtually no market for bio-based barrier materials applied to paper/board materials, though there are considerable research efforts under way. Such materials will only become commercially available if they can compete comprehensively with conventional packaging materials. Such materials may have great consumer appeal in a sustainable society scenario, but the claim of biodegradability arguably has low credibility in the absence of a waste composting industry. The impact of future policies and legislation is also largely unknown. Escalating oil prices may speed developments, but alternative raw materials to oil-based barriers, such as ethyl alcohol as a precursor to make ethylene and thereby PE- based barriers, could dampen these effects.

12.5 ACKNOWLEDGEMENT

The authors thank the Wallenberg Wood Science Centre (WWSC) for financial support.

REFERENCES

[1] G.L. Robertson (1993) *Paper and Paper-based Packaging*, Marcel Dekker, New York.

[2] J. Paltakari (2009) *Pigment Coating and Surface Sizing of Paper* Paper Engineers Association/Paperi ja Puu Oy, Finland.

[3] M. Vähä-Nissi and A. Savolainen, Dispersion barrier coating of high density base papers, in Tappi Proceedings of the Coating/Papermakers Conference, Tappi Press, Atlanta, GA, USA, pp. (1998).

[4] M. Vähä-Nissi *et al.*, *Tappi Journal* 1999, **82**, 252–256.

[5] M. Vähä-Nissi *et al.*, *Appita* 2001, **54**, 106–115.

[6] C. Andersson, *Packaging Technology and Science* 2008, **21**, 339–373.

[7] L. Stolpe, *Investigacion y Tecnica del Papel* 1996, **33**, 415–426.

[8] J.M. Krochta, Food packaging, in *Handbook of Food Engineering* (2nd Edition), D.R. Heldman and D.B. Lund (eds), CRC Press LLC, Boca Raton, FLA, pp 847–927 (2007).

[9] J.M. Krochta and C. De. Mulder-Johnston, *Food Technology* 1997, **51**, 61–74.

[10] K.S. Miller and J.M. Krochta, *Trends in Food Science and Technology* 1997, **8**, 228–237.

[11] J.J. Kester and O.R. Fennema, *Food Technology* 1986, **40**, 47–59.

[12] M.J. Kirwan and J.W. Strawbridge (2003) *Food Packaging Technology*, Blackwell Publishing, Oxford, UK.

[13] R.J. Ashley, Permeability and plastics packaging, in *Polymer Permeability* J. Comyn (ed.), Chapman & Hall, Ipswich, UK, pp 269–308 (1985).

[14] J.H. Han and A. Gennadios (2005) *Edible Films and Coatings*, Elsevier Academic Press, San Diego, US.

[15] C.J. Weber, *Biobased Food Packaging Materials for the Food Industry-Status and Perspectives*. 2000, KVL Dept. of Dairy and Food Science: Fredriksberg, Denmark.

[16] N.M. Hansen and D. Plackett, *Biomacromolecules* 2008, **9**, 1493–1505.

[17] G. Davis and J.H. Song, *Industrial Crops and Products* 2006, **23**, 147–161.

[18] S. Guilbert, B. Cuq, and N. Gontard, *Food Additives and Contaminants* 1997, **14**, 745–751.

[19] M.O. Nisperos-Carriedo, Edible coatings and films based on polysaccharides, in *Edible Coatings and Films to Improve Food Quality*, J.M. Krochta, E.A. Baldwin, and M.O. Nisperos-Carriedo (eds), Technomic, US, pp 305–336 (1994).

[20] C. Aulin, M. Gällstedt, and T. Lindström, *Cellulose* 2010, **17**, 559–574.

[21] J.J. Kester and O. Fennema, *Journal of Food Science* 1989, **54**, 1383–1389.

[22] I.K. Greener and O.R. Fennema, *Lipid Technology* 1992, **4**, 34–38.

[23] V. Morillon *et al.*, *Critical Reviews in Food Science and Nutrition* 2002, **42**, 67–89.

[24] I. Arvanitoyannis *et al.*, *Carbohydrate Polymers* 1998, **36**, 89–104.

[25] E. Psomiadou *et al.*, *Carbohydrate Polymers* 1997, **33**, 227–242.

[26] K. Petersen *et al.*, *Trends in Food Science and Technology* 1999, **10**, 52–68.

[27] A. Sorrentino, G. Gorrasi, and V. Vittoria, *Trends in Food Science and Technology* 2007, **18**, 84–95.

[28] J.W. Rhim, *Food Sci. Biotechnol.* 2007, **16**, 691–709.

[29] G.L. Robertson (1993) *Food Packaging*, Marcel Dekker, New York.

[30] R.D. Hagenmaier and P.E. Shaw, *Journal of Agricultural and Food Chemistry* 1990, **38**, 1799–1803.

274 BIOPOLYMERS

[31] J.M. Lagaron, C. R., and R. Gavara, *Materials Science and Technology* 2004, **20**, 1–7.
[32] C. Andersson, M. Ernstsson, and L. Järnström, *Packaging Technology and Science* 2002, **15**, 209.
[33] H.G. Bader and D. Göritz, *Starch/Stärke* 1994, **46**, 229–232.
[34] A. Jansson and L. Järnström, *Cellulose* 2005, **12**, 423–433.
[35] A.H. Young (1986) *Fractionation of Starch*, Academic Press Inc., Orlando.
[36] Å. Rindlav-Westling *et al.*, *Carbohydrate Polymers* 1998, **36**, 217–224.
[37] W.B. Roth and C.I. Mehltretter, *Food Technology* 1967, **21**, 72–74.
[38] B. Rioux *et al.*, *Carbohydrate Polymers* 2002, **50**, 371–378.
[39] K. Syverud and P. Stenius, *Cellulose* 2009, **16**, 75–85.
[40] H. Fukuzumi *et al.*, *Biomacromolecules* 2009, **10**, 162–165.
[41] J. Wu and Q. Yuan, *Journal of Membrane Science* 2002, **204**, 185–194.
[42] K.G. Newton and W.J. Rigg, *Journal of Applied Bacteriology* 1979, **47**, 433–441.
[43] J. Hartman *et al.*, *Journal of Applied Polymer Science* 2006, **100**, 2985–2991.
[44] J. Hartman, A.-C. Albertsson, and J. Sjöberg, *Biomacromolecules* 2006, **7**, 1983–1989.
[45] M. Gröndahl, L. Eriksson, and P. Gatenholm, *Biomacromolecules* 2004, **5**, 1528–1535.
[46] A. Höije *et al.*, *Biomacromolecules* 2008, **9**, 2042–2047.
[47] R. Sothornvit and J.M. Krochta, *Journal of Agricultural and Food Chemistry* 2000, **48**, 6298–6302.
[48] T.H. McHugh and J.M. Krochta, Permeability properties of edible films, in *Edible Coatings and Films to Improve Food Quality*, J.M. Krochta, E.A. Baldwin, and M. Nisperos-Carriedo (eds), CRC Press, US, pp 139–187 (1994).
[49] A.M. Mark *et al.*, *Food Technology* 1966, **20**, 75–77.
[50] B.L. Butler *et al.*, *Journal of Food Science* 1996, **61**, 953–955, 961.
[51] J. Lange and Y. Wyser, *Packaging Technology and Science* 2003, **16**, 149–158.
[52] M. Salame (1986) *Barrier Polymers*, John Wiley, New York.
[53] P.Y. Zhang and R.L. Whistler, *Journal of Applied Polymer Science* 2004, **93**, 2896–2902.
[54] C. Péroval *et al.*, *Journal of Agricultural and Food Chemistry* 2002, **50**, 3977–3983.
[55] J.-W. Rhim, *Critical Reviews in Food Science and Nutrition* 2007, **47**, 411–433.
[56] J. Lange and Y. Wyser, *Packaging Technology and Science* 2003, **16**, 149–158.
[57] M. Gällstedt, A. Brottman, and M.S. Hedenqvist, *Packaging Technology and Science* 2005, **18**, 161–170.
[58] M. Gällstedt and M.S. Hedenqvist, *Journal of Applied Polymer Science* 2004, **91**, 60–67.
[59] F.S. Kittur *et al.*, *Carbohydrate Polymers* 2002, **49**, 185–193.
[60] J. Vartiainen *et al.*, *Journal of Applied Polymer Science* 2004, **94**, 986–993.
[61] H. Kjellgren *et al.*, *Carbohydrate Polymers* 2006, **65**, 453–460.
[62] F. Sébastien *et al.*, *Carbohydrate Polymers* 2006, **65**, 185–193.
[63] F.S. Kittur, K.R. Kumar, and R.N. Tharanathan, *European Food Research and Technology* 2002, **206**, 44–47.
[64] J. Hosokawa *et al.*, *Industrial Engineering and Chemistry Research* 1990, **29**, 800–805.
[65] B.L. Butler *et al.*, *Journal of Food Science* 1996, **61**, 953–957.
[66] F.S. Kittur, K.R. Kumar, and R.N. Tharanathan, *Zeitschrift Lebensmittel-Untersuchung und-Forschung A* 1998, **206**, 44–47.

[67] B. Ouattara *et al.*, *International Journal of Food Microbiology* 2000, **62**, 139–148.
[68] F. Shahidi, J.K.A. Arachchi, and Y.-J. Jeon, *Trends in Food Science and Technology* 1999, **10**, 37–51.
[69] A.M. Durango, N.F.F. Soares, and S. Benevides, *Packaging Technology and Science* 2006, **19**, 55–59.
[70] I.S. Arvanitoyannis, A. Nakayama, and S. Aiba, *Carbohydrate Polymers* 1998, **37**, 371–382.
[71] A. Ebringerova and T. Heinze, *Macromolecular Rapid Communications* 2000, **21**, 542–556.
[72] A. Ebringerova, Z. Hromadkova, and T. Heinze, *Advances in Polymer Science* 2005, **186**, 1–67.
[73] T. Heinze, A. Koschella, and A. Ebringerova, in *Hemicelluloses: Science and Technology*, P. Gatenholm and M. Tenkanen (eds), American Chemical Society, Washington DC, US, pp 312–325 (2004).
[74] C. Peroval, F. Debeaufort, and S. A.-M., *Journal of Membrane Science* 2004, **233**, 129–139.
[75] M. Henriksson *et al.*, *Biomacromolecules* 2008, **9**, 1579–1585.
[76] N. Bordenave *et al.*, *Journal of Agricultural Food Chemistry* 2007, **55**, 9479–9488.
[77] E.I. Goksu *et al.*, *Journal of Agricultural Food Chemistry* 2007, **55**, 10651–10691.
[78] J.H. Briston (1988) *Plastic Films*, John Wiley and Sons Inc., New York.
[79] D.N.-S. Hon, Cellulose Plastics, in *Handbook of Thermoplastics*, O. Olabisi (ed.), Marcel Dekker Inc., NY, pp 331–349 (1997).
[80] J.J. Creely, P. Harbrink, and C.M. Conrad, *Journal of Polymer Science* 1965, **9**, 1533–1546.
[81] F. Debeaufort, *Critical Reviews of Food Products* 1998, **38**, 299–313.
[82] N. Bordenave, G. S., and V. Coma, *Biomacromolecules* 2010, **11**, 88–96.
[83] J.F. Hanlon, Films and foils, in *Handbook of Package Engineering*, J.F. Hanlon, R.J. Kelsey, and H.E. Forcinio (eds), CRC Press, US, pp 59–105 (1992).
[84] A. Sorrentino *et al.*, Barrier properties of polymer/clay nanocomposites, in *Polymer Nanocomposites*, Y.-W. May and Z.-Z. Yu (eds), Woodhead Publishing Ltd., Cambridge, UK, pp 273–296 (2006).
[85] L. Berglund, Cellulose-based nanocomposites in *Natural Fibers, Biopolymers and Biocomposites*, A.K. Mohanty, M. Misra, and L. Drzal (eds), CRC Press, US, pp 807–832 (2005).
[86] J.J. de Vlieger, Biodegradable nanocomposites, in *Proceedings of the Food Packaging Conference*, Copenhagen (2000).
[87] A. Dufresne, *Journal of Nanoscience and Nanotechnology* 2006, **6**, 322–330.
[88] E.P. Giannelis, *Advanced Materials* 1996, **8**, 29–35.
[89] G. Gorrasi *et al.*, *Polymer* 2003, **44**, 2271–2279.
[90] M. Alexandre and P. Dubois, *Materials Science and Engineering* 2000, **28**, 1–63.
[91] S.A. McGlashan and P.J. Halley, *Polymer International* 2003, **52**, 1767–1773.
[92] T. Schuman, M. Wikström, and M. Rigdahl, *Progress in Organic Coatings* 2005, **54**, 360–371.
[93] L. Ollabarrieta *et al.*, *Polymer* 2001, **42**, 4401–4408.
[94] K. Yano *et al.*, *Journal of Polymer Science Part A: Polymer Chemistry* 1993, **31**, 2493.
[95] A.J.F. De Carvalho *et al.*, *Carbohydrate Polymers* 2001, **45**, 189–194.
[96] H.-M. Park *et al.*, *Biomacromolecules* 2004, **5**, 2281–2288.
[97] S.F. Wang *et al.*, *Polymer Degradation and Stability* 2005, **90**, 123–131.

[98] M. Darder, M. Colilla, and E. Ruiz-Hitzky, *Chemistry of Materials* 2003, 15, 3774–3780.
[99] I. Siró and D. Plackett, *Cellulose* 2010, 17, 459–494.
[100] D. Klemm *et al.*, *Advances in Polymer Science* 2006, 205, 49–96.
[101] F.W. Herrick *et al.*, *Journal of Applied Polymer Science, Applied Polymer Symposium* 1983, 37, 797–813.
[102] A.F. Turbak, F.W. Snyder, and K.R. Sandberg, *Journal of Applied Polymer Science, Applied Polymer Symposium* 1983, 37, 815–827.
[103] L. Wågberg *et al.*, *Langmuir* 2008, 24, 784–795.
[104] I. Diddens *et al.*, *Macromolecules* 2008, 41, 9755–9759.
[105] M. Hubbe *et al.*, *BioResources* 2008, 3, 929–980.
[106] T. Taniguchi and K. Okamura, *Polymer International* 1998, 47, 291–294.
[107] H.L. Cox, *British Journal of Applied Physics* 1952, 3, 72–79.
[108] G. Jeronimidis and J.F.V. Vincent, *Topics in Molecular and Structural Biology* 1984, 5, 187–210.
[109] H. Yano, *Cellulose Communications* 2005, 12, 63–68.
[110] P.J. Zuraw *et al.*, *Paperboard containing microplatelet cellulose particles*. US patent 2008, MeadWestwaco Corp.
[111] L. Cousin and F. Mora, *Method of manufacture for highly loaded fiber-based composite material*. US Patent 1998, International Paper Company.
[112] Y. Matsuda, M. Hirose, and K. Ueno, *Super microfibrillated cellulose, process for producing the same and coated paper and tinted paper using the same*. US patent 2001, Tokushu Paper Mfg. Co., Ltd.

13

Agronomic Potential of Biopolymer Films

Lluís Martín-Closas and Ana M. Pelacho

Department of Horticulture, Botany and Gardening, University of Lleida, Lleida, Spain

13.1 INTRODUCTION

Polyethylene, and later other plastics, were introduced to the agricultural sector in the early 1950s in the form of films, drip-irrigation tubing and tapes, revolutionizing the commercial production of vegetable crops [1]. Different application methods were developed, depending on the spatial situation of the plastic film relative to the crop, the coverage, and the time of year. The use of these films was shown to modify the energy and the mass balance of the plant environment, thus improving production, quality, and the efficient use of resources for crop production.

Currently, the total world demand for agricultural plastic films is 3.6 million tonnes, of which 41% is used for mulch films, 40% for greenhouse covers, and 19% for silage films [2]. In Europe, the total demand is 0.7 million tonnes, with mulch and greenhouse films consumed mainly in Southern Europe, greenhouse films in Central Europe, and silage in Northern Europe. The use of various plastic films in agriculture is summarized in Table 13.1.

Agricultural film types and their characteristics vary according to application. The main polymers used are low-density polyethylene (LDPE), linear low-density

Biopolymers – New Materials for Sustainable Films and Coatings, First Edition.
Edited by David Plackett.
© 2011 John Wiley & Sons, Ltd. Published 2011 by John Wiley & Sons, Ltd.

Table 13.1   Plastic films used in agriculture

Use	Material	Crops
Greenhouse covers	LDPE, EVA, PVC	Vegetables and ornamentals
Big tunnels	LDPE, EVA, PVC	Vegetables
Low tunnels	LDPE, EVA, PVC	Strawberries, fruit vegetables
Floating covers	PP, LDPE, EVA	Vegetables
Mulches	LLDPE, LDPE, EVA, PVC	Vegetables
Silage	LLDPE, LDPE, EVA, PVC	Silage crops (Cereals, Lucerne)
Nets	LLDPE, LDPE, EVA, PVC, HDPE, PP	Vegetables, fruit crops
Windbreaks	LLDPE, LDPE, HDPE, PP	All crops
Hydroponic sacks	LLDPE, LDPE, EVA	Greenhouse crops
Blanching sheets	LLDPE, LDPE	Blanching crops
Fruit protection bags	LDPE	Fruit crops, banana
Heating tubes	LLDPE, LDPE	Greenhouse crops
Grafting strips	LDPE	Fruit crops, vegetables, vineyards
Tying tapes	LDPE, PVC	Fruit crops, vegetables, vineyards

polyethylene (LLDPE), polyvinyl chloride (PVC), ethylene–vinyl-acetate copolymers (EVA), and polypropylene (PP). Film thicknesses can range from 15 µm for mulch films to 200 µm or more for greenhouse covers. Depending on use, the films also have different types of functional additives, such as thermic fillers, UV stabilizers, antistatic and antifogging additives, and pigments. More detailed information on application of agricultural films and their features can be found in the literature [3–5].

13.2   THE POTENTIAL ROLE OF BIODEGRADABLE MATERIALS IN AGRICULTURAL FILMS

The implementation of plastics for developing agriculture has been very useful in some areas where climatic constraints limit the agricultural activity. However, after use, these plastics present a significant waste disposal issue. In Europe alone, 615 000 tonnes of agricultural plastic waste is generated every year [6]. The waste coming from greenhouse covers, silage films, pipes, fertilizer sacks, and agrochemical packaging may be recycled, although it is not always collected from the field, and sometimes economic viability is questionable.

Mulching is the most common worldwide technique applied in protected cultivation and the most common material used for mulching is polyethylene (PE). However, there is a mismatch between the long-term stability of PE and the product lifespan required in mulch applications. There have been many attempts to solve this worldwide concern [7]. Among them, the best solution is to avoid the use of PE and thereafter the waste generation (Figure 13.1), which in the case of mulching is usually not recyclable. Other alternatives have collateral drawback effects or just delay the problem. The lifetime of the mulch is expected to fit with the duration time of the crop;

Figure 13.1   Plastic waste generated from the use of polyethylene mulching after crop cycle

however, this is not the case for PE mulches, with most of the vegetable cycles lasting between two and six months and exceptionally one year. The best choice appears to be a mulch material with an outdoor service life which matches the crop duration and which would later be incorporated by the agricultural system. In agriculture, the main process for organic matter recycling is biodegradation, which can take place *in situ*, in the same agricultural soil, or in a composting facility at the farm. These options have been considered when developing plastic mulch materials as substitutes for PE.

In the context of this discussion, the use of biodegradable biopolymer films offers a very reasonable option for fitting the available natural recycling system with the material that is used. Moreover, as many biopolymers are derived from agricultural raw materials [8], the suggested solution represents a return of agricultural materials to their origins. In fact, this means that the key to the waste problem associated with conventional mulch film disposal can be found in the same agricultural system where it started, which can help to avoid more collateral drawbacks. The manufacturing of renewable biopolymers will represent a new agricultural activity, which could favour developing countries with agricultural activity at the base of their economy. Sustainability is now essential in agricultural production and use of bio-derived films in agriculture fits with this concept.

13.3 PRESENTLY AVAILABLE BIOPOLYMERS AND BIOCOMPOSITES

A summary of presently available biodegradable polymers and biocomposite materials which might be used in agriculture is presented in Table 13.2. All of these materials

Table 13.2   Biopolymers available for agricultural applications: commercial product, main raw material and resulting biopolymer, and manufacturer

Product Name	Raw material/polymer	Manufacturer
Renewable polymers or blends with oil-derived biodegradable polymers		
BioCeres®	Wheat flour/polyester	FuturMat
Biocycle®	Sucrose/(PHA/PHB)	Biocycle
Bioflex®	Starch/PLA co-polyester	FKur Kunststoff
Biograde®	Cellulose/co-polyester	FkuR Kunststoff
Biolice®	Cereal flour/co-polyester	Limagrain
Biomax TPS®	Starch/TPS-PTT	Dupont-Plantic
Biomer®	Sugars/PHB	Biomer
Bionolle®	Starch, LA, SA, Glycol/PBS,PBSA	Showa Highpolymer
Bioplast®	Starch/TPS	Biotec GmbH
Biostarch®	Starch/TPS	Biostarch
Caprowax TM	Herbal triglycerides/aliphatic polyester	Polyfea
Cardia Comp. TM	Starch/TPS, aliphatic polyester	Cardia Bioplastics
Ceresplast Comp®	Starch, Soy protein/PLA, PHA, PHBs	Ceresplast
EcolGreen	Starch/PLA	EcolBiotech
Ecovio®	Starch/PLA-Ecoflex	BASF
GS Pla®	Starch, SA, 1,4-butane-diol/PBS	Mitsubishi Chemical
Hycail	Starch/PLA	Hycail
Ingeo®	Starch/PLA	NatureWorks
Lactel®	Starch/PLA-PGA	Durect Corp.
Mater-Bi®	Starch/TPS-copolyester	Novamont
Mirel TM	Sugars/PHA	Metabolix-Telles
NaturFlex TM	Cellulose	Innovia Films
Naturplast	Aliphatic polyester (PBS) PLA, PHA	NaturePlast
NaturTec	Natural-compostable polymers blends	Nature-Tec
Plantic-films	Under development	Plantic
Solanyl®	Side stream potato starch	Rodenburg Biopolymers
Terraloy TM	Starch/TPS-PBAT-PLA-PHA	Teknor Apex
Terramac	Starch/PLA	Unitika Plastics
Composites		
AgroResin TM	Lignocellulose composite	Grenidea Tech. Ltd.
BioFibra®	Wood fiber/Polyester	FuturMat
Fibrolon®	Wood fiber/PLA	Fkur Kunststoff
Transmare®	Bamboo-Starch/Bamboo fiber-PLA	Transmare Compoud
Vegemat®	Aerial corn plants/Corn fiber-TPS-TPP	Vegeplast
Zelfo®	Cellulose fiber of different origins	Omodo
Oil-derived biodegradable materials (used in blends with renewable polymers)		
Capa®	Oil/Polycaprolactone	Perstorp
Ecoflex®	Oil/Copolyester	Basf Corp
EnPol®	Oil/Aliphatic polyester	IRE Chemical
Tone®	Oil/Polycaprolactone	Dow Plastics
Lactel®	Oil/Polycaprolactone	Durect Corp.

Source: www.ides.com, modified by the authors.

have been used or have the potential for use in agricultural films and other products. In this respect, it should be mentioned that most materials used for agricultural application are not fully of renewable origin, but in some instances can be blends of biodegradable bio-based polymers with synthetic non-renewable polymers [9]. Present examples in the market are the blends of starch-derived polymers and polylactide (PLA) with PBAT (poly-butylene adipate/terephthalate). PBAT is a biodegradable plastic produced by BASF under the trade name Ecoflex®. Starch materials have a somewhat limited range of application due to their relatively high water-absorbing properties. The use of PBAT in starch blends provides hydrophobization of the starch and thus achieves the required properties (e.g., water resistance, mechanical properties) for many applications [10,11].

One of the main goals of the biopolymer industry is to achieve the highest possible uptake of renewable materials in every product and to supply the right materials for sustainable agriculture. As a consequence, blends between different bio-based and biodegradable materials are increasingly studied [12]. Certification authorities have developed tools to establish the percentage of bio-based carbon in materials, thus allowing renewable content to be quantified. Accordingly, materials can be classified in four different groups, depending on their biobased carbon content: 20–40%, 40–60%, 60–80% and over 80% [13]. However, to date only a few materials as granulates have been certified in this way and applied in the manufacture of commercial agricultural films.

From another perspective, material biodegradability is equally or more important than the renewablility of the material in most agricultural applications. Agricultural products manufactured from renewable resources, but which do not readily biodegrade in agricultural environments are generally much less appreciated than those that biodegrade. For this reason, the possibility of commercially manufacturing PE or PVC from renewable raw materials [14] is not a solution in terms of the waste disposal issue for most agricultural applications. This is the case, even though such bio-derived PE or PVC would clearly add value in film applications where the targeted properties cannot be satisfied by use of today's biodegradable and renewable polymers. As an example, the films for greenhouse covers cannot yet be developed from renewable biodegradable polymers due to inadequate mechanical properties and the expected durability of the resulting films. Furthermore, it is accepted in this application that PE and PVC films can be recycled into other products and therefore do not create a residue.

There are tools available in the biopolymer industry for demonstrating the composting or the biodegradability of their materials. As mentioned in the introductory part of this chapter, farms have their own bio-recycling options through composting or the soil. Most raw materials in this category have obtained the certification 'OK Compost' for granules and films [13, 15], a label that recognizes the compostability. However, only some grades of a few base materials (granulates or films) are certificated 'OK Biodegradable Soil' (e.g., Dupont Biomax®, Mater-Bi®, Mirel®, and Plantic®). As most biopolymer film products for agricultural application end life in the soil, the certification 'Biodegradation in Soil' is of high value in terms of giving confidence to the converting industry and especially to end users [13]. More information on how composting and biodegradation in soil are evaluated and certified can be found in two recently published critical reviews [9, 16]. However, it has to be highlighted that, at the moment, the biodegradation in soil certification is not a globally accepted standard with unified criteria. The standards that are available are methodological, and give tools for analysing biodegradation in soil and that can be used by the different national and international certification bodies. In the near future it would be helpful to have a generally accepted international norm with specific criteria for biodegradation in soil that is suitable for real applications.

13.4   PAST AND CURRENT INTERNATIONAL PROJECTS ON BIODEGRADABLE AGRICULTURAL FILMS

The agricultural sector is a traditional one and arguably not as fully proactive in taking advantage of new equipment or materials emerging as other sectors, such as the medical or the automotive industries. This has been one of the reasons for the slow introduction of biopolymers suitable for use as films for mulching, plant pegs, plant pots and containers, trays, threads, and twines. Up to the beginning of this century, conventional plastic films produced from non-renewable fossil resources, such as PE, have been massively used and the agricultural field has invested limited efforts to studying the potential of substitutes for traditional plastics.

The consumption of fossil resources is clearly at the heart of current concerns about world climate change. Carbon from fossil resources sustained the growth of industries and societies over the last two centuries, but for obvious reasons there is now increasing attention paid to using carbon from bio-resources as part of future sustainable development on a global scale. In this context, 21st century agriculture is designed to have a leading role in minimizing the increase in CO_2 emissions through a multi-faceted approach involving recycling options and minimizing waste disposal problems. In this way, products of daily life and manufactured components should be designed in light of cradle-to-cradle (C2C) considerations, including reduced carbon footprints in a pollution-free manner. The challenge will be to maintain the high-level performance attained by fossil resource-based products and to, for example, achieve the same water-saving efficiency as obtained with conventional PE mulches that is so essential, given the global warming scenario and associated water scarcity. Increasing plant productivity per unit water consumption has long been developed through a diversity of agronomic approaches. The sustainable management of soil-related bio-resources, through the adoption of innovative agronomic techniques can greatly influence 21st century global issues such as energy demands, waste disposal, water quality, desertification and soil degradation [17].

Starting in the early 1970s, energy crisis disruptions have periodically intruded into the global economy, including agriculture. Thus, scientific, social, and political consciousness of the need for innovative new technologies and processes in the agricultural field, including alternatives to materials from non-renewable fossil resources has emerged. In addition to the 1997 Kyoto protocol, global factors which have brought this situation to a head include the steady increase of energy consumption for food production, increasing CO_2 emissions, and the increasing worldwide accumulation of plastic wastes both on land and in the oceans. In regards to the waste disposal issue, the complexity of implementing the corresponding collection systems has driven an increasing interest by various stakeholders in research aimed at the formulation, performance and utilization of biopolymer films in general, and specifically in the agri-food business.

The EU Framework Programmes (FP), as the driving force in coordinated European research, account for the most outstanding European collaborative research activities in a variety of environments and with a full range of cooperation agreements among universities, research centres, industries or technology centres (Table 13.3). The opportunity to invest in research aimed at substituting non-renewable non-biodegradable plastic films in agricultural use with other more eco-friendly materials emerged in some of the first FPs, and produced uneven results which were mostly not cost-competitive. Furthermore, industrial or social interest in such new products was irregular and discontinuous. The subject required more in-depth research and was incorporated into the 1998–2002 FP5 under the Key Action 'Sustainable agriculture, fisheries and forestry' and explicitly under Key Action 5.2: 'The integrated production and exploitation of biological materials for non-food uses'. This call included consideration of the market and the production of materials, such as bioplastics and composites, as well as the raw materials from agriculture and forestry [18]. Other areas within FP5, mainly in the materials field, also hosted projects on biopolymers for various uses, including agriculture. By that stage, industrial, environmental, and agricultural interest in biopolymers was on a rising trend which continues today. Thereafter, the 2003–2006 6th FP, with a total budget of 17 500 M€, was structured in seven thematic areas, four of which could nest research on agricultural biopolymers [19]. Presently, the 2007–2013 7th FP, has a total funding exceeding 50 000 M€ and a share of over 32 000 M€ for collaborative research. Within FP7, in addition to topics devoted to agriculture under 'Food Agriculture and Biotechnology', research on biopolymer films including those used in agriculture may also be supported in calls from several other of the 10 thematic areas [20]. The forthcoming FP8, with a focus on the 2020 European strategy, will likely bring new opportunities and open new perspectives for biopolymers in agriculture.

Using the European Union Cordis web site, a search for 'Biodegradable', 'Bioplastics' or 'Biopolymers' produced 68 FP6 results (18 NMP, 8 SME, 3 Life-SciHealth, 1 IST, 1 INCO, 1 NEST, 1 POLICIES and 35 Marie Curie Mobility actions) [19]. The same search aimed at the ongoing FP7 provided 33 entries (6 SME, 6 NMP, 6 KBBE, 1 HEALTH, 1 ENVIRONMENT, 1 ERC Starting Grant, 1 ERC Advanced Grant and 11 PEOPLE Marie Curie Actions) [20]. In both programmes the identified projects deal with a wide range of subjects, with packaging, health and use of renewable resources being the most cited. However, in spite of the tremendous environmental impact of the most conventional plastics and the enormous potential of new bio-based materials, biopolymer films remain with a few exceptions relatively unknown to society and industry at large. Knowledge is also scarce concerning the main constraints and challenges that face these bio-based materials in the agricultural sector, and which are expected to be met in the near future. These include ascertaining in-soil biodegradability and non-contaminant properties under a wide variety of conditions, implementing mechanization procedures for installation in some cases (e.g., plastic films for crop mulching), adjusting the life-cycle of each biopolymer film to the cultivation environment and to the specific characteristics of each cultivated plant species, and producing the bio-based materials at a competitive cost.

Table 13.3   Funded projects under the 5th, 6th and 7th EU Framework Programmes

Framework Programme	Acronym	Project title
FP5	BIOPLASTICS	Biodegradable plastics for environmentally friendly mulching and low-tunnel cultivation
	BIOPAL	Algae as raw material for production of bioplastics and biocomposites contributing to sustainable development of European coastal regions
	(EA*)	The use of fungal inocula in agricultural applications such as biodegradable mulch films and planting pots
	(EA)	Development of a 100% biodegradable and photo-selective mulch film for sustainable agriculture
FP6	TRIGGER	Development of pea starch film with trigger biodegradation properties for agricultural applications
	PICUS	Development of a 100% biodegradable plastic fiber to manufacture twines to stake creeping plants and nets for packaging agricultural products
	BIODESOPO**	Mechanisms of the biodegradation in soil of biodegradable polymers designed for agricultural applications
	AEROCELL	Aerocellulose and its carbon counterparts – porous, multifunctional nanomaterials from renewable resources
	BIOCELSOL	Biotechnological process for manufacturing cellulosic products with added value
FP7	HORTIBIOPACK	Development of innovative biodegradable packaging system to improve shelf life, quality and safety of high-value sensitive horticultural fresh produce
	HYDRUS	Development of crosslinked flexible bio-based and biodegradable pipe and drippers for micro-irrigation applications
	FORBIOPLAST	Forest resource sustainability through bio-based composite development
	AGROBIOFILM	Development of enhanced biodegradable films for agricultural activities

*EA = Exploratory Award,
**= Marie Curie Mobility Action.

Among ongoing and previous EU projects a significant number addressed processes for incorporating more efficient biotechnologies, targeting materials from plant origin, or using molecules such as lignin, cellulose, or other carbohydrates (e.g., starch).

The path towards increased use of renewable and biodegradable agricultural films starts with the development of the corresponding products, and there were already European projects of this type within FP5. BIOPLASTICS was one of the initial projects to address the agricultural potential of biopolymer films. This project was devoted to studying biodegradable plastic films as promising new materials providing a more convenient, environmentally friendly alternative for mulching and low-tunnel covering, and contributing to the reduction of plastic waste pollution in rural areas. The main objectives of the project were to investigate the agronomic, engineering and environmental aspects involved in the use of bioplastic films in agriculture, and to integrate these films into environmentally friendly and economically optimized low-tunnel and mulching cultivation in Europe. The European Bioplastics Association [21] supports and promotes the research and consumption of these materials. BIOPAL was another of the earlier FP5 projects, also with an emphasis on producing bioplastics for agricultural use, specifically films and clips. In this case, the focus was on using algae as the renewable source of the required compounds. Two exploratory awards are also worth mentioning: one which was aimed at combining biodegradability with photo-selectivity in mulch films, and another aimed at the control of starch-based plastic biodegradation by incorporating fungi into agricultural films and pots; these were very short projects (four and eight months respectively) and driven by SMEs.

The working field of the FP6 collaborative project TRIGGER was development of an agricultural plastic film based on pea starch which would remain functional throughout the required crop cycles. The project was also based on the idea of incorporating an on/off switching trigger into the films to initiate and control the biodegradation process at the end of the crop cycle. At this point, films could be incorporated into the soil through a tilling process, which would also activate the signalling system to start the biodegradation process. The process time span would depend upon the crop type and the soil and environmental conditions. A key consideration was to ensure that the switching system was not toxic for the soil.

There are other agricultural tools requiring somehow equivalent but different properties that are made or can be made from PE or from biopolymers. Taking this into consideration, another FP6 collaborative project (PICUS) was designed to develop a 100% biodegradable plastic fiber for specific applications, such as green-house crop twines and low-weight packaging nets. These two products would require a high-strength plastic material with high modulus and low strain, and full biode-gradability under specific conditions (e.g., composting), traits that are already available in the biopolymer industry but which are seldom combined in one material.

The implementation of more eco-friendly films in agriculture (e.g., mulches) requires investment in their formulation and adaptation to an increasing variety of cultivated species or agricultural practices, as well as their compatibility with various types of processing machinery. In addition, careful and detailed field work is required to ascertain biodegradability under a vast set of different conditions and variables, including soil types, abiotic and biotic environmental determinants, and the complex interactions between these factors. To our knowledge, the dedicated efforts in this area do not yet meet all the needs. It is noteworthy however that a mobility action under the FP6 (BIODESOPO) made one of the first moves towards this ambitious but essential goal, and generated knowledge about the mechanisms and factors influencing in-soil biodegradation of some biopolymer materials designed for agricultural applications.

Other projects did not directly deal with agricultural biopolymer films, but were designed to produce new materials from renewable resources. An example is the FP6 AEROCELL project which was concerned with producing new biodegradable materials mainly from cellulose. These materials would have new properties and therefore new applications, including plant growth supports. The area was fully new and innovative at the time of the project proposal in 2002; various aerogels have now been manufactured and show promising properties for the applications evaluated to date. Nevertheless, as noted by Innerlohinger *et al.* [22], much more research is necessary before commercialization. BIOCELSOL is another FP6 project aimed at developing a novel biotechnology-based process with a reduced environmental cost for converting cellulose into a variety of shaped items with high value.

Within FP7, HORTIBIOPACK is an ongoing SME project targeting innovative and safe biodegradable Equilibrium Modified Atmosphere Packaging (EMAP) film systems based mainly on the use of environmentally friendly biodegradable raw materials. The practical aim is improvement of the shelf-life, quality and safety of specific fresh, high-value and sensitive horticultural produce. As noted elsewhere in this book, work in this and other projects is in progress to develop biodegradable packaging materials for various applications, including food packaging. A limited portion of this research effort is dedicated to MAP packaging systems. The development of biodegradable films that are based on renewable raw materials and designed for EMAP for fresh produce, meeting the design criteria of water vapour and gas barrier properties, and ensuring safety of the produce while maintaining the original quality, firmness and colour is in progress, but is still in the relatively early stages of research and development.

In the specific area of agriculture, the EU countries are facing increasing concerns about high water consumption, particularly in irrigation. This situation is already severe for southern areas of Europe but, with climate change and global warming predictions, these concerns may spread in time to most other European countries. Consequently, multiple solutions will have to be sought. Within this perspective, HYDRUS is an FP7 SME project involving the development of plastic pipes and drippers for micro-irrigation produced with bio-based and biodegradable materials (PLA + additives). These products will need to maintain their functional properties during their service life, but also biodegrade after use without requiring their removal and disposal. The new pipes must meet the basic requirements for traditional micro-irrigation systems. When both the installed mulching and micro-irrigation systems are biodegradable, the high cost of removal at the end of service life will not be necessary and this will to some extent counteract the higher cost of these bio-based materials.

The FORBIOPLAST project is an ongoing FP7 collaborative large-scale research program which addresses improvement in biocomposite materials by using renewable forest biomass as an alternative to petro-resources. The FORBIOPLAST goals include producing bio-based products with the added value of solving petro-derived material problems and reducing waste concerns associated with disposal, polluting chemicals and hazardous substances. The base for this project is the abundance of the raw material, the avoidance of competition with food resources and the use of inexpensive bio-resources. The project takes into consideration the point that, in spite of positive experimental results, commercially successful production and marketing of materials derived from renewable resources can be difficult to achieve because of high processing costs and less than adequate properties in the final products, usually targeted to single-use sectors. The aims of the FORBIOPLAST project are the use of forest resources for the production of eco-compatible products suitable to replace other materials in many practical applications in the automotive sector, in packaging, and in the agriculture sector for mulching, greenhouses and tunnels, twines and clips, and pots. This ambitious project involves a consortium of seven RTDs, six SMEs and three INDs from nine countries and will be finished by June 2012.

AGROBIOFILM started early in 2010 and is presently the youngest of the FP7 projects investigating the agricultural use of biopolymers. This project is proof of the significant progress achieved in the production, diversification and implementation of agricultural biopolymers. The project focuses on the customization of biodegradable mulch films for specific crops and regions. Topics included in AGROBIOFILM are integrating yield enhancers for disease control and soil preparation and/or fertilization in order to increase yield and crop quality, while also reducing the costs associated with removal and disposal of plastics derived from fossil fuel feedstocks. Such products target a market worth 100 M€, accounting for 92 000 tons of plastic each year in Europe. The project managers are conscious that the envisaged solutions will need to meet the dual technical challenges of full mechanical resistance during crop cultivation and degradation and full soil absorption before a subsequent crop is planted. The latter is particularly an issue for longer cycle crops. Time will tell if the project meets expectations, but the participating SMEs are confident about the project outcomes and expect significant benefits from sales and licensing over a five year post-project period.

Results from these and other public national or privately funded projects are starting to emerge in science and in the agro-industry sectors. In terms of scientific output, a survey of papers with the topic 'mulch' in an SCI Journal revealed a low score under two from 1999 to 2004, which rose to 20–30 in 2008–2009 [23]. From 29 selected items in a Journal, organic farming and mulch are nowadays amongst the 10 most significant emerging topics. This is notwithstanding the fact that mulch is considered to be a useful, but relatively simple technology. From its high ranking position, it can be inferred that mulching is a target of increasing importance and also that there is a new wave of biopolymers for mulching that have stimulated an increasing interest in this topic. Doubtless, the development and expansion of biodegradable materials for mulching by leading companies in the life sciences and other industrial areas is connected to the second of these points, which is propelled together with the systemic stream in agriculture towards innovative practices associated with the first point.

In spite of extensive research and promising results over the last two decades, only a limited number of non-cost-competitive biopolymers are presently competing in the agricultural market with those based on petroleum. A good marketing and distribution chain, more rigorous legislation, and enforcement of up-to-date norms for the use and recycling of non-biodegradable and non-renewable agricultural plastic films in the field would undoubtedly help towards increased use of bio-based films in agriculture.

13.5   PRESENT APPLICATIONS OF BIOPOLYMER FILMS IN AGRICULTURE

13.5.1   Overview

Commercial bio-based agricultural films or film-derived commodities are just being introduced into the market. A good overview of present and future biopolymer applications can be found in the inventory of GreenPla registered products from the Japan BioPlastics Association [24]. Around 1000 products are registered in this catalogue and among them there are about 100 finished products for agricultural/ horticultural applications, mostly for protected cultivation, and occasionally for forestry. Table 13.4 shows the final product, the biopolymer used for manufacturing it, and the share of the product within the list.

The end products generated from biopolymers by the converting industry can also be certified as compostable or biodegradable. This certification is also recommended because the converter typically uses further additives for finishing the product which will be exposed to the environment. All the commodities referred to in Table 13.4 are proven to be compostable in the framework of the Japanese certification system. In Europe, there are not many certified agricultural finished products for composting, but generally speaking the applications are the same as those in Japan, that is, mainly films for mulching, but also plant pots, bedding plant trays, and clips for tutoring plants [13, 15]. In addition, there are more film-derived applications, which could be used in the farm, but which have not been certified. For example, a commercial end product certified as a film can have several agricultural applications such as mulching, soil solarization sheet, and bags for fruit protection. There are in fact no finished

Table 13.4   End-bioproducts for agricultural and forest applications registered in Green-Pla (Japan BioPlastics Association)

Product	Biopolymer material	Share (%)
Mulching	BS-LA copolymer, PBS, Starch/ Copolyester, PBAT, Fatty Ac. ester of starch/paper, PCL, PLA, CL-BS copolymer	51
Film for fumigation	PBAT, PVA, BS-LA copolymer, PLA	9
Nets, net tutoring	PBA, PLA	4
Sheets	PBS, PBAT	4
Tapes (tie, mark)	PBAT, PBS, 2CL-BS copolymer, PLA	11
Bands	PBSA	1
Wrapping films	PBAT	1
Floating covers	PLA	1
Pots	PLA, PBS, PBAT (solid), CL-BS copolymer, Starch/Copolyester, BS-LA copolymer,	11
Bedding trays	PLA (foam)	2
Planters	PLA	2

Source: www.jbpaweb.net, modified by the authors.

agricultural film products certified with the label 'Biodegradation in Soil'. The main finished product presently in the market is biodegradable mulch film.

13.5.2 Biodegradable Mulching

Film for mulching is currently the main application of biopolymer films in agriculture (see Figure 13.2). The concept for using biodegradable plastics in mulching is not new. In 1975, Wood [25], referring mainly to mulch and greenhouse films, pointed out that self-destructive biodegradable plastics could be the key to future increased use of plastics in agricultural applications. The first efforts to produce a biodegradable mulch were attempted as early as 1972 by coating plastic with paper, and testing it under field conditions on a melon crop [26]. Later on, in order to obtain a partially biodegradable mulch film, PE was blended with starch [27]. An innovative degradable mulch film was obtained by mixing conventional plant nutrients with a water-soluble polymer such as poly(vinyl alcohol), urea and starch [28]. It was likely the first biodegradable mulch as seen from literature reports. Although cellulose films (Ecopac®) remain among the mulching materials [29], newer materials considered to be 'biodegradable plastics' (Plastigone® and Biolan®) have been evaluated in the field [30]; however, it is worth mentioning that in earlier times there was some confusion between photo- and bio-degradable plastics.

Figure 13.2 Biodegradable mulching in an organic tomato commercial field in North-East Spain

In the early 1990s, studies on carbohydrate-derived thermoplastics presented new opportunities in terms of innovative biodegradable materials that could be useful for mulching [31]. New biodegradable materials based on thermoplastic starch were obtained and commercialized under the trade name Mater-Bi® [32, 33], and some grades were suggested for manufacturing mulch film [34]. At that time, not only thermoplastic starch, but also other biodegradable polymers, like PLA [35], Ecoflex® (PBAT) [10] and later Biopol® (PHA) [36], started commercial development. Other biodegradable raw materials emerged (Table 13.4) and the first commercial developments in agriculture were mulching films.

In Europe, the first reported trials with the new biodegradable materials in mulching applications were performed in the field in the late 1990s [37,38]. From then on, biodegradable plastics for mulching were tested for their agronomic behaviour, mainly in vegetable crops and in several seasons and climatic conditions. Trials included hot season crops, such as tomatoes for processing [39–41], fresh field and greenhouse tomatoes [42–45], peppers [46], melons [47–51] and cucumber [52]. Crops suited to their seasons were also tested and these included strawberries [53–56], cauliflower [57], lettuce and zucchini [58] and sweet potatoes [59]. Biodegradable mulching was evaluated for earliness in snapdragon as an ornamental species [60], and for increasing the quality of the fruits in mandarin orange groves [61]. Some preliminary unreported trials were established to improve initial crop development in cotton and for the implantation of the crop over the first two years in vineyards.

Most studies cited above were carried out with Mater-Bi® for mulching. Mater-Bi® has been improved during the last 12 years for performance as a mulch film and is nowadays amongst the best developed products. Other reported studies were performed with Ecoflex®, Biolice® and Bioflex®. These four materials are well represented as commercial mulch films in the market. Some of the above reported trials have been carried out on other polymers as mulching products, such as those shown in Table 13.4. In addition, new products are also under development, such as Mirel® [85], Bioplast®, and Ecovio®.

Paper mulching is also present in the market as a type of biodegradable mulching, although in general farmers are more reluctant to use it because of several constraints. The mechanical laying-on is more delicate than for bioplastics, which can be adjusted by adapting the plastic mulch layer and slowing the speed of installation. Additionally, paper rolls are heavier per unit length than plastic and as a result the rolls have to be changed more frequently. Both disadvantages can be overcome by using a specifically developed mulch paper layer. However, in rainy and windy regions the paper cover can blow away because of breaking of the mulch at the lateral edge in contact with the soil. On the positive side, paper mulch biodegrades very well in soil and, due to its higher puncture resistance, certain weeds are better controlled than by mulch plastics. There are some studies on how paper mulch can be improved in performance by polymerizing it with vegetable oils or by coating it with starch [62–64]. An extensive literature on paper mulching can be found in the international database.

Globally, in terms of production and quality, the field performance of bioplastic mulches is equivalent to that of PE, especially with black film. Weed control is also equivalent for both technologies, although in some long-lasting crops with a low ground cover capacity, attention has to be paid to the durability of the mulch. Most vegetable crop cycles last between two and six months and commercially available biodegradable plastics fit well with this cycle. Farmers have different options in the market for choosing the better biodegradable mulching according to their specific conditions. Crop earliness is also an important agronomic feature, mainly in the Mediterranean regions. Although in some studies PE has shown a higher earliness effect, generally biodegradable mulching has similar behaviour to PE provided that the duration of the cover is fitted with the time span of the crop cycle. It has to be stressed that biodegradable mulches are usually thinner than conventional ones. Earliness will also depend on the crop and the season in which the materials are tested. There are hardly any studies on the water-saving capacity of biodegradable mulching in the field. Some preliminary reports seem to suggest that biodegradable mulching has slightly reduced water-saving capability when compared with conventional PE [57]. Since the experimental fields where this feature has to be evaluated are usually not sensitive enough to identify differences in water saving due to material properties, it cannot yet be concluded that differences between biopolymer films and PE are significant in this respect at the farm level.

Mechanical studies have been carried out throughout the crop cycle in order to determine the degradation rate of Mater-Bi® mulches [55, 65, 66], Bioflex® and Biofilm® [66], and Ecoflex® [67] during its functional service life. All the materials tested had adequate mechanical properties for mechanical laying-on (Figure 13.3). A decrease in mechanical properties was observed in the first week and continued into the fourth or fifth week depending on the material and testing conditions. After this period, stabilization was noticed. The decay in mechanical properties did not affect the mulch soil covering and the mechanical properties were satisfactory from planting

Figure 13.3 Mechanical laying on of a biodegradable mulch film in a commercial field

to harvesting. The direct effect of UV solar radiation is the main cause for this decay [68, 69]; although, for some materials, water can also be a significant factor. Biodegradable black films for mulching show a very high absorbance of UV radiation, as is the case for PE [66]. As black films are opaque to photosynthetically active radiation (PAR), the loss of PAR transparency is not as relevant as in clear translucent films, where mechanical property decay is accompanied by a significant decrease in transmission of PAR [70]. Depending on the crop and crop environment, the loss of PAR transmission can modify crop earliness and productivity. Mater-Bi® black and transparent mulch films (30–60 µm) PAR and infrared transmission coefficients are given by Vox and Schettini [71], while Martin-Closas et al. have reported the infrared transmission characteristics of black Mater-Bi®, Bioflex® and Biofilm® mulches (15–17 µm) [66].

After harvesting, the mulch film is ploughed into the soil and after a period between three months to over a year it has visually disappeared. At field level only fragmentation and visual presence of the material can be tracked, and effective biodegradation is very difficult to ascertain. The degree of above-soil degradation will mainly depend on the time frame of exposure to radiation, rainfall, and crop covering. In-soil degradation depends mainly on soil chemical features and on its biological activity (environmentally dependent). Qualitative scales on the remaining covered surface and in-soil mulching fragmentation have usually been applied for evaluating mulch degradation in the field. Field degradation of a PE and a biodegradable mulch film above- and in-soil are presented in Figure 13.4. Kapanen et al. [72] reported less than 4% of the initial weight of the Mater-Bi® film remaining in the soil 12 months after

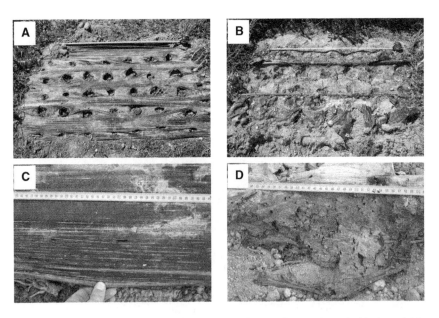

Figure 13.4   Above (A, B) and in-soil (C, D) degradation of PE (A, C) and a biodegradable mulch (B, D) at the end of a 115 days crop cycle in North-East Spain

tillage. Laboratory experiments simulating the soil environment appear to be the best way to guarantee in-soil biodegradation under several different conditions [9]. However, there are hardly any reports on in-soil mulch biodegradation, which is more frequently carried out in compost [73].

A new market demand is for long-lasting biodegradable mulching films in applications requiring functionality for more than one year. This kind of mulching will be applied to long-lasting vegetable or ornamental crops, for implantation of vineyards or orchards, and also for establishing forest covers [74]. In certain crops such as melon, biodegradable mulching can stick to the fruit if it is in direct contact with the mulch. This inconvenience also has to be overcome if biodegradable mulching is to be better accepted by the farmer. Detailed studies on the environmental impact of biodegradable mulching by life cycle analysis (LCA) are also needed. Some studies have applied LCA to conventional plastics (e.g., PE and EVA) [6], but did not include the biodegradable options. Other studies with biodegradable and conventional multilayer films found that biodegradable films offered lower global warming potential and fossil energy depletion impacts than multilayer films [75].

13.6 POTENTIAL USES: CURRENT LIMITATIONS AND FUTURE APPLICATIONS

In addition to mulching, new agricultural applications for biodegradable films are being developed. Starting from the mulch experience, soil solarization and low-tunnel developments are at an advanced stage but are not yet fully commercially developed.

13.6.1 Solarization with Biodegradable Films

Solarization is an agronomic technique applied for the purpose of soil disinfestation after intensive periods of culture. The procedure consists of laying a transparent plastic sheet on bare, humid soil during the hot season in order to increase the soil temperature and to thereby disinfest it from pathogens, pests and weeds [76]. Candido et al. [47] found equivalent effects for Mater-Bi® and PE sheets in terms of soil temperature increases through solarizing. When biodegradable sheets were compared with EVA films, the latter produced a higher soil temperature [77, 78]. No difference in disinfestation activity was found when soil nematodes were monitored in melon [77] or eggplant crops [79], but EVA films provided much better disinfestation activity against soil fungi [78] in cucurbitaceous crops. Thus, there is still a way to go in terms of improving the potential of biodegradable films in solarization, specifically their radiometric characteristics and life span [78].

13.6.2 Biodegradable Low Tunnels

Low-tunnel crop is a protected cultivation technology mainly used for obtaining early crop production (Figure 13.5). Since residual low-tunnel films are usually accepted by the recycling industry, the introduction of biodegradable film technology in this case is

Figure 13.5   Biodegradable Mater-Bi® low tunnel. Courtesy of D. Briassoulis [65, 68, 81]

not as pressing as for mulch films. Nevertheless, mulching and low-tunnel technologies are usually used together and therefore the idea of a totally biodegradable system, which will avoid the removal of plastics from the field, is very attractive.

Briassoulis [80] details the mechanical requirements for low-tunnel biodegradable films. The analysis of the mechanical behaviour of biodegradable low-tunnel films proves that rather good mechanical performance is possible and that stabilization schemes used with conventional PE films may not be suitable for biodegradable films [81]. Scarascia-Mugnozza *et al.* [70] studied the effect of solar radiation on the radiometric properties of biodegradable films for low tunnels. After field exposure of experimental films (UV-stabilized or non-UV- stabilized Mater-Bi®) in a strawberry crop, these authors found a decrease in the PAR transmission coefficient and less change in long-wavelength infrared transmission. The early production of the plants under the UV-stabilized biodegradable films was higher than that obtained by the plants under PE low tunnels, and final production was similar. These results were explained by the lower infrared transmission of biodegradable films compared with PE which allowed higher temperatures inside the biodegradable film tunnels. The PAR transmission coefficient of a low-tunnel biodegradable film (40 µm) was lower than that of PE (86 vs 90%), but the diffuse PAR was much higher in the biodegradable film than in the PE (79 vs 27%) film. This means a more homogenous light environment inside the biodegradable film tunnels. The infrared transmission coefficient was between 10 and 16% for the biodegradable cover and ca. 40% for the PE cover [71]. The results obtained for low tunnel application are encouraging; however the tested biodegradable materials require improved transmission of PAR radiation.

Figure 13.6 Biodegradable Mater-Bi® fruit protecting bag in use for table grapes. Courtesy of S. Guerrini, Novamont S.p.A

13.6.3 Fruit Protecting Bags

Fruit protecting bags (Figure 13.6) are used for various fruit crops such as apples, pears, peaches, mangoes, table grapes and bananas. Their function is to protect the fruit from predators and atmospheric agents, and to favour more homogeneous maturation and quality. Biodegradable plastics and paper bags are adequate for this application because they allow a higher transpiration of the fruit inside the bag than conventional PE bags, and also a higher exchange of humidity between the inside and the outside of the bags, thus effectively decreasing the humidity inside the bag and therefore reducing crop losses. This product is already on the market, mainly as paper or wax-paper bags, although improvements are needed to reduce the influx of water from the outside to the inside of the bag during rainfall.

13.6.4 Future Biodegradable Film Applications

One of the potential applications of biodegradable materials in agriculture is non-woven floating covers. Floating covers are installed over a crop to protect it from atmospheric agents and pests and to improve the crop environment microclimate, thus increasing earliness and quality. As in the case of mulching, after use the floating cover residues are dirty with soil and, although recovery from the field is easier than for mulches, recycling is not always feasible. There are already some commercial products on the market, mainly manufactured with PLA, and also some patents on thermoplastic starch. Some preliminary attempts have been made in Europe by designing a biodegradable floating cover for winter and summer crops using Bionolle® from Showa Denko [82].

Other applications that are under development are all the agricultural goods which are obtained by cutting the films and which will end their life cycle in the soil or in a composting environment, or that are very difficult to recover from the field. Examples are biodegradable plastic strips for grafting and plastic tapes for tutoring, tying or marking plants. These products are intensively used in greenhouse crops, in orchards, and in vineyards. In the case of grafting strips a very elastic biodegradable material is required, and strength resistance is expected in the case of tapes. Further biodegradable film applications under development are in hydroponic sacks for soilless cultivation and tree protection cylinders.

Aside from biodegradable films, other agricultural biodegradable products are being developed. Among them plant pots are the most frequently considered (Table 13.4), but bedding trays, bands, strings and nets for tutoring and pheromone supports may also be mentioned. In addition, some preliminary work has been done to develop irrigation tapes [83].

13.7   CONCLUDING REMARKS

Biopolymer films have a significant presence in today's agricultural market as mulch films, although according to a recent survey in Europe, their use remains less than 10% of the total [84], which is very low when compared with the use of PE mulching. A contributing factor is the relatively high cost of biodegradable mulch films. In France and Italy the PE mulch in current use is thicker than that used in other countries (e.g. Spain) and thus in these two countries there is more possibility for biodegradable mulch films to become competitive. The main reasons for using biodegradable films are environmental and for reducing labour costs, and the main constraint is the higher price. Not surprisingly, a significant increase in biodegradable mulch films was noticed in regions where their usage was subsidized.

Currently, a significant number of new applications are being developed and will likely appear commercially in the near future. The main factors for this development will be not only higher cost-competitiveness, but also research-based innovation and more efficient use of currently available biopolymers. In situations where the biodegradable polymers are bio-derived, these developments will be fully in line with the increasing need for sustainable industry and society in the 21st century.

13.8   ACKNOWLEDGEMENTS

The authors are grateful for financial support received from the Spanish Ministry of Science and Innovation (Grant Nr. AGL2008-03733) and the University of Lleida, Spain (Sustainable agriculture grant 2009 P09011).

REFERENCES

[1] W. J. Lamont, *HortTech.* 1996, 6, 150–154.
[2] Reynolds, A., Market Overview. Agricultural film markets, trends, and business development, in *Proceedings of Agricultural Films 2009*, AMI Ltd., Barcelona Spain (2009).
[3] López, J. C., Pérez-Parra, J., and Morales M.A. (2009). *Plastics in Agriculture. Application and usages handbook.* CEPLA-Plastics Europe. Almería, Spain.
[4] E. Espí, A. Salmerón, A. Fontecha, *et al.*, *J. Plast. Film Sheet.* 2006, 22, 85–102.
[5] Papaseit P., Badiola, J., and Armengol E. (1997). *Plastics and Agriculture.* Ediciones de Horticultura S.L., Reus, Spain.
[6] U. Bos, C. Makishi, and M. Fischer, *Acta Hort.* 2008, 801, 341–349.
[7] W. J. Lamont, *HortTech.* 2005, 15, 477–481.
[8] A. Rouilly, L. Rigal, *J. Macromol. Sci. Part C – Polym. Rev.* 2002, C42, 441–479.
[9] D. Briassoulis, C. Dejean, *J. Polym. Environ.* 2010, 18, 384–400.
[10] Yamamoto, M., Witt, U., Skupin, G., *et al.* Biodegradable aliphatic-aromatic polyesters: "Ecoflex®", in Y. Doy and A. Steinbüchel (eds), Biopolymers, Volume 4, Polyesters III - Applications and Commercial Products, Wiley, Weinheim, Germany, pp 299–305 (2002).
[11] S. V. Malhotra, V. Kumar, A. East, *et al.*, *The Bridge* 2007, 37, 17–24.
[12] L. Yu, K. Dean, and L. Li, *Prog. Polym. Sci.* 2006, 31, 576–602.
[13] http://www.okcompost.be Accessed on September 2010.
[14] Anonymous *Chem. Eng. Prog.* 2007, 103, 15–15.
[15] http://www.dincertco.de Accessed on September 2010.
[16] D. Briassoulis, C. Dejean, and P. Picuno, *J. Polym. Environ.* 2010, 18, 364–383.
[17] R. Lal, *Agron. Sustain. Dev.* 2008, 28, 57–64.
[18] http://cordis.europa.eu/fp5 Accessed on September 2010.
[19] http://cordis.europa.eu/fp6/dc/index.cfm?fuseaction=UserSite.FP6HomePage Accessed on September 2010.
[20] http://cordis.europa.eu/fp7/home_en.html Accessed on September 2010.
[21] http://www.european-bioplastics.org Accessed on September 2010.
[22] J. Innerlohinger, H. K. Weber, and G. Kraft, *Lenzinger Berichte*, 2006, 86, 137–143.
[23] E. Lichtfouse, M. Hamelin, M. Navarrete, *et al.*, *Agron. Sustain. Dev.* 2010, 30, 1–10.
[24] http://www.jbpaweb.net Accessed in September 2010.
[25] A. S. Wood, *Modern Plastics* 1975, 52, 37–39.
[26] Lipe, W. N., Martin, C., Eddins, R., *et al.* A biodegradable plastic-coated paper mulch for cantaloupe production, *Progress Report*, Texas Agricultural Experiment Station 3081, 5 pp (1972).
[27] C. L. Swanson, R. P. Westhoff, W. M. Doane, *et al.* *Am. Chem. Soc. Polym. Preprints, Division Polym. Chem.* 1987, 28, 105–106.
[28] S. M. Lahalih, S. A. Akashah, and F. H. Al-Hajjar, *Ind. Eng. Chem. Res.* 1987, 26, 2366–2372.
[29] L. Boldrin, *Plasticulture* 1989, 83, *51–52*, 54.
[30] C. W. Marr, W. J., Jr. Lamont, *HortScience* 1990, 25, 1661.
[31] G. Mantovani, G. Vaccari, *Industria Saccarifera Italiana* 1991, 84, 171–179.
[32] Bastioli, C., Belloti, V., and Gilli, G., The use of Agricultural Commodities as a source of new plastic material, in *Proceedings of Biodegradable Packaging and Agricultural Films, APRIA Conference*, Paris, France pp 1–36 (1990).

[33] C. Bastioli, V. Belloti, L. del Giudice, et al., J. Environ. Polym. Degrad. 1993, 1, 181–191.

[34] C. Bastioli, Polym. Degrad. Stabil. 1998, 59, 263–272.

[35] Gruber, P., and O'Brien, M., Polylactides. NatureWorks™ PLA, in Y. Doy, and A. Steinbüchel (Eds.), Biopolymers, Volume 4, Polyesters III – Applications and Commercial Products, Wiley, Weinheim, Germany, pp 235–247 (2002).

[36] Asrar, J., and Gruys, K. J., Biodegradable Polymer (Biopol®), in Y. Doy, and A. Steinbüchel (Eds.), Biopolymers, Volume 4, Polyesters III – Applications and Commercial Products, Wiley, Weinheim, Germany, pp 53–72 (2002).

[37] C. Weber, KTBL-Arbeitspapier, 1998, 251, 78–83.

[38] C. Manera, S. Margiotta, and P. Picuno, Colture Protette, 1999, 28, 59–64.

[39] L. Martin-Closas, J. Soler, and A. M. Pelacho, KTBL-Schrift, 2003, 414, 78–85.

[40] R. Armendariz, J. I. Macua, I. Lahoz, et al., Acta Hort. 2006, 724, 199–202.

[41] L. Martin-Closas, M. A. Bach, and A. M. Pelacho, Acta Hort. 2008, 767, 267–274.

[42] V. Candido, V. Miccolis, D. Castronuovo, et al., Acta Hort. 2006, 710, 415–420.

[43] M. M. Moreno, A. Moreno, Sci. Hortic. 2008, 116, 256–263.

[44] M. Ngouajio, R. Auras, R. T. Fernandez, et al., HortTechnology, 2008, 18, 605–610.

[45] M. M. Moreno, A. Moreno, and I. Mancebo, Span. J. Agric. Res. 2009, 7, 454–464.

[46] J. K. Olsen, R. K. Gounder, Aust. J Exp. Agric. 2001, 41, 93–103.

[47] V. Candido, V. Miccolis, G. Gatta, et al., Acta Hort. 2001, 559, 705–712.

[48] A. González, J. A. Fernández, P. Martín, et al., KTBL-Schrift 2003, 414, 71–77.

[49] G. Incalcaterra, A. Sciortino, F. Vetrano, et al., Options Mediterraneennes Serie A, Seminaires Mediterraneens, 2004, 60, 181–184.

[50] J. López, A. González, J. A. Fernández, et al., Acta Hort. 2007, 747, 125–130.

[51] F. Vetrano, S. Fascella, G. Iapichino, et al., Acta Hort. 2009, 807, 109–114.

[52] C. Weber, Gemüse, 2000, 36, 30–32.

[53] F. Lieten, Fruit Belge, 2002, 497, 95–96.

[54] C. A. Weber, HortTech. 2003, 13, 665–668.

[55] G. Scarascia-Mugnozza, E. Schettini, G. Vox, et al., Polym. Degrad. Stabil. 2006, 91, 2801–2808.

[56] A. P. Bilck, M. V. E. Grossmann, and F. Yamashita, Polym. Test. 2010, 29, 471–476.

[57] G. Magnani, F. Filippi, A. Graifenberg, et al., Colture Protette, 2005, 34, 59–68.

[58] G. Minuto, L. Pisi, F. Tinivella, et al., Acta Hort. 2008, 801, 291–298.

[59] L. JoonSeol, J. KwangHo, K. HagSin, et al., Korean J. Crop Sci. 2009, 54, 135–142.

[60] E. Schettini, G. Vox, and B. de Lucia, Sci. Hortic. 2007, 112, 456–461.

[61] Y. Tachibana, T. Maeda, O. Ito, et al., Int. J. Mol. Sci. 2009, 10, 3599–3615.

[62] R. L. Shogren, J. Sustain. Agriculture, 2001, 16, 33–47.

[63] R. L. Shogren, R. C. Hochmuth, Hort. Science, 2004, 39, 1588–1591.

[64] Y. Zhang, J. H. Han, and G. N. Kim, Commun. Soil Sci. Plant Anal. 2008, 39, 1026–1040.

[65] D. Briassoulis, Polym. Degrad. Stabil. 2006, 91, 1256–1272.

[66] L. Martin-Closas, A. M. Pelacho, P. Picuno, et al., Acta Hort. 2008, 801, 275–282.

[67] T. Kijchavengkul, R. Auras, M. Rubino, et al., Chemosphere 2008, 71, 942–953.

[68] D. Briassoulis, Polym. Degrad. Stabil. 2007, 92, 1115–1132.

[69] T. Kijchavengkul, R. Auras, M. Rubino, et al., Polym. Degrad. Stabil. 2010, 95, 99–107.

[70] G. Scarascia-Mugnozza, E. Schettini, and G. Vox, Biosyst. Eng. 2004, 87, 479–487.

[71] G. Vox, E. Schettini, Polym. Test. 2007, 26, 239–251.

[72] A. Kapanen, E. Schettini, G. Vox, et al., J. Polym. Environ. 2008, 16, 109–122.

[73] T. Kijchavengkul, R. Auras, M. Rubino, et al., Chemosphere, 2008, 71, 1607–1616.

[74] S. Cao, L. Chen, Z. Liu, *et al.*, *J. Arid Environ*. 2008, **72**, 1374–1382.

[75] R. Vidal, P. Martínez, E. J. Mulet, *et al.*, *J. Polym. Environ*. 2007, **15**, 159–168.

[76] J. I. Stapleton, J. E. Devay, *Crop Protection*, 1986, **5**, 190–198.

[77] D. Castronuovo, V. Candido, S. Margiotta, *et al.*, *Acta Hort*. 2005, **698**, 201–206.

[78] G. Russo, A. Candura, and G. Scarascia-Mugnozza, *Acta Hort*. 2005, **691**, 717–724.

[79] V. Candido, V. Miccolis, M. Basile *et al.*, *Acta Hort*. 2005, **698**, 195–199.

[80] D. Briassoulis, *Biosyst. Eng*. 2004, **87**, 209–223.

[81] D. Briassoulis, *J. Polym. Environ*. 2006, **14**, 289–307.

[82] Siwek, P., Libik, A., Gryza, I., *et al.*, Physico-mechanical properties and utility of melt-blown biodegradable nonwovens, in *Proceedings of the CIPA Congress. Session IV. Conference II*, Almería, Spain pp 11 (2009).

[83] D. Briassoulis, M. Hiskakis, and E. Babou, *Acta Hort*. 2008, **801**, 373–380.

[84] Ziermann, A. Diploma Thesis: 'Framework conditions for the use of biodegradable mulch films', University of applied Science, Weihestephan, Germany, (2010). Unpublished.

[85] Krishnaswani, R., Kelly, P., and Schwier, C.E., The effectiveness of Biodegradable Poly(Hydroxy Butanoic acid) copolymers in Agricultural mulch film application, in *Proceedings of the Plastic Encounter ANTEC*. Society Plastic Engineer, Milwaukee, Wisconsin, USA pp 1–5 (2008).

14

Functionalized Biopolymer Films and Coatings for Advanced Applications

David Plackett and Vimal Katiyar
Risø National Laboratory for Sustainable Energy, Technical University of Denmark, Roskilde, Denmark

14.1 INTRODUCTION

As outlined in previous chapters, there is now growing interest in sustainable biopolymers, particularly in commodity applications such as packaging, edible films and coatings, paper and paperboard, and agronomy (e.g., agricultural mulch films). There have also been numerous developments in medical uses of biopolymers, including, for example, polylactide (PLA)- or polylactide-*co*-glycolide (PLGA)-based devices for bone fracture fixation, suture materials, sheets for preventing adhesion, blood vessel prostheses and drug delivery [1]. In the future, it is likely that the use of functionalized biopolymer films or coatings will be extended to other specialized high-performance applications and, on the basis of current literature, the introduction of bio-based polymeric materials in optoelectronics (e.g., solar cells), chemical or biological sensors and various advanced coating systems can be envisaged. The driving forces for this trend will be the need for sustainably derived and, in some

Biopolymers – New Materials for Sustainable Films and Coatings, First Edition.
Edited by David Plackett.
© 2011 John Wiley & Sons, Ltd. Published 2011 by John Wiley & Sons, Ltd.

instances, degradable polymers and/or the unique characteristics which certain bio-polymers can bring to these devices and applications. An additional concern driving such interest is the potential increased cost of using petroleum-derived polymers in the future. This chapter is designed to give the reader a flavor of recent research in this field and an indication as to where future opportunities and challenges may lie.

In the context of this introduction, it is worth noting that bio-based nanocompo-sites in particular offer new opportunities as ecological, bio-inspired and functional hybrid materials with applications suggested in electrochemical and potentiometric sensors, electrode materials, enzymatic sensors, and nanoreactors [2]. The possibility to couple biopolymers with magnetic or photoactive nanoparticles, which in turn may be amenable to further assembly into bioactive species, has also been raised and, as discussed later, the concept of flexible magnetic cellulose films or nanopaper has quite recently been investigated [3].

Bionanocomposites have gained increasing attention as an important sub-set in the broader nanocomposite field, which was launched by researchers at Toyota over two decades ago [4]. Research on the topic of nanocomposites has increased dramatically in subsequent years and, in addition to various commercial applications based on conventional polymers, there is now interest in biopolymers as nanocomposite matrices (Figure 14.1). The result has been the development of biomaterials in which properties introduced through nanoreinforcement are combined with the often unique biocompatibility and degradability characteristics of individual biopolymers.

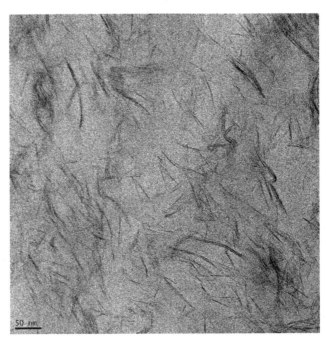

Figure 14.1 A transmission electron microscopy (TEM) image of an extruded bio-nanocomposite film showing the dispersion of organomodified montmorillonite (MMT) clay platelets in a polylactide (PLA) matrix. Scale bar = 50 nm

One of the key incentives for research in this area has been the perceived need for a new generation of 'green' sustainable materials which also biodegrade under controlled conditions and could therefore potentially help to alleviate the growing global plastic waste problem. Biocompatibility, as defined by the lack of injurious effects in living organisms, will be important in some anticipated uses and, in this respect, the characteristics of both the biopolymer and nanofiller need to be considered. As reflected in the literature, a continued increase in research on bionanocomposites can be expected because of the wide availability of biopolymers from renewable resources and their potential synergy with nanofillers. The investigation of bionanocomposite technologies is discussed at intervals in this chapter and it is worth mentioning here that a diversity of advanced applications such as electroluminescent displays, carriers for influenza vaccines, DNA vectors and sensor devices, and electroanalytical instruments have been proposed [5–7]. This chapter reviews some of the recent research on bio-based polymer films and coatings which may be functionalized chemically or physically in order to open up a new generation of high-value devices for specalized applications in such fields as optoelectronics and sensor technologies.

14.2   OPTOELECTRONICS

14.2.1   Photovoltaics

The investigation and development of organic photovoltaics (OPVs) is now an increasingly important topic in sustainable energy research. This situation has arisen because polymer-based solar cells, although at present still in the development and pre-commercial phase, offer a potentially cheaper and in some respects a more technically convenient alternative to conventional silicon solar cells for various energy needs. As has been demonstrated, rapid bulk production of flexible solar cell films using roll-to-roll printing technology is feasible (Figure 14.2).

Figure 14.2   Post-production processing of polymer solar cells involving lamination on a PET substrate between barrier foils in order to ensure maximum service life (Source: Solar Energy Programme, Risø DTU)

The state of the art in OPVs has been detailed in two recent books, covering the principles and practice of organic solar cell technology and encompassing materials, design and manufacture [8,9]. Strategies for inclusion in garments and textiles have been discussed as one example of the many possible applications of OPVs [10]. Although technical advances are required before a wide range of OPVs can be adopted on a large scale (i.e., greater conversion efficiency and stability), the rate of OPV technology development is now such that the first significant commercial applications are likely within the next decade.

In parallel with the development of OPVs, there has also been some interest in utilizing renewable materials where possible in photovoltaic devices. As one example, chitosan has been studied as a binder because of its abundant –OH and –NH$_2$ groups, excellent compatibility, nontoxicity, relative cheapness, ease of handling, conductivity and good mechanical properties [11]. The first bismuth-based solar cell using a chitosan matrix embedded with bismuth oxyiodide (BiOI) nanoplate microspheres was recently reported and was shown to have promising photochemical activity [12]. The complex electrical conductivity mechanism of chitosan thin film has been studied and its suitability for photovoltaics and electrochromic devices discussed by a number of authors [13–15]. As an example, Mohamad *et al.* [13] prepared polymer electrolyte films from a blend of polyethylene oxide (PEO) and chitosan with the addition of ammonium iodide. Devices were obtained by sandwiching a film of this type with the highest measured room temperature conductivity between indium–tin oxide plates with a zinc telluride semiconductor. The PEO–chitosan blend hosting an I—I^{3-} redox couple was shown to operate as a junction solar cell.

Conventional polymers such as polyethylene terephthalate (PET) and polyethylene naphthenate (PEN) have been used as sheets for encasing and protecting polymer solar cell active components because of their flexibility and thermal properties. This is the case even though these polymers do not provide the significant barrier to oxygen and water vapour permeability which would be valuable in some of these devices. The environmental impact of solar cells throughout their life cycle could therefore be reduced by substituting PET or PEN with bio-derived alternatives, providing that the sustainability of these alternatives could be confirmed and suitable barrier properties could be introduced. With this in mind, Strange *et al.* [16] investigated commercially available poly-L-lactic acid (PLLA) as a substrate and included a PLLA/clay nanocomposite film in the study. The conclusion was that further development was necessary, including the need for an increased glass transition temperature (T_g) as well as nanocomposite films with better clay dispersion. Recent developments reported in the literature suggest that there may be technical solutions to the issue of thermal stability through the use of PLA stereocomplexes [17] although, to the authors' knowledge, this idea has yet to be pursued, possibly for cost reasons. In a recent study, Levy [18] examined biodegradable solar cell backsheets, including examples of corn-derived PLA, cellulose from cotton, and nylon obtained from castor beans. Although PLA was considered suitable in view of its transparency and UV stability, its weak moisture resistance was a drawback. Similarly, Levy concluded that the use of cellulose would require its conversion to more hydrophobic material through chemical modification. Nylon 11 was viewed as an attractive option because, although not biodegradable, it was derived from a renewable resource and also showed good moisture resistance.

The introduction of the first fully sustainable solar cell biopolymer technology was claimed by the company BioSolar, Inc. (Santa Clarita, CA, USA) in 2009. Literature from this company points out that traditional bioplastics have been unreliable in PV applications in the past because of their chemical characteristics with regard to environmental exposure, as well as insufficient stability when exposed to high temperatures and fragility during processing. The innovation developed by BioSolar is a protective backing for solar cells derived from cotton and castor beans. It is claimed that the resulting bioplastic significantly reduces manufacturing costs in comparison with petroleum-based alternatives and, for example, is expected to cost 25% less than Tedlar®, the polyvinyl fluoride (PVF) film produced by DuPont, which is used today by most silicon solar cell producers. BioSolar has announced a two-phase strategy to disseminate its bioplastic technology, involving first-generation films using a robust PLA as superstrate, substrate and backsheet films and second-generation films which will employ cellulose-based resins for non-roll-to-roll applications [19].

Dye-sensitized solar cells (DSCs) are a class of thin film devices and have been under development since their invention in the early 1990s [20]. As discussed by Gonçalves et al., there are some expectations that this type of solar cell will become commercial in the near future [21]. Considering the use of bio-materials in DSCs, Kaneko et al. fabricated a solid-state dye-sensitized solar cell using an inexpensive κ-carrageenan-based gel containing an iodide redox electrolyte as the active organic medium [22]. Carrageenans are a family of gel-forming, viscosifying polysaccharides obtained commercially by extraction of certain red seaweeds and are composed of a galactose backbone with various degrees of sulfation. The most important commercial types are called ι-, κ-, and λ-carrageenans and are defined according to their chemical structures (i.e., number and location of ester sulphate groups on the repeating galactose units) and gelation properties. The DSC device made by Kaneko et al. [22] gave an incident photon-to-electricity conversion efficiency of more than 7%, and a small prototype solar cell based on this approach provided sufficient electricity to run low energy demand devices.

A solar-based LCD screen has been developed using power from a transparent solar cell plate made of zinc oxide and copper aluminum oxide in combination with a biopolymer thin film-based secondary battery to prolong the running time. This biopolymer transparent battery was composed of microbatteries for storing electrical energy generated by photoelectric conversion [23]. More recently, Singh et al. produced electrically conductive chitosan composite membranes doped with an ionic liquid, and a DSC using this new biopolymer electrolyte showed promising performance [24].

The idea of using an ionic liquid in combination with a biopolymer to create a so-called ion jelly has been reported by Vidinha et al. [25]. The objective of this work was to produce polymeric conducting materials which could be used in electrochemical devices such as batteries, fuel cells, electrochromic windows or photovoltaics. The research focused particularly on gelatin because it is well characterized, widely available, and relatively inexpensive. Gelatin is prepared by thermal denaturation after acid or alkaline pretreatment of collagen. Dissolution of gelatin in an ionic liquid produces a viscous solution which gels below 35 °C and combines flexibility with conductivity. Both the ionic liquid and gelatin can potentially be functionalized and the combination offers the chance to tailor ion jelly properties according to need.

14.2.2   Other Optoelectronic Devices

In general, optoelectronics can be described as the study and application of electric devices that source, detect and control light, and is usually considered to be a sub-field of photonics. Optoelectronic devices are electrical-to-optical or optical-to-electrical transducers. As well as the use of these devices to explore biological environments, a field known as biooptoelectronics, there is now considerable interest in development of bio-based optoelectronic materials. In this category, one particularly interesting direction involves the fabrication of electronic and photonic devices using the unique properties of DNA thin films. This has spawned a new area of research within the broader field of natural or organic photonics [26]. In recent years, such investigations have focused on the idea of DNA as a natural biopolymeric material, with unique properties for applications such as molecular electronic devices, nanoscale robotics and DNA-based computation. For example, the potential use of DNA as a marine waste product from salmon processing has been studied [27,28] (Figure 14.3).

Thin film fabrication based on supramolecular self-assembly has been carried out using DNA–quaternary ammonium surfactant (e.g., cetyltrimethylammonium chloride or CTMA) complexes. These complexes are thermostable up to 230 °C, highly transparent from 350 to ~1700 nm and have been successfully used in organic light emitting diodes (OLED), cladding and host materials in nonlinear optical (NLO) devices, and organic field-effect transistors (OFET). Using a thin film of DNA–CTMA-based biopolymer as the gate insulator and pentacene, a polycyclic aromatic hydrocarbon consisting of five linearly fused benzene rings, as the organic semiconductor, a bio-organic field-effect transistor or BioFET was obtained in which the current was modulated over three orders of magnitude using gate voltages less than 10 V. Given the possibility to custom-functionalize the DNA film for specific purposes (e.g., biosensing), DNA–CTMA with its unique structural, optical and electronic properties appears to have a number of future applications [29]. Along similar lines, Popescu et al. [30] prepared thin films of DNA complexed with CTMA as well as collagen and found an optical damage threshold which was about an order of magnitude higher for these biopolymers than for a number of conventional synthetic polymers [30]. Sarma et al. also investigated a DNA-based biopolymer as a gate dielectric for fabrication of organic field-effect transistors [31]. In their approach, doping DNA with a conductive polymer led to a significant decrease in the overall blend resistivity with the potential for use in low-cost OFET back planes in new-generation flexible displays and other microelectronic applications. Whitten et al. used a free-standing film of a biopolymer incorporating carbon nanotubes (CNTs) as a material with good mechanical strength and specific capacitance suitable for use in free-standing electrodes [32]. Grote [28] summarized the work of the US Air Force Research Laboratory, Materials and Manufacturing Directorate in organizing an international team of government, industrial and university investigators, which has examined bio-based materials for electronic and photonic devices and it is apparent from this description that DNA biopolymers have been a particular focus.

OFETs have been fabricated on resorbable biomaterial substrates. An example is the work of Bettinger and Bao [33]. These authors mentioned the potential development of biodegradable electronic devices suitable for use in biomedical or

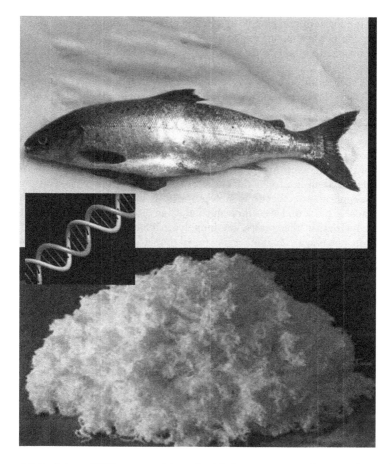

Figure 14.3   Salmon DNA extracted from food industry waste products can be used to create thin films with advanced functionality

environmental applications. The objective of their research was to prepare an organic thin film transistor using a small-molecule semiconductor in combination with a biodegradable polymeric substrate and dielectric. Devices were found to be stable after exposure to water and could be considered resorbable in an *in vitro* degradation environment.

Considering other initiatives, an optically transparent gelatin-based anti-reflective film for contact lenses was prepared by Mitsuo [34]. Preparation of this type of film involved dispersion of Ti, Sn, Zn, Al and Si oxides or hydroxides bound to gelatin. This process ensured nanoscale dispersion and good optical properties. Optical phase conjugation can be measured in gelatin films through the use of laser radiation ($\lambda = 632.8$ nm) generated by a He–Ne laser [35]. As shown in this research, phase conjugate reflectivity as a function of gelatin concentration, backward and forward beam intensity, probe beam intensity, mean pumping beam intensity and the angle between the forward pumping beam and the probe beam could be measured using degenerate four-wave mixing experiments.

Mitaji invented a visual device which was considered suitable as an artifical retina with chitosan as one component [36]. Optical devices using nanopatterned biopolymer films such as those based on silk, chitosan, collagen, gelatin or agarose, with inherent optical functions as well as electrical conductivity, have also been developed [37].

A further route to new optoelectronic devices is to prepare block copolymers in which one of the polymers can be removed so as to generate structured films. An example is poly(3-hexyl thiophene)-block-polylactide in which the lactide polymer is selectively etched away using sodium hydroxide solution [38,39]. The structured films can then be used in idealized bulk heterojunctions in hybrid energy and solar devices. This approach should be applicable to other degradable polymers as long as it is possible to etch away the degradable block and leave the semiconducting block with unimpaired optoelectronic properties. In these cases, the biopolymer component clearly contributes a fabrication advantage even if other functional properties are not relevant.

Bacterial cellulose has been used to provide flexible films for fabrication of OLEDs [40]. Although visible light transmission in these films can be as low as 40%, this property may be modified by adjusting the size of the dispersed cellulose. Cellulose-ester-based films have also been used for preparation of direct backlights and polarizers. Light utilization efficiency in these cases depends upon degree of esterification and applications in optoelectronics, including solar cells, have been suggested. Methods such as chemical vapour deposition, electron beam evaporation and ion beam sputter deposition have been used as ways to produce optical coatings with different reflectance and/or transmittance characteristics. A holographic technique, based on use of dichromated gelatin as one of the best materials for phase-hologram recording, has also been proposed as an alternative to these classical approaches [41]. Dynamic holographic recording using low-intensity red light from a He–Ne laser was achieved by Korchemskaya et al. through use of labelled fluoro-bacteriorhodopsin-gelatin films [42]. Pre-illumination of the films with blue light significantly increased the diffraction efficiency and hence the films could also be used in optical memory devices.

A recent review by d'Ischia et al. discusses melanins, which are the key components of the human pigmentary system, and refers to their unexplored potential as bio-optolectronic materials [43]. Various groups [44–46] have reported thin film preparation from eumelanins and the first device-quality spun-cast optoelectronic eumelanin films were recently reported [47].

14.3   SENSORS

14.3.1   Chemical Sensors

The investigation of electrochemical sensors based on biopolymers containing intercalated clays or layered double hydroxides LDHs (Figure 14.4) has become an important research area in the broader field of clay-modified sensors and biosensors [48]. One of the first references to this topic was published in 2003, when potentiometric sensors based on chitosan incorporating cationic montmorillonite

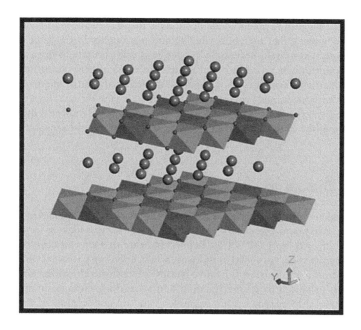

Figure 14.4 Schematic of a layered double hydroxide (LDH) showing tetrahedrally coordinated metal hydroxide layers hosting two metal cations, for example Mg^{2+} and Al^{3+} or Mg^{2+} and Fe^{3+}, and interlayers of inorganic anions (e.g., CO_3^{2-}) and water (depicted in this illustration as spheres located between the mixed metal hydroxide layers). Under the right pH conditions, LDHs can be organomodified so that organic anions, including complex biomolecules such as DNA and negatively charged polysaccharides such as chitosan, replace inorganic anions and water in the interlayer spaces

(MMT) clays were reported [49]. Depending upon process conditions, it is possible to intercalate chitosan molecules in clay galleries in monolayer or bilayer configurations. In the bilayer case, the protonated amine groups which do not interact with the negatively charged MMT surfaces are available as anion exchange sites. The improved mechanical properties of bionanocomposites are also an issue when considering their suitability as electrochemical sensor devices. The conversion of MMT from a material with cation-exchange capability to one with anion-exchange properties allows, for example, monovalent acrylate ions to be retained and then polymerized which, in turn, produces a photostable, water-superabsorbent material [50]. The same principle has been applied to sepiolite, a fibrous clay mineral, through the use of chitosan to reverse the starting cation-exchange behavior [51]. The use of LDHs in combination with negatively charged polysaccharides has resulted in conversion from anion- to cation-exchange behavior, allowing incorporation in sensors such as those used for detecting calcium ions.

There have been a number of studies on gelatin in combination with layered silicates, leading to the preparation of gelatin-based nanocomposites [52,53]. The use of layered perovskite, a calcium titanium oxide mineral, in particular has resulted in self-supporting films which have functionality related to the dielectric properties of the filler. These materials can be easily handled and otherwise processed as thin films and coatings, suggesting potential applications in the microwave industry or in high-frequency devices. Other materials which gain functionality through the characteristics of perovskites could potentially be designed and this would open up further practical opportunities related to optical, optoelectronic, superconducting or ferroelectric properties. Fernandes *et al.* reported a variety of gelatin micro- and nanocomposites with functional properties but found that, of the various clays, it was only MMT that displayed intercalation of gelatin [54]. Adding pH indicator dyes to these bionanocomposite films allows their potential use as optical pH sensors.

14.3.2   Biosensors

The bionanocomposite approach described above may also be applied to the development of enzymatic biosensors by incorporating actives such as proteins and enzymes in layered inorganic solids. Reports indicate that enzymes can retain their activity even when encapsulated in this way, which opens up pathways for the use of bionanocomposites as carriers for active phases in biosensors. In addition to cationic clays, LDHs produced by co-precipitation or a re-stacking mechanism may also be used to immobilize enzymes. One example is the immobilization of urease with potential use of the resulting biohybrids in capacitance biosensors [55].

Hydrophobized chitosan films have been employed as sensors to quantify glucose dehydrogenase/glucose systems using a rotating disc electrode [56]. The immobilization of enzymes in these systems depends on the extent and type of chitosan functionalization. An amperometric urea sensor has been produced by immobilizing urease and glutamate dehydrogenase on a chitosan film containing superparamagnetic iron oxide, which is then deposited on an indium–tin oxide-coated glass plate electrode. Using this device, peak current was linearly dependent on urea concentration. The iron oxide nanoparticles increased the active surface area available for enzyme immobilization, enhanced the electron transfer process and lengthened the working life of the electrode. Other researchers have shown that thin films composed of chitosan in combination with graphene and gold nanoparticles or CNTs with silver nanoparticles showed good electrocatalytic activity towards H_2O_2 and O_2 as well as an electrochemical response to glucose [57,58]. Similarly, Luo *et al.* demonstrated that chitosan/glucose oxidase/nickel ferrite composite films laid down on a glassy carbon electrode provided a rapid means for detection of glucose [59]. Dubas *et al.* employed a layer-by-layer deposition technique for preparation of polyelectrolyte multilayer (PEM) thin films which were sensitive to ethanol content in water [60]. In this case, PEM films were composed of a functionalized cationic chitosan combined with an anionic dye (Nylosan®) deposited on a glass slide. These thin films responded linearly with ethanol content in the range 10–45%.

Functionalized biopolymer nanocomposites have been used as films or coatings which can transmit an electrical current between source and drain electrodes as a function of volatile chemical concentrations. Applications of these types of chemical sensors using single wall CNTs functionalized with a biopolymer can include determination of the freshness of foods, wastewater analysis, detection of explosives, identification of spilled chemicals or biowarfare agents, and diagnosis of disease states through sampling of patient odours, tissues and body fluids.

The inherent piezoelectric properties of chitosan make it suitable as a sensor in biomedical applications and as a mechanical strain sensor. The latter idea was pursued by Richter *et al.* who prepared a device using a thin film of chitosan reinforced with chitin nanowhiskers [61]. This film was bound to a cantilever beam with terminals connected to a charge amplifier. Data closely matched those collected by conventional means. Furthermore, the use of chitin nanowhiskers provided mechanical strength while maintaining material homogeneity.

A hybrid chitosan coating with electrically conducting polyaniline was prepared on an indium–tin oxide electrode by Khan and Dhayal [62]. Strong covalent bonding between chitosan and polyaniline layers provided low charge transfer resistance. As an example, interaction between ochratoxin-A, one of the most significant food-contaminating mycotoxins, and a rabbit antibody immobilized on this type of coating gave increased charge transfer due to the higher proportion of carboxylic and hydroxyl functionalities at the chitosan–polyaniline-functionalized matrix interface with a linear response of current to ochratoxin-A concentration in the electrolyte.

The cationic nature of chitosan makes it a useful material as a conducting fiber as well as a functionalized wire for production of unique biosensing devices. An example is the detection of glucose by use of chitosan-coated wires functionalized with glucose oxidase. For structural reasons, several biopolymers exhibit high-shear piezoelectric effects, although physical properties are often inadequate for sensing applications. One route to overcome this problem has been through formation of biopolymer–CNT composite materials, as exemplified by the work of Lovell *et al.* on CNT–polypeptides [63].

14.4   MISCELLANEOUS APPLICATIONS

As outlined by Olsson *et al.* [3], there are interesting possibilities for introducing renewable character into magnetic nanomaterials, which have already been investigated for biomedical, data storage and device applications [64,65]. The key concept is to use nanocellulose (e.g., derived from bacteria) to create magnetic aerogels which are dry, lightweight, higly porous and flexible. Such gels also have the property of absorbing water and then releasing it on compression. Potential applications could include microfluidics devices and electronic actuators. The same principle allows the formation of magnetic nanopaper or nanofilms based on bacterial cellulose in combination with magnetic nanoparticles. This research, targeted at new high-value applications, is one of the many fascinating new initiatives in the rapidly expanding field of nanocellulose research and development [66].

The surface protection of historic monuments and buildings can be included amongst the diverse potential uses for biopolymer films and coatings. In one study on this type of application, various coatings based on zein, chitosan, PLA or polyhydroxybutyrate (PHB) were applied to marble as a form of protection against sulphur dioxide. High-molecular weight PLA provided significant protection to the marble and reduced gypsum formation in polluted environments [67].

The various positive properties of chitosan, especially its chemical resistance, mechanical strength, antimicrobial properties and thermal stability, have also been exploited in corrosion inhibition of mild steel. For example, an electrophoretically deposited hydroxyapatite–chitosan nanocomposite provided better corrosion resistance to stainless steel [68]. Good anti-corrosion performance was also obtained when zinc–chitosan coatings were electrically deposited on mild steel [69]. Anti-corrosive coatings based on epoxy resin emulsions modified with chitosan and including red ferric oxide as a pigment were prepared by Chen and Wang [70]. The resulting formulations had good emulsion stability, colloid protection, and antibacterial/mildew-proofing performance, as well as weathering resistance.

In another study of relevance to construction, Ishihara *et al.* prepared spherical cellulose particles with average grain size of 0.5–50 μm which were dispersed into polyvinyl alcohol (PVOH) to produce a coating which had a contact angle of 5° when applied on heat exchanger fins [71]. The result of this hydrophilic coating was improved heat transfer and therefore improved heat exchanger performance. The possibility to functionalize the cellulose particles in order to support bactericides, fungicides, deodorants and fragrances was also raised in this context. Lin *et al.* developed an aqueous coating of chitosan and PVOH and demonstrated both air purification properties and the inherent antimicrobial properties of chitosan [72].

Films prepared from cellulose or PLA have been examined as battery separators. Cellulose films regenerated from *N*-methyl morpholine oxide (NMMO) solutions performed better than PLA films in this respect. Bele *et al.* suggested that graphite anodes pretreated with gelatin solution would have prospects for use in lithium ion batteries [73]. As little as 1.75% gelatin was required to produce identical electrode conditions as with conventional polyvinylidene fluoride (PVDF) or polytetrafluoro-ethylene (PTFE) binders at concentrations up to 5%. The production of polymer platforms for fundamental studies in cell biology has been investigated by Ma *et al.* [74]. These researchers used variants of soft lithography such as microcontact printing to create non-fouling topography features of a comb polymer with demarcated cell adhesive regions.

In an example from the world of cosmetics, soluble derivatives of chitin/chitosan with aldoses/ketoses showed excellent gel- and film-forming capability with moisturizing and antibacterial properties [75].

The biocompatibility, bioactivity and ready availability of chitosan are features which have attracted enormous interest in its use as a medical antimicrobial coating for enhanced wound healing [76]. One example from many such studies is the improved performance of orthopedic and craniofacial implant devices. In one such case, Bumgardner *et al.* used coatings made from 90% deacetylated chitosan bound to titanium-grafted silane glutaraldehyde molecules and found a good growth rate of osteoblast cells relative to unmodified titanium [77].

14.5   CONCLUDING REMARKS

As discussed in this chapter, films and coatings based on biopolymers derived from renewable plant or animal origins have the potential to play an important role in future optoelectronic devices and sensor technologies as well as a variety of other high-value applications. The examples discussed here, including research reports and patents from the recent literature, cover a wide range from new sustainable substrates for solar cells through to applications based on the unique electrical characteristics offered by thin DNA complex films. As could be expected for such a diverse set of topics, reports vary in nature from fundamental studies through to commercial issues. Broadly speaking, new developments are being driven either by the increasing need for industry to invest in sustainable materials or by the advantageous properties offered by individual biopolymers and, in that respect, the reports on chitosan- or DNA-based films mentioned here are of particular note. Other examples could have been cited but, as discussed, the objective was to provide the reader with a flavor of the relevant literature. If supported by suitable R & D investment, there seems to be considerable promise for the future of biopolymer films and coatings in high-value areas such as optoelectronics and sensor technologies, as well as in the commodity uses described in other chapters.

REFERENCES

[1] K. Letchford, A. Södergaard, D. Plackett, *et al.*, 2011. Lactide and glycolide polymers. Chapter 9 in A. Domb, N. Kumar, A. Azra (eds), *Biodegradable Polymers in Clinical Use and Clinical Development*, John Wiley and Sons, Chichester, West Sussex, England.

[2] M. Darder, P. Aranda, and E. Ruiz-Hitzky, *Adv. Mater*, 2007, **19**, 1309–1319.

[3] R.T. Olsson, M.A.S. Azizi Samir, G. Salazar-Alvarez, *et al.*, *Nature Nanotech*. 2010, **5**, 584–588.

[4] Y. Kojima, A. Usuki, M. Kawasumi, *et al.*, *J Mater. Res,*. 1993, **8**, 1185–1189.

[5] H. Keishin, *Petrotech*. 2007, **30**, 106–109.

[6] E. Ruiz-Hitzky, M. Darder, P. Aranda, *et al.*, *Adv. Mater.*, 2009, **21**, 4167–4171.

[7] M. Darder, M. Colilla, E. Ruiz-Hitzky, *et al.*, *Appl. Clay Sci*. 2005, **28**, 199–208.

[8] Krebs, F.C. (2008) Polymer photovoltaics: A practical approach, SPIE Press, Bellingham, WA, USA.

[9] Krebs, F.C. (2010) Polymeric solar cells: Materials, design, manufacture, DEStech Publications, Lancaster, PE, USA.

[10] F.C. Krebs, M. Biancardo, B. Winther-Jensen, *et al.*, *Sol. Energy Mater. Sol. Cells* 2006, **90**, 1058–1067.

[11] Q. Li, E.T. Dunn, E.W. Grandmaison, *et al.*, *J. Bioact. Compat. Polym*. 1992, **7**, 370–397.

[12] K. Zhao, X. Zhang, and L. Zhang, *Electrochem. Commun*. 2009, **11**, 612–615.

[13] S.A. Mohamad, R. Yahya, Z.A. Ibrahim, *et al.*, *Sol. Energy. Mater. Sol. Cells* 2007, **91**, 1194–1198.

[14] M.Z.A. Yahya, A.K. Arof, *Carbohydr. Polym*. 2004, **55**, 95–100.

[15] M.H. Buraidah, L.P. Teo, S.R. Majid, *et al.*, *Int. J. Photoen*. 2010, 1–7.

[16] M. Strange, D. Plackett, M. Kaasgaard, *et al.*, *Sol. Energy Mater. Sol. Cells* 2008, **92**, 805–813.

[17] H. Tsuji, *Macromol. Biosci.*, 2005, **7**, 569–597.

[18] S. B. Levy. Bio-based backsheet, in Proceedings of SPIE *Reliability of Photovoltaic Cells, Modules, Components, and Systems*, SPIE Volume 7048, G.D. Neelkanth (ed.), SPIE, Bellingham, WA, USA (2008).

[19] S.B. Levy, *Solar Industry Magazine*, 2009, **2**, 68–72.

[20] M. Grätzel, *J. Photochem. Photobiol. C. Photochem Rev.*, 2003, 4, 145–153.

[21] L.M. Gonçalves, V. de Zea Bermudez, H.A. Ribeiro, *et al.*, *Energy Environ. Sci.* 2008, **1**, 655–667.

[22] M. Kaneko, T. Hoshi, *Chem. Lett.*, 2003, **32**, 872–873.

[23] F. Xu (2005) Self-powered screen, Chinese Patent 101169529A.

[24] P.K. Singh, B. Bhattacharya, R.K. Nagarle, *et al.*, *Synth. Metals* 2010, **160**, 139–142.

[25] P. Vidinha, N.M.T. Lourenco, C. Pinheiro, *et al.*, *Chem. Commun.* 2008, **44**, 5842–5844.

[26] A. Steckl, *Nature Photonics*, 2007, **1**, 3–5.

[27] T.B. Singh, N.S. Sariciftci, and J.G. Grote, *Adv. Polym. Sci.* 2010, **223**, 189–212.

[28] J.G. Grote, *J. Nanophot.* 2008, **2**, 1–5.

[29] E. Jones (2010) Development of biopolymer-based resonant sensors. PhD Thesis, University of Dayton, Dayton, Ohio, USA.

[30] R. Popescu, C. Pirvu, M. Moldoveanu, *et al.*, *Mol. Cryst. Liq. Cryst.* 2010, **522**, 229–237.

[31] K.R. Sarma, S. Dodd, C. Chanley, *et al.*, Proceedings of SPIE *Optical Materials in Defence Systems Technology*, SPIE Volume 7118, J.G.Grote, F. Kajzar, M. Lindgren (eds), SPIE, Bellingham, WA, USA (2008).

[32] P.G. Whitten, A.A. Gestos, G.M. Spinks, *et al.*, *J. Biomed. Mater. Res. Part B: Appl. Biomater.* 2007, **82**, 37–43.

[33] C.J. Bettinger, Z. Bao, *Adv. Mater.* 2009, **21**, 1–5.

[34] S. Mitsuo. Japanese Patent 20070926 (2007).

[35] Q.M. Ali, P.K. Palanisamy, *Optic. Laser. Technol.* 2007, **39**, 1262.

[36] S. Mitaji. Visual sense device and device for measuring optical characteristics thereof, Japanese patent JP 2005241620 (2005).

[37] D. Kaplan, F. Omenotto, B. Lawrence, *et al.*, Nanopatterned biopolymer optical device, 2008 World patent WO/2008/127404.

[38] I. Botiz, S. Darling, *Macromol.* 2009, **42**, 8211–8217.

[39] I. Botiz, A.B. Martinson, and S. Darling, *Langmuir*, 2010, **26**, 8756–8761.

[40] C. Legnani, C. Vilani, V.L. Calil, *et al.*, *Thin Sol. Film.* 2008, **517**, 1016–1020.

[41] P. Sharlandjiev, B. Markova, *J. Mater. Sci: Mater. Electr.* 2003, **14**, 863–864.

[42] E. Korchemskaya, N. Burykin, A. de Lera, *et al.*, *Photochem. Photobiol.* 2005, **81**, 920–923.

[43] M. d'Ischia, A. Napolitano, A. Pezzella, *et al.*, *Angew. Chem. Int. Ed.* 2009, **48**, 3914–3921.

[44] P. Meredith, S. Subianto, and G. Will (2005) Process for the production of thin films of melanin and melanin-like molecules by electrosynthesis. World patent (WO/2005/026216).

[45] P. Diaz, Y. Gimeno, P. Carro, *et al.*, *Langmuir*, 2005, **21**, 5924–5930.

[46] J.E. deAlbuquerque, C. Giacomantonio, A.G. White, *et al.*, *Eur. Biophys.* 2006, **35**, 190–195.

[47] J.P. Bothma, J. de Boor, U. Divakar, *et al.*, *Adv. Mater.* 2008, **20**, 3539–3542.

[48] C. Mousty, *Appl. Clay Sci.*, 2004, **27**, 159–177.

[49] M. Darder, M. Colilla, and E. Ruiz-Hitzky, *Chem. Mats.* 2003, **15**, 3774–3780.

[50] H. Qiu, J. Yu, and J. Zhu, *Polym. Polym. Comps.* 2005, **13**, 167–172.

[51] M. Darder, M. Lopez-Blanco, P. Aranda, *et al.*, *Chem. Mats.* 2006, **18**, 1602–1610.

[52] J.F. Martucci, A. Vazquez, and R.A. Ruseckaite, *J. Therm. Anal. Cal.* 2007, **89**, 117–122.

[53] M. Darder, A.I. Ruiz, P. Aranda, *et al.*, *Curr. Nanosci.* 2006, **2**, 231–241.

[54] M.F. Fernandes, A.I. Ruiz, M. Darder, *et al.*, *J. Nanosci. Nanotech.* 2009, **9**, 221–229.

[55] C. Forano, S. Vial, and C. Mousty, *Curr. Nanosci.* 2006, **2**, 283–294.

[56] K. Sjoholm, S. Minteer, Proceedings of the 44th ACS Midwest Regional Meeting (2009).

[57] C. Shan, H. Yang, D. Han, *et al.*, *Biosens. Bioelectr.* 2010, **25**, 1070–1074.

[58] J. Lin, C. He, Y. Zhao, *et al.*, *Sens. Act. B: Chem.* 2009, **137**, 768–773.

[59] X.L. Luo, J.-J. Xu, Q. Zhang, *et al.*, *Biosens, Bioeletron.* 2005, **21**, 190–196.

[60] S.T. Dubas, C. Iamsamai, and P. Potiyaraj, *Sens. Act.B: Chem.* 2006, **113**, 370–375.

[61] H. Richter, F.G. Torres, and J. Sanchez. Proceedings of the World Forum on Smart Materials and Smart Structures Technology, Chongqing and Nanjing, China, May 22–27, (2008), 1.

[62] R. Khan, M. Dhayal, *Biosens. Bioelectr.* 2009, **24**, 1700–1705.

[63] C. Lovell, E. Worthington, T.J. Deming, *et al.*, *Polym. Prepr.* 2005, **46**, 802–803.

[64] S.A. Majetichin *Nanostructured Materials; Processing, Properties and Applications* (ed. Koch, C. C.) pp 439–485, William Andrew Publishing, 2007.

[65] A. Millan, A., Palacio, F., Snoeck, *et al.*, in *Polymer Nanocomposites* (eds Mai, Y. W. and Yu, Z.-Z.) pp 441–484, Woodhead Publishing, 2006.

[66] D. Klemm, D. Schumann, F. Kramer, *et al.*, Nanocelluloses as innovative polymers in research and application. In *Advances in Polymer Science*, Volume 205, Polysaccharides II, pp. 49–96 (2006).

[67] Y. Ocak, A. Sofuoglu, F. Tihminlioglu, *et al.*, *Prog. Org Coat.* 2009, **66**, 213–220.

[68] X. Pang, I. Zhitomorsky, *Mater Charact.* 2007, **58**, 339–348.

[69] K. Vathsala, T. Venkatesha, B.M. Praveen, *et al.*, *Engineering*, 2010, 580–584.

[70] Y. Chen, Q. Wang (2008), Chinese Patent 101148557.

[71] S. Ishihara, K. Kobayashi, and M. Ikeda. (2003), Japanese Patent 2003128977 A.

[72] J. Lin, C. He, Y. Zhao, *et al.*, *Sens. Act. B: Chem.* 2009, **137**, 768–773.

[73] M. Bele, M. Gaberscek, R. Dominko, *et al.*, *Carbon* 2002, **40**, 1117–1122.

[74] H. Ma, J. Hyun, Z. Zhang, *et al.*, *Adv. Funct. Mater.* 2005, **15**, 529–540.

[75] C. Juneau, A. Georgalas, and R. Kapino, *Cosm. Toilet*, 2001, **116**, 73–75.

[76] J.S. Boateng, K.H. Matthews, H.N.E. Stevens, *et al.*, *J. Pharm. Sci.* 2007, **97**, 2892–2923.

[77] J.D. Bumgardner, R. Wiser, P.D. Gerard, *et al.*, *J. Biomat. Sci, Polym Ed.* 2003, **14**, 423–438.

15

Summary and Future Perspectives

David Plackett

Risø National Laboratory for Sustainable Energy, Technical University of Denmark, Roskilde, Denmark

15.1 INTRODUCTION

The chapters in this book provide a technical update on a number of key biopolymers or types of biopolymer which are now commercially available, under development, or are promising candidates for future film and coating applications. The approach to each chapter has varied and has naturally reflected the interests and expertise of each author. As a result of the combination of chapters in two sections addressing, first, basic knowledge on production/synthesis of individual biopolymers and, second, practical issues for their use in key films and coatings applications, it is hoped that the reader has now gained useful knowledge of this rapidly developing field. Since each of the chapters in this book has covered key points in the production, characterization and use of biopolymers in films and coatings, this exercise will not be repeated here. Rather, by way of a summary, this concluding chapter describes some of the latest developments and advances reported in the very recent literature with reference to bioplastics and bio-thermoset resins as well their nanocomposites. These examples, in the author's opinion, are useful in that they reflect some of the most important issues and exciting challenges in this expanding area of polymer science and technology.

Biopolymers – New Materials for Sustainable Films and Coatings, First Edition.
Edited by David Plackett.
© 2011 John Wiley & Sons, Ltd. Published 2011 by John Wiley & Sons, Ltd.

15.2   BIOPLASTICS

Although, as outlined earlier in Chapters [2–4], bioplastics such as thermoplastic starch, polylactides (PLAs) and polyhydroxyalkanoates (PHAs) have been investigated and under commercial development for years, if not decades, it seems clear that there is now a 21st century resurgence of interest in these bio-derived polymers. This situation has come about because of a combination of factors, most notably the increasing profile of the 'green' agenda as an issue in industry, governments and societies in general, as well as a realization that polymers from fossil fuels can now realistically be replaced by bio-alternatives in certain markets, albeit usually at a price penalty. Consequently, companies are continuing to make manufacturing and processing investments and research aimed at new, better-performing bio-based materials continues to receive funding globally. The way in which bioplastics are being adopted has been summarized in a very recent industry magazine article [1].

In addition to the 'established' bioplastics, and as indicated in Chapters [5–8], biopolymers such as chitosan, some proteins (e.g., wheat gluten, corn zein), hemicelluloses and cellulosics also offer opportunities for new bio-based films and coatings, even if chemical or physical modification may often be required to meet the needs of particular applications. Furthermore, as pointed out in Chapters [10–13], although packaging is the dominant commodity market for bioplastics, there are important and significant outlets for these materials in markets for edible films and coatings, paper and paperboard coatings, and agricultural films. As the reader will realize from Chapter 12, there is also a clear connection between developments aimed at biopolymer coatings for paper and paperboard and the principal market for such products in the packaging arena. Looking ahead still further it is argued in Chapter 14 that there could be increasing interest in 'green' polymer films and coatings for high-performance materials in optoelectronics and sensor technologies, driven in part by the unique chemical and physical properties of biopolymers such as chitosan and DNA, to give two examples.

In the present environment, the continued development and commercialization of bioplastics will be driven by several key issues. First, technologies that can reduce product costs (e.g., through processing efficiencies or strength enhancements leading to use of thinner films) will clearly be beneficial in terms of closing the price gap with conventional synthetic polymers. Second, new ways of using presently undervalued and/or under-utilized industry by-products or co-products for commercial manufacturing of biopolymers will be in focus, with a special emphasis on directions which will avoid conflict with agricultural food production. Third, development and use of sustainable bio-based materials in the context of reduced greenhouse gas emissions and the cradle-to-cradle concept is likely to gain more attention worldwide. Fourth, as pointed out at intervals during the preceding chapters, wider adoption of bioplastics could be boosted if characteristics such as the mechanical, barrier and thermal properties can be improved. The latter issue represents the primary goal in much of the current technical research at the interface of biopolymer science and nanotechnology, as it relates to commodity products. Of the four issues, the second and fourth in particular are the subject of increased interest worldwide, as suggested by very recent reports in the literature and discussed in the following paragraphs.

Taking PHAs derived from bacteria as a first example, and as outlined by Keshavarz and Roy [2], the likely future trend with these bacterial polyesters will involve greater attention to efficient and economical processes, which will particularly mean advances in polymer isolation and purification. Furthermore, as suggested by the work of Jian Yu and his group at the Hawaii Natural Energy Institute, the use of waste streams (e.g., from the food industry) as feedstock for PHA synthesis offers an interesting example of synergy in respect to PHA production and the interest of other local industries [3,4]. As mentioned in a recent article in *The Economist* magazine, researchers are now once again studying plastics obtained from casein as an output from the dairy industry [5] and, similarly, the feasibility of producing PHAs from dairy industry residues has recently been demonstrated. Interestingly, this research showed that, unlike other reports in the literature involving use of pure cultures, this PHA production method from whey in the presence of an enriched biomass allowed direct use of lactose under non-sterile fermentation conditions and without the need for pH control [6]. The link between agricultural by-products and PHA production has also been recently explored with reference to using olive oil industry wastes. Apart from adding value to these waste materials, their use in PHA synthesis also provides a potential way of reducing environmental risks associated with waste transportation and storage. In this work, specific strains of *Azotobacter* were used used in combination with olive mill wastes to produce polyhydroxy-butyrate (PHB) under anaerobic conditions [7].

In addition to a focus on the use of waste products to create bioplastics there are many reports on studies which are aimed at greater fundamental knowledge about processes and products. A good recent example is the work of Lagrain *et al.* which was aimed at a better understanding of wheat gluten protein at the molecular level, with a view to its use as a bio-based material [8]. As explained by these authors, significant knowledge gaps still exist in respect to the native structure and functionality of gluten proteins. Improved understanding of structure–property relationships in gluten proteins would be valuable in terms of elevating processing research to the next level. The conclusions of these researchers, which might equally apply to other biopolymers, point to a need for more knowledge about the chemical reactions which occur during processing as well as the mechanical performance of gluten-based bioplastic end-products. These suggestions emphasize the value of a multidisciplinary approach involving chemical, biological and processing expertise to solve problems and enhance the properties of biopolymer products.

As stated by several contributors to this book, one way to address new product opportunities is through development of biopolymer blends. This is now a very active area of research in which a rational scientific approach is needed to complement and reinforce empirical studies. An example from the very recent literature is reported by Gonzalez-Gutierrez *et al.* [9] in which albumen protein was combined with glycerol-plasticized starch to derive highly transparent films. Tests with these biodegradable films revealed strength values at low deformation which were comparable to those of commodity plastics. This research is also an example of how adjustments in processing parameters may lead to quite different products and introduce new versatility in biomaterials manufacturing. As in many cases, the need for new 'green' materials in the packaging and container industries was cited as the driving force.

15.3   BIO-THERMOSET RESINS

In contrast to the bioplastics field, developments in bio-based thermoset resins, especially those based fully on renewable resources, have been relatively limited. A search of the literature using the term 'bio-thermoset' proves to be relatively unproductive, although this may be because this specific term is not widely used at present. In this book, we cover one aspect of this very broad field in Chapter 9 which provides an up-to-date summary of furan polymers, derivable from furfural and furfuryl alcohol, which may be obtained from plant hemicelluloses. As the reader will appreciate, a rich portfolio of furan polymers and derivatives has been investigated and the basis for new industrial initiatives has now become available.

At present, the Belgian company TransFurans Chemicals BVBA (TFC) (http://www.transfurans.be) may be the most advanced in respect to commercial development of furan resins, offering renewable thermosets in the form of Biorez® resins for moulding or water-soluble impregnating resins or as reinforced Furolite® resins for uses where stability against heat, fire or corrosive environments is needed. TFC has played an active role in the EU Sixth Framework BIOCOMP project (http://www.biocomp.eu.com) as well as the current EU Seventh Framework WOODY project (http://www.woodyproject.eu), both of which have been aimed at practical adoption of bio-resins in advanced composites.

The recent literature contains a state-of-the-art review on thermosetting biomaterials which emphasizes that, in addition to furans, bio-thermosets can also include phenolics, epoxies, polyurethanes and polyesters as well as other varieties such as protein-based thermosets [10]. An enormous potential market for bio-thermosets exists in the form of adhesives and composite building materials as well as in speciality markets such as high-performance bicycle frames and other vehicle parts. Looking further ahead, bio-based thermosets may offer an alternative option as a matrix for the rapidly developing wind turbine blade industry and a source of new film and coatings materials for advanced applications in electronic devices.

15.4   NANOCOMPOSITES BASED ON INORGANIC NANOFILLERS

Improvement in the properties of bioplastic films and coatings through the formation of nanocomposites incorporating inorganic nanofillers, especially montmorillonite (MMT) clays, continues to inspire new research activities. A review of literature in this field indicates studies covering the full range from fundamental work aimed at understanding the properties of solvent-cast films through to more practically oriented work on extrusion processing. The research literature is based on a very wide spectrum of biopolymer/clay combinations and a feature of this field has been realization that the theoretically achievable enhancement in mechanical and barrier properties has often remained quite elusive. This observation may generally be attributed to two factors, the distribution and morphology of clay fillers during and after high-shear processing and/or adverse affects of processing on individual biopolymers (e.g., changes in crystallinity, molecular weight reduction). The biopolyesters (i.e., PLAs and PHAs) are particularly notable in this respect since, in addition to thermophysical effects, even trace quantities of water can cause polymer molecular weight reductions with further implications in terms of reduced crystallinity, barrier and mechanical characteristics. Researchers worldwide are, however, rising to these challenges and, as a result, nanocomposite technology transfer into the packaging field in particular is already under way. This trend should continue in the coming years and scientists are also now addressing a broad set of questions relating to the safety and possible environmental consequences of nanoparticulates in food packaging. This concern is motivated, for example, by possible migration of clays from packaging films, although this may arguably be unlikely in many cases, and also by potential migration of the clay organomodifiers typically employed to ensure clay/polymer compatibility during melt processing.

Although packaging films remain a main focus of attention in research on bio-polymer/clay nanocomposites, as shown in a recent review article many other exciting applications exist in the medical and environmental fields [11]. The review mentions layered double hydroxides (LDHs), anionic clays typically synthesized in the labor-atory, and their suitability for applications such as delivery of entrapped drug molecules. In terms of commodity biopolymer film applications, LDHs have been less well studied than the MMT clays and, as revealed in recent PLA processing research, may in some cases present a slightly greater challenge in respect to avoiding degradation of biopolyesters during melt processing [12].

15.5   NANOCOMPOSITES BASED ON CELLULOSE NANOFILLERS

As mentioned and well illustrated in Chapters 8 and 12, the development of new biopolymer-based films and coatings reinforced with nanocellulose, either in the form of cellulose nanowhiskers or microfibrillated cellulose (MFC), is now widely recognized as a valuable new route towards property-enhanced biomaterials. Research on MFC was recently reviewed [13] and a comprehensive review of nanocellulose as a new source of materials has been published [14]. There is a broad panoply of potential uses for the three main types of nanocellulose (microfibrillated cellulose generated by high-shear homogenization of wood pulps, nanocrystalline cellulose produced after acid hydrolysis of plant-derived cellulose and bacterial nanocellulose), which covers such diverse areas as paper, biodegradable films and coatings for packaging, various technical films and membranes and medical applications in wound dressings, implants and cardio-vascular grafts.

Research activity in nanocellulose goes back some decades but, with regard to practical use in new bio-based films and coatings, now appears to have reached a critical juncture at which scale-up to industrially relevant production volumes is being tackled. Present initiatives of this type in Sweden and Canada are especially note-worthy [15,16]. Successful developments in this direction should attract increased commercial interest and catalyze future technology transfer.

In addition to nanocellulose production activities, a survey of research reported in 2009–2010 gives an insight into the breadth of interest in biopolymer/nanocellulose combinations and an improved understanding of their basic properties. Recent examples include the work of Azeredo *et al.* [17] who investigated the mechanical and water vapour properties of films based on chitosan in combination with Avicel® cellulose nanofibers and glycerol and that of Khan *et al.* [18] who addressed use of nanocellulose provided by the Canadian research organization FP Innovations Paprican to reinforce methylcellulose films. A further development reflected in the recent literature is the use of a range of new non-wood fibers to produce nanocellulose. The research of Sousa *et al.* on Curava fibers and the work of Cherian *et al.* on nanocellulose isolated from pineapple leaf fibers by steam explosion may be mentioned as examples [19,20]. Finally, a new approach to scaled-up production of bacterial cellulose sheets was recently reported, pointing to a future in which this valuable form of nanocellulose may be produced in semi-continuous fashion at a lower cost than is currently feasible [21].

15.6   CONCLUDING REMARKS

This final chapter provides an insight into some of the latest biopolymer developments reported in the literature and in popular articles, including both advances in basic scientific knowledge as well as exciting new materials opportunities. Through this summary, as well as the individual chapters, it is hoped that the reader can better appreciate the raw biomaterials and their processing as well as the main markets into which sustainable biopolymer films and coatings may increasingly be introduced. There are many sources for further information on sustainable bio-based films and coatings through individual biopolymer manufacturers or by way of industry associations such as the European Bioplastics Association (EBA) (http://www.european-bioplastics.org). The Bioplastics magazine published by the EBA and the biopolymer.net website (http://www.biopolymer.net) can also be excellent sources of information in terms of new industrial developments based on bioplastics.

REFERENCES

[1] A.P. Ambekar, P. Kukade, and V. Mahajan, *Popular Plastics and Packaging*, September 2010, 30–32.
[2] T. Kershavarz, I. Roy, *Curr. Op. Microbiol.* 2010, **13**, 321–326.
[3] G. Du, J. Yu, *Environ. Sci. Technol.* 2002, **36**, 5511–5516.
[4] G. Du, L.X. Chen, and J. Yu, *J. Polym. Environ.* 2004, **12**, 89–94.
[5] Plastics: There and back again, *The Economist*, 28 October 2010.
[6] F. Bosco, F. Chiampo, *J. Biosci. Bioeng.* 2010, **109**, 418–421.
[7] F. Cerrone, M.d. Mar Sanchez-Peinado, B. Juarez-Jimenez, *et al.*, *J. Microbiol. Biotechnol.* 2010, **20**, 594–601.
[8] B. Lagrain, B. Goderis, K. Brijs, *et al.*, *Biomacromol.* 2010, **11**, 533–541.
[9] J. Gonzalez-Gutierrez, P. Partal, M. Garcia-Morales, *et al.*, *Biores. Tech.* 2010, **101**, 2007–2013.
[10] J.-M. Raquez, M. Deléglise, M.-F. Lacrampe, *et al.*, *Progr. Polym. Sci.* 2010, **35**, 487–509.
[11] E. Ruiz-Hitzky, P. Aranda, M. Darder, *et al.*, *J. Mater. Chem.* 2010, **20**, 9306–9321.

[12] V. Katiyar, N. Gerds, C.B. Koch, *et al.*, *J. Appl. Polym. Sci.*, 2011, accepted for publication.

[13] I. Siró, D. Plackett, *Cellulose* 2010, **17**, 459–494.

[14] D. Klemm, D. Schumann, F. Kramer, *et al.*, Nanocelluloses as innovative polymers in research and application. In *Advances in Polymer Science*, Volume 205, Polysaccharides II, pp. 49–96 (2006).

[15] M. Ankerfors, *Beyond*, 2010, **2**, 8.

[16] C. McCormick, *Pulp and Pap. Can.* 2010, July/August issue, 15–16.

[17] H.M.C. Azeredo, L.H.C. Mattoso, R.J. Avena-Bustillos, *et al.*, *J. Food Sci.* 2010, **75**, N1–N7.

[18] R.A. Khan, S. Salmieri, D. Dussault, *et al.*, *J. Agric. Food Chem.* 2010, **58**, 7878–7885.

[19] S.F. Souza, A.L. Leao, J.H. Cai, *et al.*, *Mol. Cryst. Liq. Cryst.* 2010, **522**, 42–52.

[20] B.M. Cherian, A.L. Leao, S.F. de Souza, *et al.*, *Carbohydr. Polym.* 2010, **81**, 720–725.

[21] D. Kralisch, N. Hessler, D. Klemm, *et al.*, *Biotech. Bioeng.* 2010, **105**, 740–747.

Index

Page numbers marked in *italics* indicate figures; page numbers marked in **bold** indicate tables.

AC *see* acrylic copolymers
acetyl coenzyme A (Ac-CoA), 68
acetyl tributyl citrate (ATBC), 77–8
AcGGM *see* O-acetylgalactoglucomannan
acid-binding properties, 96
acrylic copolymers (AC), 164–5, *164*
active edible films/coatings, 247
active packaging, 88, 91–4, 213–14,
  223–4
additive/ingredient carriers, 235
AEROCELL programme, **284**, 286
aerogels, 154, 311
AFM *see* atomic force microscopy
agar, 241, 242
AGROBIOFILM programme, **284**, 287
agronomy, 3, 277–99
  applications of biopolymer films, 277–8,
    288–96, **288**
  biodegradability, 278–9, *279*, 281–7,
    289–93, 296
  chitosan, 94
  current biopolymers and
    biocomposites, 279–82, **280**
  fruit protecting bags, 295, *295*
  future developments, 293–6
  international projects, 282–7, **284**
  low tunnels, 293–4, *294*

  mulching, 278–9, *279*, 287,
    289–93, *289*, *291–2*
  plastic films used in agriculture, 277–8,
    278
  regulatory frameworks, 282–3,
    284
  solarization, 293
alginates, 241, 242, 245–6
alkaline extraction, 142
all-cellulose composites, 154
amylopectin
  paper/board applications, 258–9
  starch-based polymers, 16, *17*, **18**, 22, 26,
    34, 39
amylose
  paper/board applications, 258–9
  starch-based polymers, 16, *17*, **18**, 22,
    25–6, 34, 39
anhydride-modified cellulose, 170–1
animal-based proteins, 113–17
anionic polymerization, 187
anti-corrosion properties, 312
antimicrobial properties
  chitosan, 88, 91–6, 95
  edible films/coatings, 245
  food packaging applications, 224
  proteins, 117

Biopolymers – New Materials for Sustainable Films and Coatings, First Edition.
Edited by David Plackett.
© 2011 John Wiley & Sons, Ltd. Published 2011 by John Wiley & Sons, Ltd.

antioxidant properties
    chitosan, 96–7
    edible films/coatings, 244
    proteins, 117
ATBC *see* acetyl tributyl citrate
atom transfer radical polymerization
    (ATRP), 169, 172
atomic force microscopy (AFM), *157*
ATRP *see* atom transfer radical
    polymerization

β-glucans, 135, 137, 140
bacterial biosynthesis, 68–71, *70*, 82
bacterial cellulose (BC), 153, 156–9,
    157–9, 163, *163–4*
barrier properties
    cellulose, 165–7, *166*, *168–70*
    chitosan, 94, *95*, 98
    edible films/coatings, 234–5, *234–5*,
        242, 244, 247
    food packaging applications, 214–18,
        216, 223
    paper/board applications, 255–6,
        262–5, 267, 271
    polylactides, 57–9
    proteins, 117, 125
BC *see* bacterial cellulose
biaxially oriented PP (BOPP) films, 7
bio-thermoset resins, 317, 320
BIOCELSOL programme, **284**, 286
biocomposites
    agronomy, 279–82, *280*
    polyhydroxyalkanoates, 79–81
    see also bionanocomposites
biodegradability, 9
    agronomy, 278–9, *279*, 281–7,
        289–93, 296
    chitosan, 88
    commercial bio-derived polymer
        production, 9–10, 12
    food packaging applications, 218–19,
        224–7
    furans, 205–6
    low tunnels, 293–4, *294*
    mulching, 289–93
    paper/board applications, 256–7
    polyhydroxyalkanoates, 75–6, 82
    simulation tests, 32, 39
    solarization, 293
    starch-based polymers, 23, 27, 32–9,
        33–4, 36–8
BIODESOPO programme, **284**, 285–6

biofuels
    hemicelluloses, 146
    proteins, 108
biomass production methods, 12–13
biomedical applications
    cellulose, 173
    chitosan, 88, 90–1
    hemicelluloses, 146–7
bionanocomposites
    functionalized biopolymer films/
        coatings, 302, *302*
    future developments, 317, 320–2
    paper/board applications, 267–71, 272
    polyhydroxyalkanoates, 80–1
BIOPAL programme, **284**, 285
bioplastics
    commercial bio-derived polymer
        production, 11–12
    food packaging applications, 214, 218–19
    future developments, 317, 318–19
    paper/board applications, 257
    polyhydroxyalkanoates, 71–2
BIOPLASTICS programme, **284**, 285
biosensors, 95, 310–11
black films, 290–2
block copolymers
    polyhydroxyalkanoates, 66–7, *66*, 73
    polylactides, 51–2, *52*
BOPP *see* biaxially oriented PP
bulk polycondensation, 46–7, 50

C2C *see* cradle-to-cradle
carbon dioxide barriers, 58, 220
carbon nanotubes (CNT), 306, 310
carboxymethyl cellulose (CMC), 265
carboxymethylation, 144–5
carrageenans, 241, 242
casein, 114–15, 239
cationic polymerization, 187–9, *188–91*
cellophane, 266–7
cellulose, 3, 151–78
    all-cellulose composites, 154
    biomedical applications, 173
    bionanocomposites, 321–2
    cellulose-based composites, 162–72
    composites with pristine fibres, 162–5,
        163–4
    coupling agents, 165–6, *165–6*
    crystallinity, 153, 155, 266
    derivatives, 173–4
    edible films/coatings, 234, 240
    electronic applications, 172–3

food packaging applications, 222
functionalized biopolymer films/
    coatings, 312
hierarchical morphology, 152, *153*
miscellaneous chemical
    modifications, 171–2
model cellulose films, 159–60
nanofillers, 321–2
nano-objects, 154–9, *155–9*
nanoparticle-coupled, 172
new materials from pristine
    cellulose, 154–60
paper/board applications, 258, 265–7
polymer grafting, 167–71
production and properties of
    biocomposites, 79–80
solvent/solution properties, 160–2, *161–2*
sources and characteristics, 151–3
structure, 152, *152*
superficial fibre modification, 164–72
surface hydrophobization, 165–7, *166*,
    168–70
surface-initiated polymerization, 167–71
cellulose nanocrystals (CN), 154–5, *156*
cellulose whiskers, 80, *156*
cellulose-based composites, 162–72
chain extension, *46*, 48
chain-growth polymerization, 186–9
chemical sensors, 308–10, *309*
chitin, 87–8
chitosan, 3, 87–105
    active packaging applications, 88, 91–4
    agricultural applications, 94
    antimicrobial properties, 88, 91–6, *95*
    applications and properties, 88, 89–97
    barrier properties, 94, *95*, 98
    biodegradability, 88
    biomedical applications, 88, 90–1
    biosensors, *95*
    cellulose-based composites, 163, *163–4*
    cosmetics applications, 90
    degree of N-acetylation, 88–9
    edible films/coatings, 240, 242, 244–5
    fat trapping agents, 90
    food industry applications, 88, 91–4, 95–7
    functionalized biopolymer films/
        coatings, 309–13
    future developments, 318
    molecular weight distribution, 89
    paper/board applications, 259, 262–3, 269
    physical and chemical
        characterization, 88–9

plasticization, 93
polymer blends, 94, 97–8
processing, 97–8
solvent and solution properties, 89
sources and characteristics, 87–8
waste water treatment, 88, 90
clarifiers, 96
click chemistry, 171
CMC *see* carboxymethyl cellulose
CN *see* cellulose nanocrystals
CNT *see* carbon nanotubes
co-extrusion, 78–9
collagen
    edible films/coatings, 234, 238, 245
    sources and characteristics, 115
compatibilizers, 23, 25
compostability, 10
    agronomy, 288
    starch-based polymers, 23
compression moulding, 121–5, *122–4*
construction applications, 312
controlled drug delivery
    chitosan, 90–1
    functionalized biopolymer films/
        coatings, 301
    hemicelluloses, 147
copolymerization
    cellulose, 164–5, *164*
    chitosan, 94, 97–8
    furans, 186
    polyhydroxyalkanoates, 66–7, **66**,
        69–70, 73
    polylactides, 51–2, *52*
corn zein, 109–10, *110*, 234, 238
cosmetics applications, 90, 312
cottonseed, 112–13
coupling agents, 165–6, *165–6*
CPVC *see* critical pigment volume
    concentrations
cradle-to-cradle (C2C) concept, 9, 13,
    282, 318
critical pigment volume concentrations
    (CPVC), 255–6
cross-linking agents
    chitosan, 93–4
    furans, 192, 197, 205
    hemicelluloses, 145
    proteins, 127–8
crystallinity
    cellulose, 153, 155, 266
    polyhydroxyalkanoates, 73, 74
    polylactides, 56–7

DA *see* Diels–Alder
debranching enzymes, **33**
deep fat frying foods, 246–7, **249**
degree of polymerization (DP)
   cellulose, 152, 160
   furans, 189
   hemicelluloses, 133–4
   polylactides, 52
delignification, 142–3
denaturation temperature, 119
Diels–Alder (DA) reactions, 201–5, *201*
differential scanning calorimetry (DSC)
   polylactides, 53
   starch-based polymers, 19, 22
DMA *see* dynamic mechanical analysis
DNA thin films, 306–7, *307*, 312, 318
DP *see* degree of polymerization
dry foods, 221–2, 246, **249**
dry forming, 121–8, **122–4**, **126**
DSC *see* differential scanning calorimetry;
   dye-sensitized solar cells
dye-sensitized solar cells (DSC), 305
dynamic mechanical analysis (DMA), 53

economic factors, 11–12
edible films/coatings, 3, 233–54
   additive/ingredient carriers, 235
   applications, 243–7
   composite/multilayer films, 242–3
   deep fat frying foods, 246–7, **249**
   dry foods, 246, **249**
   fruit and vegetables, 243–4, **248**
   lipids, 241–2, 246
   mass transfer control, 234–5, *234*, 246–7
   materials, 236–43, *237*
   meat and poultry, 244–6, **248–9**
   paper/board applications, 257–8
   physical and mechanical protection, 235–6
   plasticization, 243, 246
   polysaccharides, 234, 239–41
   properties and characteristics, 233–6
   proteins, 234, 237–9, 245
   research directions, 247, 250
   safety issues, 244
   sensorial properties, 236, 244, 250
   stability, 247, 250
   surface condition control, 235, *235*
egg white, 115
electronic applications, 172–3
emulsifiers, 96, 243
emulsion films/coatings, 244, 257
endoamylases, 33

engineered nanomaterials (ENM), 225
ENM *see* engineered nanomaterials
environmental factors
   agronomy, 277, 282–3
   food packaging applications, 218–19
   see also biodegradability
enzymatic degradation, 32–5, 33–4, 36–8, 39
enzyme immobilization, 95
EO *see* essential oils
essential oils (EO), 244
esterification, 143–4
ethanol biomass production, 12–13
etherification, 144–5
ethylene-vinyl alcohol (EVOH), 94, 98
exfoliated nanoclays, 267–8, *268*
exoamylases, 33
extraction methodologies, 141–3
extrusion moulding, 125–8, **126**

FA *see* furfuryl alcohol
fat trapping agents, 90
fatty acids, 241
FCM *see* Food Contact Material
field assays, 32
floating covers, 295–6
Food Contact Material (FCM)
   regulations, 216–18, 225
food industry applications
   chitosan, 88, 91–4, 95–7
   proteins, 121, 128–9
   see also edible films/coatings
food packaging applications, 213–32
   biodegradability, 218–19, 224–7
   biopolymer applications, 220–2
   chitosan, 88, 91–4
   degradation and shelf-life, 214–16,
      215–16, 220–2
   environmental factors, 218–19
   functional properties, 214–16, *215–16*,
      222–3
   integrated approach, 225–7
   long shelf-life/dry or liquid foods, 221–2
   material specifications, 214–19
   modified atmosphere packaging, 225–7,
      226
   paper/board applications, 256, 272
   regulatory framework, 216–19,
      225
   research directions, 222–7
   safety issues, 216–18, 224–5
   short shelf-life/fresh foods, 220–1
   see also active packaging

food simulant liquids (FSL), 225
FORBIOPLAST programme, **284**, 286–7
Fourier transform infrared spectroscopy
  (FTIR), 19, *92*
FP *see* Framework Programmes
fractional extraction, 142
Framework Programmes (FP), 283–7, **284**
free radical polymerization, 186–7
fresh foods, 220–1
fruit protecting bags, *295, 295*
FSL *see* food simulant liquids
FTIR *see* Fourier transform infrared
  spectroscopy
functional food processes, 95
functional properties
  edible films/coatings, 243
  food packaging applications, 214–16,
    215–16, 222–3
functionalized biopolymer films/
  coatings, 301–15
  biosensors, 310–11
  chemical sensors, 308–10, *309*
  optoelectronic applications, 301, 303–8,
    303, 307
furans, 3, 179–209
  anionic polymerization, 187
  biodegradability, 205–6
  cationic polymerization, 187–9, *188–91*
  chain-growth polymerization, 186–9
  Diels–Alder reactions, 201–5, *201*
  free radical polymerization, 186–7
  future developments, 320
  linear polymerizations, 202–3, *203*
  miscellaneous polymers, 199–201,
    200–1, 205
  nonlinear polymerizations, 203–4, *204*
  polyamides, 194–5, *194–5*
  polycondensation of furfuryl
    alcohol, 189–93, *190–1*
  polycondensation of 5-methylfurfural,
    193–4, *194*
  polyesters, 195–7, *196–7*
  polymers, 183–206
  polyurethanes, 198–9, *198–9*
  precursors and monomers, 181–3,
    181–2, 184–5
  reversible polymer cross-linking, 205
  sources and characteristics, 179–81
  step-growth polymerization, 189–201
  structure, *180*
furfuryl alcohol (FA), 183, 189–93, *190–1*
2-furyl oxirane, 187

gas chromatography-mass spectrometry
  (GC-MS), 52
GCV *see* gross calorific values
gel permeation chromatography (GPC), 53
gelatin
  edible films/coatings, 238
  functionalized biopolymer films/
    coatings, 305, 307
  sources and characteristics, 115–16, *116*
gelatinization process, 18–22
genetically modified organisms
  (GMO), 71–2
glass transition temperatures
  polyhydroxyalkanoates, 74
  proteins, 117
  starch-based polymers, 19–20, *20,*
    22, 27
glycerol plasticization
  edible films/coatings, 246
  future developments, 319
  proteins, 109, 112, 120, *125*
glycerol–water systems, 19, *20*
GMO *see* genetically modified organisms
GPC *see* gel permeation chromatography
grafting-from anionic polymerization, 171
grain sorghum, 111
green nanocomposites, 80–1, *82*
gross calorific values (GCV), 219

HA *see* hydroxyapatites
heat fractionation, 143
HEC *see* hydroxyethyl cellulose
hemicelluloses, 3, 133–50
  applications, 146–7
  co-constituents, 141
  degradation, 145–6
  distribution within plants, 140–1
  esterification, 143–4
  etherification, 144–5
  extraction methodologies, 141–3
  modifications, 143–6, *144*
  paper/board applications, 263–5
  principal sugar units, *134,* 135
  sources and characteristics, 133–4,
    134, 137–41, **138–9**, 147–8
  species composition, 138–40
  structures, 134–7, *135–6*
high performance liquid chromatography
  (HPLC), 52, 54
HMF *see* hydroxymethylfurfural
HORTIBIOPACK programme, **284,**
  286

HPC *see* hydroxypropylcellulose
HPLC *see* high performance liquid
  chromatography
HPMC *see* hydroxypropylmethylcellulose
hydrocolloids, 247
hydrolytic degradation
  polyhydroxyalkanoates, 74–5
  polylactides, 56
hydroxyapatites (HA), 80
hydroxyethyl cellulose (HEC), 13, 163
hydroxymethylfurfural (HMF), 181–3
hydroxypropylcellulose (HPC), 266
hydroxypropylmethylcellulose
  (HPMC), 265
HYDRUS programme, **284**, 286
hygroscopicity
  chitosan, 93–4, *95*, 98
  food packaging applications, 221–2
  proteins, 118, 120–1

IL *see* ionic liquids
industrial membrane bioreactors, 95
ingredient/additive carriers, 235
injection moulding, 125–8
inorganic nanofillers, 320–1
intelligent packaging, 213–14
intercalated nanoclays, 267–8, *268*
ionic liquids (IL), 160–1, *161–2*, 305
isocyanate-modified cellulose, 170–1

kafirin, 111
keratin, 115

lactic acid bacteria (LAB), 45
lactic acid (LA), 43–7, *44*, 52–3
Langmuir–Blodgett (LB) films, 159–60
layered double hydroxides (LDH), 80,
  308–9, *309*, 310, 321
LB *see* Langmuir–Blodgett
LCA *see* life cycle assessment
LCC *see* lignin-carbohydrate complexes
LDH *see* layered double hydroxides
life cycle assessment (LCA), 9, 13, 293
lignin-carbohydrate complexes (LCC), 141
lignocellulosic fibres, 138, **138**, 142–3
linear polymerizations, 202–3, *203*
lipids
  edible films/coatings, 241–2, 246
  oxidation, 245
  paper/board applications, 257
  starch-based polymers, 17–18
liquid foods, 221

low tunnels, 293–4, *294*
lupin, 111–12

mannans, 135, 136–40, 146
MAP *see* modified atmosphere packaging
Mark–Houwink constants, 53
market for biopolymers, 4–8, *4*, *5*
  biomass production methods, 12–13
  bioplastics, 11–12
  coatings, 7–8
  commercial bio-derived polymer
    production, 9–12
  plastic films, 5–7, *6*, *7*
mass transfer control, 234–5, *234*, 246–7
mechanical properties
  agronomy, 291–2, 294
  chitosan, 94
  edible films/coatings, 235–6
  paper/board applications, 264–5, **264**,
    267, 269–71
  polylactides, 58
  proteins, 122, **123–4**, 125
medical *see* biomedical; pharmaceutical
melt-flow index (MFI), 54
membrane bioreactors, 95
5-methylfurfural, 193–4, *194*
MFC *see* microfibrillated cellulose
MFI *see* melt-flow index
microfibrillated cellulose (MFC), 154–6,
  157, 163–5, 173, 269–71, *270*, 321
microsphere formulations, 147
milk proteins, 239
mineralization rates, 27, 32
MMT *see* montmorillonite
model cellulose films, 159–60
modified atmosphere packaging
  (MAP), 225–7, *226*
molecular weight distribution
  chitosan, 89
  polylactides, 50
montmorillonite (MMT), 267–8, 302,
  308–10, 320–1
mulching, 278–9, *279*, 287, 289–93, *289*,
  291–2
multilayer co-extrusion, 78–9
multilayer films
  edible films/coatings, 242–3
  production and usage, 6–7
multiphase materials, 77–81
multiwalled carbon nanotubes
  (MWCNT), 205
myofibrillar proteins, 116–17

nanocellulose, 269–71, 270, 311, 321–2
nanoclays, 267–9, 268, 272
nanocomposites see bionanocomposites
nanofillers, 303, 320–1
nano-objects, 154–9, 155–9
nanoparticle-coupled cellulose, 172
NLO see nonlinear optical
NMR see nuclear magnetic resonance
non-woven floating covers, 295–6
nonlinear optical (NLO) devices, 306
nonlinear polymerizations, 203–4, 204
nuclear magnetic resonance (NMR)
    spectroscopy, 22, 26, 52–4

O-acetylgalactoglucomannan
    (AcGGM), 265
oat avenin, 111, 112
OFET see organic field-effect transistors
OLED see organic light emitting diodes
OML see overall migration limit
OMMT see organically modified
    montmorillonite
OP see oxygen permeability
optical properties, 59, 271
optoelectronic applications, 301,
    303–8, 303, 307
OPV see organic photovoltaics
organic field-effect transistors
    (OFET), 306–7
organic light emitting diodes (OLED), 306,
    308
organic photovoltaics (OPV), 303–4
organically modified montmorillonite
    (OMMT), 80
OTR see oxygen transmission rate
overall migration limit (OML), 217–18
oxygen barriers, 58
oxygen permeability (OP)
    food packaging applications, 215–16,
        220, 223
    paper/board applications, 259, 260–1,
        262–5, 263, 271
    proteins, 120, 125
oxygen transmission rate (OTR), 215–16

packaging applications, 3
    chitosan, 88, 91–4
    production and usage, 6
    see also food packaging applications
paper mulching, 290
paper/board applications, 3, 255–76
    biodegradability, 256–7

bionanocomposites, 267–71, 272
biopolymer films/coatings, 257–67,
    260–2, 263
    cellulose, 258, 265–7
    chitosan, 259, 262–3
    future developments, 318
    hemicelluloses, 263–5
    mechanical properties, 264–5, 264,
        267, 269–71
    nanocellulose, 269–71, 270
    nanoclays, 267–9, 268, 272
    objectives and methods, 255–7
    polysaccharides, 256, 257
    proteins, 120–1
    starch-based polymers, 258–9
PAR see photosynthetically active radiation
PCL see polycaprolactone
PE see polyethylene
peanut protein, 113
pectins, 138, 141
PEM see polyelectrolyte multilayer
PEN see polyethylene naphthenate
PEO see polyethylene oxide
permeability see barrier properties
PET see polyethylene terephthalate
pH sensitivity, 118
PHA see polyhydroxyalkanoates
pharmaceutical applications
    chitosan, 90–1
    proteins, 116
    see also biomedical applications
photosynthetically active radiation
    (PAR), 292, 294
photovoltaics (PV), 303–5, 303
physical properties, 235–6
PICUS programme, 284, 285
piezoelectric properties, 311
pigment concentrations, 255–6
PLA see polylactides
plant-based ethanol, 12–13
plant-based proteins, 108–13
plastic films, 5–7, 6, 7
    see also bioplastics
plasticization
    chitosan, 93
    edible films/coatings, 243, 246
    future developments, 319
    paper/board applications, 265, 266
    polyhydroxyalkanoates, 76–8
    polylactides, 57
    proteins, 109, 112, 119–20, 122, 125
    starch-based polymers, 19–22, 20–1

PLGA *see* polylactide-*co*-glycolide
polyamides, 194–5, *194–5*
polycaprolactone (PCL), 26–7, 35, 171
polycondensation
  furans, 189–94, *190–1*
  polylactides, 46–7, 50
polyelectrolyte multilayer (PEM) thin
  films, 310
polyesters, 195–7, *196–7*
polyethylene naphthenate (PEN), 304
polyethylene oxide (PEO), 304
polyethylene (PE)
  agronomy, 277–9
  biomass production methods, 12–13
  production and usage, 5, **6**
polyethylene terephthalate (PET), 304
polyhydroxyalkanoates (PHA), 3, 10, 65–86
  bacterial biosynthesis, 68–71, **70**, 82
  biocomposites, 79–81
  biodegradability, 75–6, 82
  chemical synthesis, 72
  crystallinity and characteristic
    temperatures, 73, 74, 79
  degradation, 74–7
  food packaging applications, 222
  future developments, 318–19
  genetically modified organisms, 71–2
  hydrolytic degradation, 74–5
  mechanical properties, 72–3, **73**
  multilayer co-extrusion, 78–9
  multiphase materials, 77–81
  plasticization, 76–8
  polymer blends, 78
  production and commercial
    products, 70–1, 81–2, **81**
  properties, 72–4
  sources and characteristics, 65–7
  structures, **66**, 67, 72–3
  synthesis, 67–72
  thermal degradation, 76–7
polylactide-*co*-glycolide (PLGA), 301
polylactides (PLA), 3, 10, 43–63
  applications, 59–60
  barrier properties, 57–9
  bulk polycondensation, 46–7, 50
  chain extension, 46, 48
  characterization, 52–4
  chemical properties, 49–54
  conversion and yields, 50–1
  copolymerization, 51–2, *52*
  food packaging applications, 217,
    219, 221

functionalized biopolymer films/
  coatings, 301, 304–5, 311
future developments, 318
hydrolytic stability, 56
lactic acid, 43–7, *44*, 52–3
mechanical and thermo-mechanical
  properties, 58
molecular weight distribution, 50
optical properties, 59
plasticization, 57
printability, 59
processability, 54–5
production methods, 44–9, *46*
properties, 54–9
ring-opening polymerization, 43–6, 48–9,
  50–1, 53
sources and characteristics, 43–4
starch-based polymers, 26–7, 35
tacticity, 49–50
thermal stability, 55
thermal transitions and crystallinity, 56–7
polymer blends
  chitosan, 94, 97–8
  future developments, 319
  polyhydroxyalkanoates, 78
  starch-based polymers, 23–7, *24–6*, 27,
    28–31, 35–9, *36–8*
polymer grafting, 167–71
polymer scaffolds, 91
polypropylene (PP), 5, **6**
polysaccharides
  edible films/coatings, 234, 239–41
  paper/board applications, 256, 257
  starch-based polymers, 22
polystyrene (PS), 5, **6**
polyurethanes, 198–9, *198–9*
polyvinyl butyrate (PVB), 5, **6**
polyvinyl chloride (PVC), 5, **6**
post-metering, 256
potassium sorbate, 19–20
PP *see* polypropylene
printability, 59, 255
processability, 54–5
proteins, 3, 107–32
  animal-based, 113–17
  applications, 110, 116, 120–1
  barrier properties, 117, 125
  casein, 114–15
  collagen, 115
  compression moulding, 121–5, **122–4**
  corn zein, 109–10, *110*
  cottonseed, 112–13

dry forming of protein films, 121–8, 122–4, 126
edible films/coatings, 234, 237–9, 245
egg white, 115
extrusion and injection moulding, 125–8, 126
gelatin, 115–16, 116
hygroscopicity, 118, 120–1
kafirin/grain sorghum, 111
keratin, 115
lupin, 111–12
myofibrillar proteins, 116–17
oat avenin, 111, 112
peanut protein, 113
pH sensitivity, 118
plant-based, 108–13
plasticization, 109, 112, 119–20, 122, 125
rapeseed, 108, 108
rice bran protein, 111
solution casting, 117–21, 128
solvent selection, 119
sources and characteristics, 107–17
soy protein, 110–11, 127–8
temperature sensitivity, 119
viscosity, 118
wheat gluten, 109, 109, 128
whey protein, 113–14, 114
PS see polystyrene
PV see photovoltaics
PVB see polyvinyl butyrate
PVC see polyvinyl chloride
pyrroles, 180

radical polymerization, 186–7
RAFT see reversible addition-fragmentation transfer
rapeseed, 108, 108
RBP see rice bran protein
recycling, 8, 281, 282
regulatory frameworks
  agronomy, 282–3, 284
  food packaging applications, 216–19, 225
relative humidity (RH)
  food packaging applications, 220
  paper/board applications, 265–7, 271
resins, 241–2
retrogradation process, 22–3
reversible addition-fragmentation transfer (RAFT), 169
reversible polymer cross-linking, 205
RH see relative humidity
rice bran protein (RBP), 111

ring-opening polymerization (ROP)
  polyhydroxyalkanoates, 72
  polylactides, 43–6, 48–9, 50–1, 53

safety issues
  edible films/coatings, 244
  food packaging applications, 216–18, 224–5
SBC see Sustainable Biomaterials Collaborative
scanning electron microscopy (SEM), 157–8, 164, 166
seaweed extracts, 241, 242, 245–6
SEM see scanning electron microscopy
sensorial properties, 236, 244, 250
SFPI see sunflower protein isolate
shelf-life, 214–16, 215, 220–2
silver nanoparticles, 172
simulation tests, 32, 39
SML see specific migration limit
sodium hypochlorite, 142
softgel capsules, 116
solarization, 293
solid state polycondensation (SSPC), 47
solution casting, 117–21, 128
solvent-assisted polycondensation, 47
solvent/solution properties
  cellulose, 160–2, 161–2
  chitosan, 89
  proteins, 119
soy protein, 110–11, 127–8
soy protein isolate (SPI), 239, 242, 247
specific migration limit (SML), 217–18
SPI see soy protein isolate
SSPC see solid state polycondensation
starch-based polymers, 3, 15–42
  biodegradability, 23, 27, 32–9, 33–4, 36–8
  edible films/coatings, 240
  future developments, 318
  gelatinization process, 18–22
  mechanical properties, 28–31
  paper/board applications, 258–9
  plasticization, 19–22, 20–1
  polymer blends, 23–7, 24–6, 27, 28–31, 35–9, 36–8
  retrogradation process, 22–3
  sources and characteristics of starch, 15–18, 16–18, 18, 20–1, 21
  structure, 16–18
starch-converting enzymes, 33
starch-derived polymers, 281
step-growth polymerization, 52–3, 189–201

stress–strain curves, 21–2, *21*
sulphated hemicelluloses, 145, 147
sunflower protein isolate (SFPI), 127
superficial fibre modification, 164–72
surface condition control, 235, *235*
surface hydrophobization, 165–7, *166*,
    168–70
surface-initiated polymerization, 167–71
sustainability, 8–9
Sustainable Biomaterials Collaborative
    (SBC), 9

tacticity, 49–50
tactoids, 268, *268*
TEM *see* transmission electron microscopy
thermal degradation
    polyhydroxyalkanoates, 76–7
    polylactides, 55
thermal transitions, 56–7
thermo-mechanical properties, 58
thermoplastic elastomers, 199
thermoplastic films
    paper/board applications, 257, 269
    production and usage, 5–7, *6*
    see also starch-based polymers
thermoset resins, 317, 320
thiophenes, 180
tissue engineering, 91
titania nanoparticles, 172
transferases, 33
transmission electron microscopy
    (TEM), 155, *156*, *270*, *302*
TRIGGER programme, 284, *285*

triglycerides, 17–18, 241
twin-screw extruders, 23, 25, 26, 128

viscosity, 118

waste products, 319
waste water treatment, 88, 90
water vapour permeability (WVP)
    edible films/coatings, 242, 244, 247
    food packaging applications, 215, 222
    paper/board applications, 257–9, **262**,
        263, 265
    polylactides, 58
    proteins, 125
water vapour transmission rate
    (WVTR), 215
waxes, 241, 243
wheat gluten
    edible films/coatings, 238–9
    future developments, 319
    sources and characteristics, 109, *109*, 128
whey protein, 113–14, *114*
whey protein isolate (WPI), 239, 242, 247
worldwide markets *see* market for
    biopolymers
WPI *see* whey protein isolate
WVP *see* water vapour permeability
WVTR *see* water vapour transmission rate

X-ray diffraction (XRD), 16, *18*
X-ray scattering, 16
xylans, 135–6, 139–40, 142–3, 146
xyloglucans, 135, 137, 140, 146–7

Printed and bound by CPI Group (UK) Ltd, Croydon, CR0 4YY

16/04/2025

14658545-0005